T0206888

Advanced R Statistical Programming and Data Models

Analysis, Machine Learning, and Visualization

Matt Wiley

Joshua F. Wiley

Apress®

Advanced R Statistical Programming and Data Models: Analysis, Machine Learning, and Visualization

Matt Wiley
Columbia City, IN, USA

Joshua F. Wiley
Columbia City, IN, USA

ISBN-13 (pbk): 978-1-4842-2871-5
https://doi.org/10.1007/978-1-4842-2872-2

ISBN-13 (electronic): 978-1-4842-2872-2

Library of Congress Control Number: 2019932986

Managing Director, Apress Media LLC: Welmoed Spahr
Acquisitions Editor: Steve Anglin
Development Editor: Matthew Moodie
Coordinating Editor: Mark Powers

Cover designed by eStudioCalamar

Cover image designed by Freepik (www.freepik.com)

Distributed to the book trade worldwide by Springer Science+Business Media New York, 233 Spring Street, 6th Floor, New York, NY 10013. Phone 1-800-SPRINGER, fax (201) 348-4505, e-mail orders-ny@springer-sbm.com, or visit www.springeronline.com. Apress Media, LLC is a California LLC and the sole member (owner) is Springer Science + Business Media Finance Inc (SSBM Finance Inc). SSBM Finance Inc is a Delaware corporation.

For information on translations, please e-mail editorial@apress.com; for reprint, paperback, or audio rights, please email bookpermissions@springernature.com.

Apress titles may be purchased in bulk for academic, corporate, or promotional use. eBook versions and licenses are also available for most titles. For more information, reference our Print and eBook Bulk Sales web page at http://www.apress.com/bulk-sales.

Any source code or other supplementary material referenced by the author in this book is available to readers on GitHub via the book's product page, located at www.apress.com/9781484228715. For more detailed information, please visit http://www.apress.com/source-code.

Printed on acid-free paper

Table of Contents

TABLE OF CONTENTS

TABLE OF CONTENTS

About the Authors

Matt Wiley is a tenured, associate professor of mathematics with awards in both mathematics education and honors student works. He earned degrees in pure mathematics, computer science, and business administration through the University of California and Texas A&M University Systems. He serves as director of quality enhancement at Victoria College, facilitating comprehensive assessment programs, key performance indicator dashboards and one-click reports, and data consultation for campus stakeholders. Outside academia, he is managing partner at Elkhart Group LLC, a statistical consultancy. With experience in programming R, SQL, C++, Ruby, Fortran, and JavaScript, he has always found ways to meld his passion for writing with his joy of logical problem solving and data science. From the boardroom to the classroom, Matt enjoys finding dynamic ways to partner with interdisciplinary and diverse teams to make complex ideas and projects understandable and solvable.

Joshua F. Wiley is a lecturer in the Monash Institute of Cognitive and Clinical Neurosciences and School of Psychological Sciences at Monash University. He earned his PhD from the University of California, Los Angeles, and completed his postdoctoral training in primary care and prevention. His research uses advanced quantitative methods to understand the dynamics between psychosocial factors, sleep, and other health behaviors in relation to psychological and physical health. He develops or codevelops a number of R packages including varian, a package to conduct Bayesian scale-location structural equation models, and MplusAutomation, a popular package that links R to the commercial Mplus software, and miscellaneous functions to explore data or speed up analysis in JWileymisc.

About the Technical Reviewer

Andrew Moskowitz is an analytics and data science professional in the entertainment industry focused on understanding user behavior, marketing attribution and efficacy, and using advanced data science concepts to address business problems. He earned his PhD in quantitative psychology at the University of California, Los Angeles, where he focused on hypothesis testing and mixed effects models.

Acknowledgments

To our dear family, who may not always understand everything we write, yet are nevertheless content to place our books on fireside mantels and coffee tables.

Introduction

This book shows how to conduct data analysis using the popular R language. Our goal is to provide a practical resource for conducting advanced statistical analyses using R. As this is an advanced book, the reader is assumed to have some background in using R, including familiarity with general data management and the use of functions.

Because the book is primarily practical, we do not provide in-depth theoretical or conceptual introductions to the various statistical models discussed. However, to aid understanding and their appropriate application, we do provide some conceptual background on each analytic technique discussed.

Conventions

Bold lowercase letters are used to refer to a vector, for example, **x**. Bold uppercase letters are used to refer to a matrix, for example, **X**. Generally, the Latin alphabet is used for data and the Greek alphabet is used for parameters. Mathematical functions are indicated with parentheses, for example, $f(\cdot)$.

In the text, reference to R code or function will be in monospaced font `like this`. R function names have parentheses included to help indicate it is a function, such as `mean()` to indicate the mean function in R.

Package Setup

Throughout the book, we will make use of many different R packages that make tasks easier or provide more robust or sophisticated graphing and analysis options.

Although not required for readers, we make use of the `checkpoint` package to help ensure the book is reproducible [23]. If you do not care about reproducibility and are happy to take your chances that our code that worked with one version of R and packages also works with whatever versions you have, then you can just skip reading this section. If you want reproducibility, but do not care why or how it works, then just create R scripts for the code for each chapter, save them, and then run the `checkpoint` package at the beginning. If you care and want to know why and how it all works, read on the next few paragraphs.

Details on Reproducibility

The many additional packages available for R are one of its greatest strengths. However, they also create some challenges. For example, as a reader, suppose that on your computer, you have R v3.4.3 installed and as part of that in January you had installed the ggplot2 package for graphs. By default, you will have whatever version of ggplot2 was available in January when you installed it. Now in one chapter, we tell you that you need both the ggplot2 and cowplot packages. Because you already had ggplot2 installed, you do not need to install it again. However, suppose that you did not have the cowplot package installed. So, whenever you happen to be reading that chapter, you attempt to install the cowplot package, let's say it's in April. You will now by default get the latest version of cowplot available for that version of R as of April.

Now imagine a second reader comes along and also had R v3.4.3 but had neither the ggplot2 nor the cowplot package installed. They also read the chapter in April, but they install both packages in April, so they get the latest version of both packages available in April for R v3.4.3.

Even though both you and the other reader had the same version of R installed, you will end up with different package versions from each other, and likely different versions yet from whatever versions we used to write the book.

The end result is that different people, even with the same version of R, very likely are using different versions of different packages. This can pose a major challenge for reproducibility. If you are reading a book, it can be a frustration because code does not seem to work as we said it would. If you are using code in production or for scientific research or decision-making, nonreproducibility can pose an even bigger challenge.

The solution to standardize versions across people and ensure results are fully reproducible is to control not only the version of R but also the version of all packages. This requires a different approach to package installation and management than the default system, which uses the latest package versions from CRAN. The checkpoint package is designed to solve this challenge. It does require some extra steps and processes to use, and at first may seem a nuisance, but the payoff is that you can be guaranteed that you are not only using the same version of R but also the same version of all packages.

To understand how the checkpoint package works, we need a bit more background regarding how R's libraries and package system work.

Mainstream R packages are distributed through CRAN. Package authors can submit new versions of their packages to CRAN, and CRAN updates nightly. For some operating systems, CRAN just stores the package source code, such as for Linux machines. For others, such as Windows operating systems, CRAN builds precompiled package binaries and hosts those. CRAN keeps old source code but generally not old binary packages for long. On a local machine, when `install.packages` is run, R goes online to a repository, by default CRAN, finds the package name, downloads it, and installs it into a local *library*. The local *library* is basically just a directory on your own machine. R has a default location it likes to use as its local library, and by default when you install packages, they are added to the default library. Once a package is installed, when it is loaded or opened using `library()`, R goes to its default library, finds a package with the same name, and opens it.

The `checkpoint` package works by creating a new library on the local machine, for a particular version of R for a particular date. Then it scans all the R script files in R's current working directory—you can identify this using the `getwd()` function—and identifies any calls to the `library()` or `require()` functions. Then it goes and checks whether those packages are installed in the local library. If they are not, it goes to a snapshot of CRAN taken by another server setup to support the `checkpoint` package. That way, `checkpoint` can install the version of the package available from a specific date. In that way, the `checkpoint` package can ensure that you have the same specific version of R and specific version of all packages that we used when writing the book. Or if you are trying to re-run some analysis from a year ago, you can get the same version of those packages on a new computer.

Assuming that you have the following code in an R script, you can use the `checkpoint` package to read the R script and find the call to `library(data.table)`, and it will install the `data.table` package, which is a great package for data management [29]. If you do not want `checkpoint` to look in the current working directory, you can specify the project path, as we do to the book in this example. You can also change where `checkpoint` sets its library to another folder location, instead of the default location, which we also do. We accomplish both of these using variables set as part of our R project, `book_directory` and `checkpoint_directory`. If you are using checkpoint on your own machine, set those variables to the relevant directories, for example, as `book_directory <- "path/to/your/directory"`. Note that whatever folder you choose, R will need read and write privileges for that folder.

```
library(checkpoint)
checkpoint("2018-09-28", R.version = "3.5.1",
  project = book_directory,
```

```
  checkpointLocation = checkpoint_directory,
  scanForPackages = FALSE,
  scan.rnw.with.knitr = TRUE, use.knitr = TRUE)

library(data.table)

options(
  width = 70,
  stringsAsFactors = FALSE,
  digits = 2)
```

Data Setup

One of the datasets we will use throughout this book is a longitudinal study, the Americans' Changing Lives (ACL) [45]. This is publicly available data and can be downloaded by going to http://doi.org/10.3886/ICPSR04690.v7.

The Americans' Changing Lives (ACL) is a longitudinal study with five waves of data, shown in Table I-1.

Table I-1. *ACL Study Collection Waves*

Wave	Year
W1	1986
W2	1989
W3	1994
W4	2002
W5	2011

All we need is the data file in R format, which should be called 04690-0001-Data.rda. You may also find it helpful to download the PDF documentation of the dataset for more details about the study. After you have downloaded the data, you should extract the zip folder.

After setting up our R session and necessary libraries, we load the data. You will need to adjust the path to wherever you saved the data file after extracting it from the zip folder. Because it is a RDA file, loading it loads an R object into the workspace. Next we convert to a data table, select just the variables we are going to use for this book, and change the variable names to be a bit more intuitive. The suffix (e.g., "W1") indicates which wave the variable comes from. Finally, we convert some variables to factor class and then save the dataset using the saveRDS() function with compression. This will allow us to read our cleaned dataset back into R in later chapters with ease and to assign it to any object name we wish, rather than being stuck with the object name in the RDA file.

```
load("../ICPSR_04690/DS0001/04690-0001-Data.rda")
ls()
```

```
## [1] "book_directory"          "checkpoint_directory"
## [3] "da04690.0001"            "render_apress"
```

```
acl <- as.data.table(da04690.0001)
acl <- acl[, .(
  V2, V1801, V2101, V2064,
  V3007, V2623, V2636, V2640,
  V2000,
  V2200, V2201, V2202,
  V2613, V2614, V2616,
  V2618, V2681,
  V7007, V6623, V6636, V6640,
  V6201, V6202,
  V6613, V6614, V6616,
  V6618, V6681
)]

setnames(acl, names(acl), c(
  "ID", "Sex", "RaceEthnicity", "SESCategory",
  "Employment_W1", "BMI_W1", "Smoke_W1", "PhysActCat_W1",
  "AGE_W1",
  "SWL_W1", "InformalSI_W1", "FormalSI_W1",
  "SelfEsteem_W1", "Mastery_W1", "SelfEfficacy_W1",
  "CESD11_W1", "NChronic12_W1",
  "Employment_W2", "BMI_W2", "Smoke_W2", "PhysActCat_W2",
  "InformalSI_W2", "FormalSI_W2",
  "SelfEsteem_W2", "Mastery_W2", "SelfEfficacy_W2",
  "CESD11_W2", "NChronic12_W2"
        ))
```

INTRODUCTION

```
acl[, ID := factor(ID)]
acl[, SESCategory := factor(SESCategory)]
acl[, SWL_W1 := SWL_W1 * -1]

saveRDS(acl, "advancedr_acl_data.RDS", compress = "xz")
```

CHAPTER 1

Univariate Data Visualization

Most statistical models discussed in the rest of the book make assumptions about the data and the best model to use for them. As data analysts, we often must specify the distribution that we assume the data come from. Anomalous values, also called extreme values or outliers, may also have undue influence on the results from many statistical models. In this chapter, we examine visual and graphical approaches to exploring the distributions and anomalous values for one variable at a time (i.e., univariate). The goal of this chapter is not specifically to create beautiful or publication quality graphs nor to show results, but rather to use graphs to understand the distribution of a variable and identify anomalous values. This chapter focuses on univariate data visualization, and the next chapter will employ some of the same concepts but applied to multivariate distributions and cover how to assess the relations between variables.

```
library(checkpoint)
checkpoint("2018-09-28", R.version = "3.5.1",
  project = book_directory,
  checkpointLocation = checkpoint_directory,
  scanForPackages = FALSE,
  scan.rnw.with.knitr = TRUE, use.knitr = TRUE)

library(knitr)
library(ggplot2)
library(cowplot)
library(MASS)
library(JWileymisc)
library(data.table)

options(width = 70, digits = 2)
```

© Matt Wiley and Joshua F. Wiley 2019
M. Wiley and J. F. Wiley, *Advanced R Statistical Programming and Data Models*,
https://doi.org/10.1007/978-1-4842-2872-2_1

The `ggplot2` package [109] creates elegant graphs, and the `cowplot` package is an add-on that makes graphs cleaner [117]. The `MASS` package provides functions to test how well different distributions fit data [98]. The `JWileymisc` package is maintained by one of this text's authors and provides miscellaneous functions that allow us to focus on the graphics in this chapter [114]. The `data.table` package will be used a lot for data management [29].

1.1 Distribution

Visualizing the Observed Distribution

Many statistical models require that the distribution of a variable be specified. Histograms use bars to graph a distribution and are probably the most common graph used to visualize the distribution of a single variable. Although relatively rare, stacked dot plots are another approach and provide a precise way to visualize the distribution of data that shows the individual data points. Finally, density plots are also quite common and are graphed by using a line that shows the approximate density or amount of data falling at any given value.

Stacked Dot Plots and Histograms

Dot plots work by plotting a dot for each observed data value, and if two dots would fall on top of each other, they are stacked up [118]. Compared to histograms or density plots, dot plots are unique in that they actually plot the raw individual data points, rather than aggregating or summarizing them. This makes dot plots a nice place to start looking at the distribution or spread of a variable, particularly if you have a relatively small number of observations.

The granular approach, plotting individual data points, is also dot plots limitation. With even moderately large datasets (e.g., several hundred), it becomes impractical to plot individual values. With thousands or millions of observations, dot plots are even less effective at visualizing the overall distribution.

We can create a plot easily using `ggplot2`, and the results are shown in Figure 1-1.

```
ggplot(mtcars, aes(mpg)) +
  geom_dotplot()

## 'stat_bindot()' using 'bins = 30'. Pick better value with 'binwidth'.
```

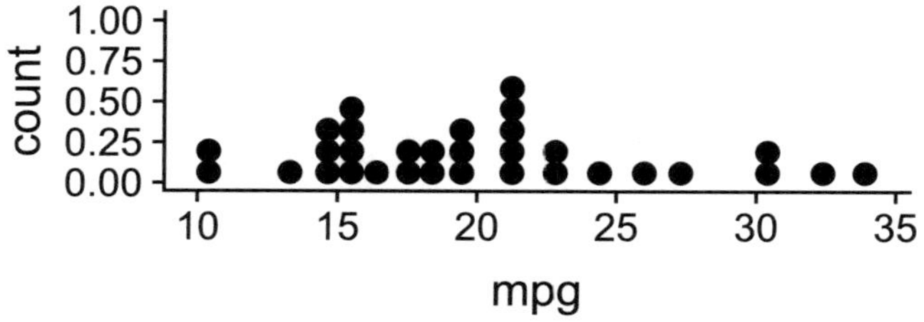

Figure 1-1. *Stacked dot plot of miles per gallon from old cars*

As a brief aside, much of the code for `ggplot2` follows the format shown in the following code snippet. In our case, we wanted a dot plot, so the geometric object, or "geom", is a dot plot (`geom_dotplot()`). Many excellent online tutorials and books exist to learn how to use the `ggplot2` package for graphs, so we will not provide a greater introduction to `ggplot2` here. In particular, Hadley Wickham, who develops `ggplot2`, has a recently updated book on the package, *ggplot2: Elegant Graphics for Data Analysis* [109], which is an excellent guide. For those who prefer less conceptual background and more of a cookbook, we recommend the *R Graphics Cookbook* by Winston Chang [20].

```
ggplot(the-data, aes(variable-to-plot)) +
  geom_type-of-graph()
```

Unlike a dot plot that plots the raw data, a histogram is a bar graph where the height of the bar is the count of the number of values falling within the range specified by the width of the bar. You can vary the width of bars to control how many nearby values are aggregated and counted in one bar. Narrower bars aggregate fewer data points and provide a more granular view. Wider bars aggregate more and provide a broader view. A histogram showing the distribution of sepal lengths of flowers from the famous iris dataset is shown in Figure 1-2.

```
ggplot(iris, aes(Sepal.Length)) +
  geom_histogram()

## 'stat_bin()' using 'bins = 30'. Pick better value with 'binwidth'.
```

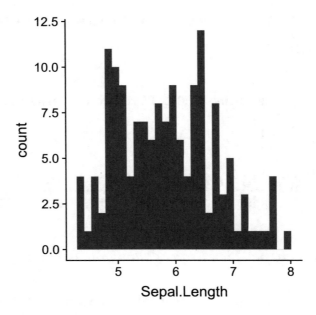

Figure 1-2. *Histogram of sepal length from the iris data*

If you know the shape of a distribution (e.g., a normal distribution), you can examine whether the histogram for a variable looks like the shape of a distribution you recognize. In the case of the sepal length data, they appear approximately normally distributed, although they are clearly not perfect.

If data do not appear to follow the distribution we hoped for (e.g., normal), it is common to apply a transformation to the raw data. Again, histograms are a useful way to examine how the distribution looks after transformation. Figure 1-3 shows a histogram of annual Canadian lynx trappings. From the graph we can see the variable is positively skewed (has a long right tail).

```
ggplot(data.table(lynx = as.vector(lynx)), aes(lynx)) +
  geom_histogram()

## 'stat_bin()' using 'bins = 30'. Pick better value with 'binwidth'.
```

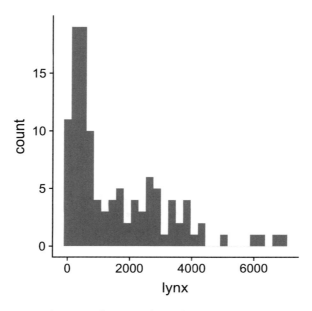

Figure 1-3. *Histogram of annual Canadian lynx trappings*

For positive skew, a square root or log transformation can help to reduce positive skew and make variables closer to a normal distribution, assuming that there are no negative values. This histogram of lynx trappings after a natural log transformation is shown in Figure 1-4.

```
ggplot(data.table(lynx = as.vector(lynx)), aes(log(lynx))) +
  geom_histogram()
```

```
## 'stat_bin()' using 'bins = 30'. Pick better value with 'binwidth'.
```

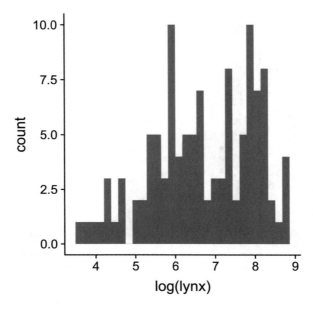

Figure 1-4. *Histogram of annual Canadian lynx trappings after a natural log transformation*

Density Plots

Another common tool to visualize the observed distribution of data is by plotting the empirical density. The code for `ggplot2` is identical to that for histograms except that `geom_histogram()` is replaced with `geom_density()`. The code follows and the result is shown in Figure 1-5.

```
ggplot(iris, aes(Sepal.Length)) +
  geom_density()
```

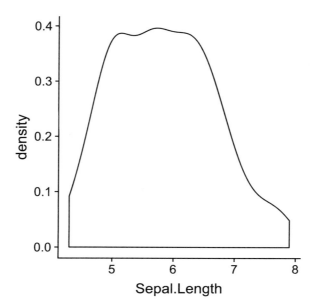

Figure 1-5. *This is the density plot for our sepal lengths*

Empirical density plots include some degree of smoothing, because with continuous variables, there is never going to be many observations at any specific value (e.g., it may be that no observation has a value of 3.286, even though there are values of 3.281 and 3.292). Empirical density plots show the overall shape of the distribution by applying some degree of smoothing. At times it can be helpful to adjust the degree of smooth to see a coarser (closer to the raw data) or smoother (closer to the "distribution") graph. Smoothing is controlled in ggplot2 using the adjust argument. The default, which we saw in Figure 1-5, is adjust = 1. Values less than 1 are "noisier" or have less smoothing, while values greater than 1 increase the smoothness. We compare and contrast noisier in Figure 1-6 vs. very smooth in Figure 1-7.

```
ggplot(iris, aes(Sepal.Length)) +
  geom_density(adjust = .5)

ggplot(iris, aes(Sepal.Length)) +
  geom_density(adjust = 5)
```

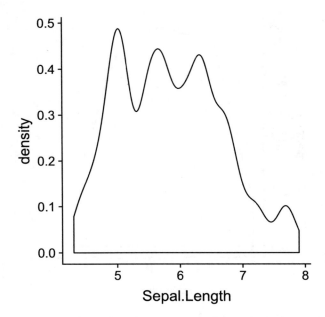

Figure 1-6. *A noisy density plot*

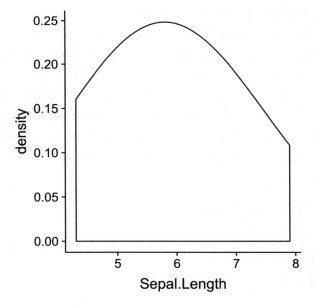

Figure 1-7. *A very smooth density plot*

Comparing the Observed Distribution with Expected Distributions

Although it is helpful to examine the observed data distribution, often we are examining the distribution to see whether it meets the assumption of the statistical analysis we hope to apply. For instance, linear regression assumes that data are (conditionally) normally distributed. If the empirical distribution appeared sufficiently not normal or was closer to a different distribution, then it may be inappropriate to use normal linear regression.

Q-Q Plots

To assess whether data fit or are close to a specific expected distribution, we can use a quantile-quantile or Q-Q plot. A Q-Q plot graphs the observed data quantiles against theoretical quantiles from the expected distribution (e.g., normal, beta, etc.). Q-Q plots can be used to examine whether data come from almost any distribution. Because the normal or Gaussian distribution is by far the most common, the default function in ggplot2 for making Q-Q graphs defaults to assuming a normal distribution. With a Q-Q plot, if the data perfectly match the expected distribution, the points will fall on a straight line. The following code creates Figure 1-8.

```
ggplot(iris, aes(sample = Sepal.Length)) +
  geom_qq()
```

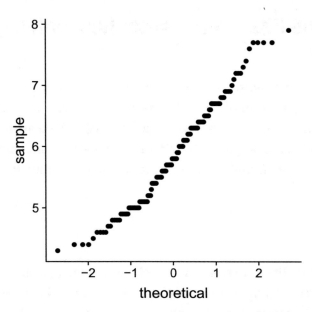

Figure 1-8. *Normal data look like a straight line. Sepal.Length seems fairly normal*

To better understand how geom_qq() works, we can make one ourselves. R has basic functions for many probability distributions built in. These allow one to generate random numbers (e.g., rnorm()), obtain the probability that an observation from a given distribution falls above or below a certain value (e.g., pnorm()), calculate the quantile from a distribution (e.g., qnorm()), and get the density of a distribution at a particular value (e.g., dnorm()). By convention, these are named r, p, q, d, followed by the distribution name (or an abbreviation of that, like "norm" for normal). Applying this knowledge, we obtain the following quantile in which 10% (.10) of the values of a normal distribution with mean = 0 and standard deviation of 1 would lie, using qnorm():

```
qnorm(p = .1, mean = 0, sd = 1)
```

```
## [1] -1.3
```

It is straightforward how to apply this to normal distributions with different means or standard deviations. Suppose we have three data points. If they are normally distributed, we might expect the middle point to fall at a probability of .5 for a normal distribution, and the other two to fall about half between 0 and .5 or .5 and 1 (i.e., .25 and .75). We can easily obtain the normal quantiles for these probabilities.

```
qnorm(p = c(.25, .50, .75), mean = 0, sd = 1)
```

```
## [1] -0.67 0.00 0.67
```

To aid coming up with appropriately spaced probabilities, we could use the ppoints() function.

```
ppoints(n = 3, a = 0)
```

```
## [1] 0.25 0.50 0.75
```

Rather than perfectly spaced, ppoints() defaults to a small adjustment. For ten or fewer data points, (1:N - 3/8)/(n + 1 - 2 * 3/8), and for more than ten data points, (1:N - 1/2)/(n + 1 - 2 * 1/2). Either way, the idea is the same.

```
ppoints(n = 3)
```

```
## [1] 0.19 0.50 0.81
```

All that remains is to sort our data and plot against the theoretical normal quantiles. Adding the mean and standard deviation is not technically necessary; they are a linear adjustment. Either way, the points should fall in a straight line, but using them makes the theoretical quantiles have the same mean and scale as our raw data.

We use the qplot() function from the ggplot2 for graphing here. Note that the q in qplot() stands for "quick" as it is shorthand for the longer specification using the ggplot() function used for more advanced graphs. All this is to say the q is unrelated to quantiles. Lastly, just to help with visualization, we add a straight line with a slope of 1 and intercept of 0 using geom_abline(). This function is named for the common equation of a line as a function of x:

$$b*x+a \tag{1.1}$$

where the parameters are named intercept (a) and slope (b). We show the result in Figure 1-9.

```
qplot(
  x = qnorm(
    p = ppoints(length(iris$Sepal.Length)),
    mean = mean(iris$Sepal.Length),
    sd = sd(iris$Sepal.Length)),
  y = sort(iris$Sepal.Length),
  xlab = "Theoretical Normal Quantiles",
  ylab = "Sepal Length") +
  geom_abline(slope = 1, intercept = 0)
```

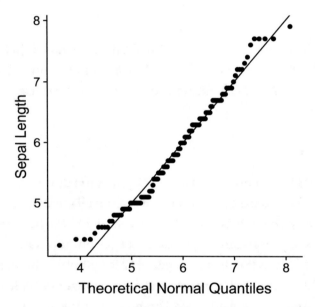

Figure 1-9. *Shows theoretical norms on the x axis (based on predictions from mean and standard deviation)*

In this case, we can see that the data are reasonably normally distributed as all points fall fairly closely to the normal line and are roughly symmetric around this line.

Although testing whether data are consistent with a normal distribution is common, real data may be closer to many other distributions. We can make a Q-Q plot using geom_qq() by specifying the quantile function for the expected distribution. For instance, going back to the lynx trapping data, Figure 1-10 evaluates the fit of the raw lynx data with a log-normal distribution (qlnorm()).

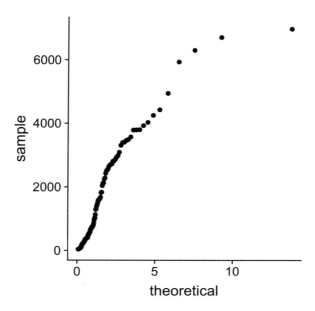

Figure 1-10. *Testing whether lynx data are consistent with a log-normal distribution*

When using less common distributions, sometimes ggplot2 does not know how to pick the parameters for the distribution by default. If required, we can pass parameters for the expected distribution as a named list to the dparams argument. The following example does this to test whether the lynx data are consistent with a Poisson distribution in Figure 1-11.

```
ggplot(data.table(lynx = as.vector(lynx)), aes(sample = lynx)) +
  geom_qq(distribution = qlnorm)

ggplot(data.table(lynx = as.vector(lynx)), aes(sample = lynx)) +
  geom_qq(distribution = qpois, dparams = list(lambda = mean(lynx)))
```

13

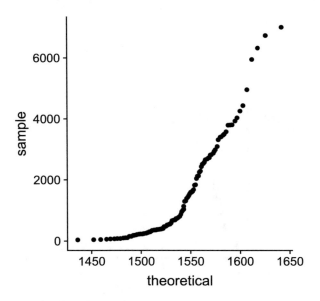

Figure 1-11. *Testing whether lynx data are consistent with a Poisson distribution*

Density Plots

Another way to examine whether the observed distribution appears consistent with an expected distribution is to plot the empirical density against the density for the expected distribution. To do this, we can use geom_density() to plot the empirical density and then the stat_function() function, which is a generic way to plot any function. If we plot the function dnorm(), it will plot a normal density. We need only specify that the mean and standard deviation of the normal distribution should be based on our data. The results are shown in Figure 1-12. Again the data appear to be close to a normal distribution, although not perfect.

```
ggplot(iris, aes(Sepal.Length)) +
  geom_density() +
  stat_function(fun = dnorm,
                args = list(
                  mean = mean(iris$Sepal.Length),
                  sd = sd(iris$Sepal.Length)),
                colour = "blue")
```

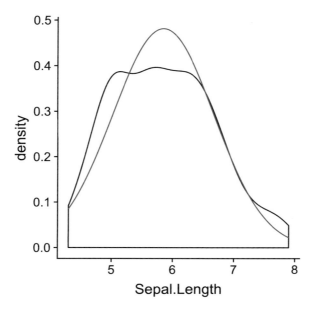

Figure 1-12. *A normal curve and our density plot (with default smoothness of 1)*

Although plotting the empirical and expected densities does not provide any information not captured by the Q-Q plot, it can be more intuitive to "see" the distributions on top of each other, rather than see the two distributions plotted against each other.

Fitting More Distributions

With either Q-Q plots or observed and expected density plots, we can evaluate many different distributions. However, for distributions outside of the normal distribution, it is often required to specify their parameters, to get the quantiles or densities. It is possible to calculate the parameters by hand and pass them to the appropriate quantile or density function, but using the `fitdistr()` function from the `MASS` package, we can fit many distributions and have R estimate the parameters by just specifying the name of the distribution. Currently, `fitdistr()` supports the following distributions:

- Beta
- Cauchy
- Chi-squared
- Exponential
- F

- Gamma

- Geometric

- Log-normal

- Logistic

- Negative binomial

- Normal

- Poisson

- t

- Weibull

Although this is far from an exhaustive list of statistical distributions, it is more than adequate for nearly all statistics used and covers all of the distributions used in the statistical analyses discussed in this book.

To see how to use `fitdistr()`, we make up some random data from a beta distribution. Beta distributions are useful for proportions, as the beta distribution is bounded by 0 and 1. We use `set.seed()` to make our example reproducible.

```
set.seed(1234)
y <- rbeta(150, 1, 4)
head(y)
```

```
## [1] 0.138 0.039 0.111 0.099 0.377 0.384
```

The `fitdistr()` function takes the data, a single variable character string indicating the distribution name, and the starting values for the parameters of the distribution as a list.

```
y.fit <- fitdistr(y, densfun = "beta",
                  start = list(shape1 = .5, shape2 = .5))
```

From `fitdistr()` we can get the estimated parameters of the distribution. We extract those explicitly.

```
y.fit
```

```
##     shape1      shape2
##     1.08        4.28
##    (0.11)      (0.52)
```

```
y.fit$estimate["shape1"]
```

```
## shape1
##     1.1
```

```
y.fit$estimate["shape2"]
```

```
## shape2
##     4.3
```

We can also extract the log likelihood—often abbreviated LL—which tells us about how likely the data are to come from that distribution with those parameters. A higher likelihood indicates a closer match between the tested distribution and the data. As a note, in addition to the log likelihood, or LL, it is also common to report $-2 * LL$ often written as simply –2LL. Finally, more complex models often (though not always) provide at least a slightly better fit to data. To account for this, you can evaluate the degrees of freedom used for a given likelihood. The logLik() function extracts both the log likelihood and the degrees of freedom.

```
logLik(y.fit)
```

```
## 'log_Lik.' 95 (df=2)
```

Although likelihood values are not easy to interpret individually, they are extremely useful for comparison. If we fit two distributions, the one that provides the higher (log) likelihood is a better fit for the data. We can use fitdistr() again to fit a normal distribution and then compare the LL from a beta distribution to the LL from a normal distribution.

```
y.fit2 <- fitdistr(y, densfun = "normal")
logLik(y.fit2)
```

```
## 'log_Lik.' 67 (df=2)
```

With the same degrees of freedom, the LL is higher for the beta (LL = 95.4) than the normal distribution (LL = 67.3). These results suggest we should pick the beta distribution for these data.

The JWileymisc package provides an automatic way to fit many distributions and automatically view a density or Q-Q plot via the testdistr() function. Very little R code is required to accomplish this for a normal distribution, with the results shown in Figure 1-13.

```
testdistr(y)
```

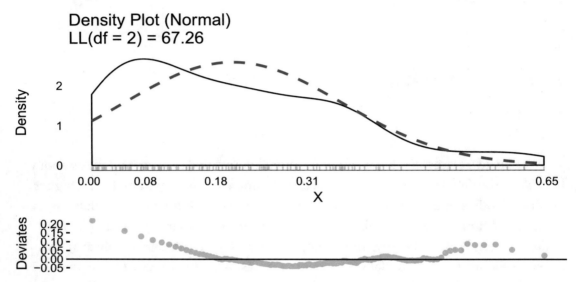

Figure 1-13. *Density plot with superimposed normal distributions and normal Q-Q plot*

To compare the fit of a normal vs. beta distribution, we can fit each and plot them the two graphs together as a panel of graphs using the plot_grid() function from the cowplot package. The results in Figure 1-14 show a good concordance of the data with a beta distribution (although hardly surprising given we generated the data from one!) and a poor fit with the normal. Note that warning messages from the density function are common and not necessarily cause for concern in this instance.

```
test.beta <- testdistr(y, "beta",
                       starts = list(shape1 = .5, shape2 = .5),
                       varlab = "Y", plot = FALSE)
```

```
## Warning in densfun(x, parm[1], parm[2], ...): NaNs produced
## Warning in densfun(x, parm[1], parm[2], ...): NaNs produced
## Warning in densfun(x, parm[1], parm[2], ...): NaNs produced
## Warning in densfun(x, parm[1], parm[2], ...): NaNs produced
```

18

```
test.normal <- testdistr(y, "normal", varlab = "Y", plot = FALSE)

plot_grid(
    test.beta$DensityPlot, test.beta$QQPlot,
    test.normal$DensityPlot, test.normal$QQPlot,
    ncol = 2)
```

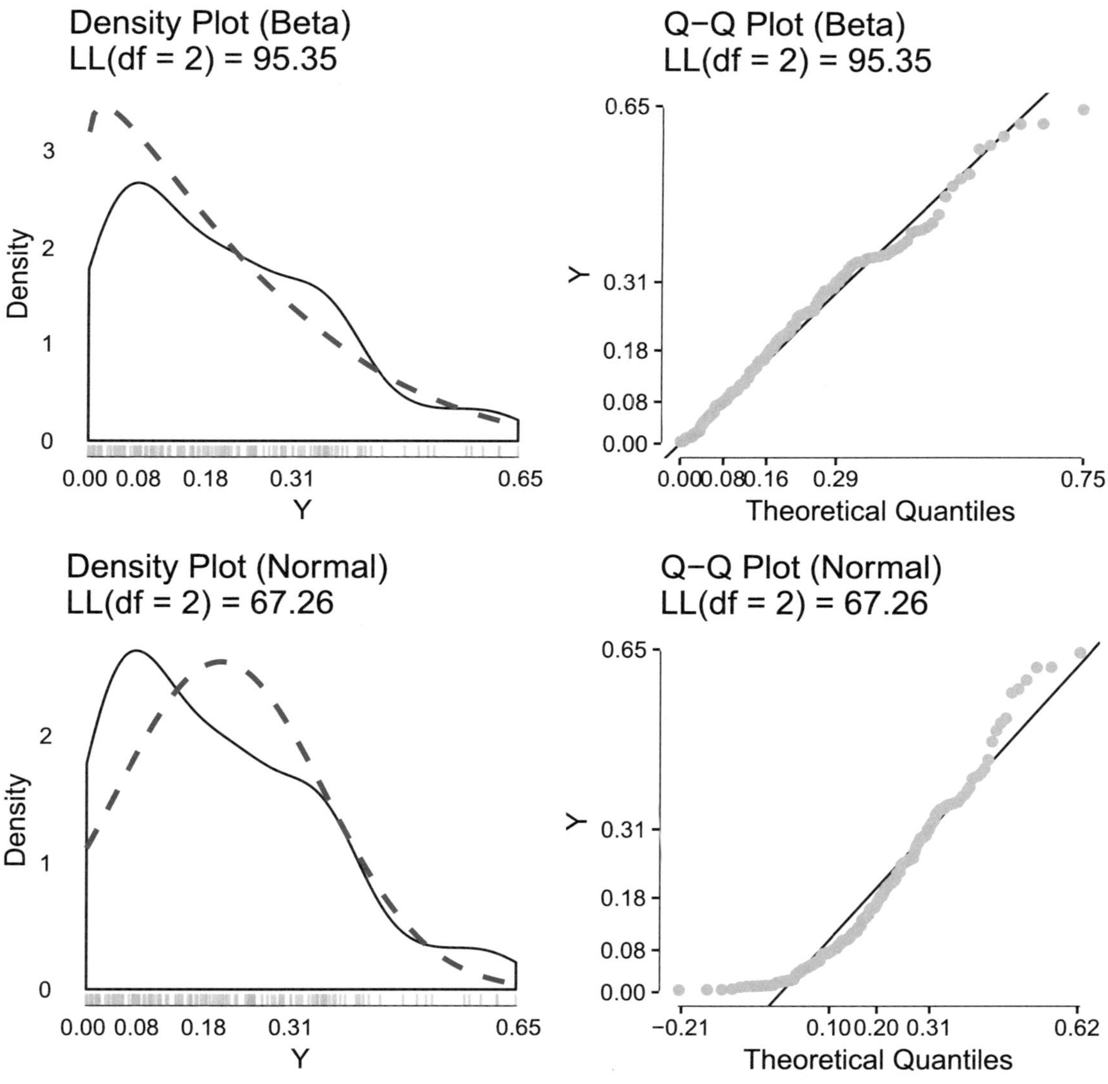

Figure 1-14. *Shows density plot with superimposed beta or normal distributions along with Q-Q plot fits*

For discrete distributions, such as for counts, `testdistr()` makes a slightly different type of plot, designed to better show the observed proportions vs. the probability mass function from the theoretical distribution. Specifically, the density values are the observed proportions and then the expected probabilities of each value for a given distribution.

To look at an example, first we simulate some data from a negative binomial distribution and then plot the results assuming a Poisson distribution in Figure 1-15 and the results assuming a negative binomial distribution in Figure 1-16. This comparison shows both in terms of the log likelihood and the deviates from expected values that the negative binomial is a better fit to the data than is the Poisson.

```
set.seed(1234)
y <- rnbinom(500, mu = 5, size = 2)
testdistr(y, "poisson")

testdistr(y, "nbinom")
```

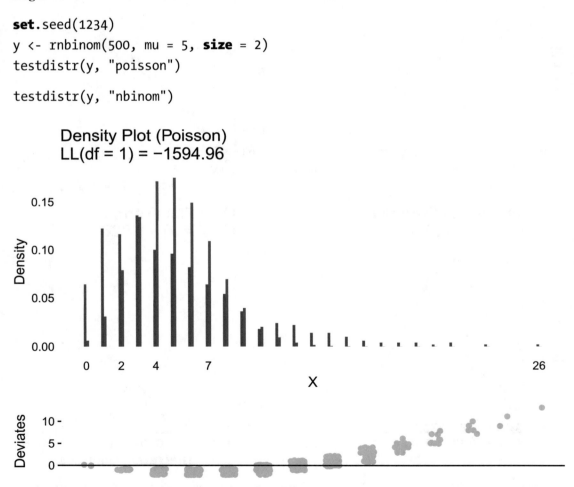

Figure 1-15. *Discrete observed proportions with the theoretical probabilities from a Poisson plotted in blue*

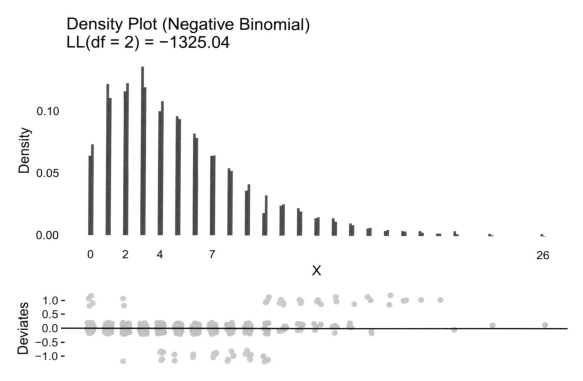

Figure 1-16. *Discrete observed proportions with the theoretical probabilities from a negative binomial distribution plotted in blue*

1.2 Anomalous Values

Anomalous values are values that differ from the rest of the values or are in some way not standard or atypical. Anomalous values are also often called outliers or extreme values. It is difficult to precisely define what qualifies as an anomalous value, but generally they are data points that appear in some way incongruent with the majority, typically in a relatively extreme way.

For data from a normal distribution, a common threshold for outliers is any values that are outside of z-scores of ±3. These thresholds are based off of the probability assuming a normal distribution. Specifically, if the data are normally distributed, about 0.10% of the data will fall below a z-score of −3 and about 0.10% will fall above a z-score of +3. The exact probabilities are as follows, using the pnorm() function.

```
pnorm(c(-3, 3))
```

```
## [1] 0.0013 0.9987
```

Because these thresholds are based off of a normal distribution, they do not necessarily apply meaningfully to data that are not normally distributed. Although many statistical analyses, such as linear regression, assume that the outcome is (conditionally) normally distributed, few distributional assumptions are made about predictors. Nevertheless, anomalous values on predictor variables, particularly if they are at the extremes, can strongly influence results.

It is often easier to visually identify anomalous values than it is to define them using quantitative criteria. For example, Figure 1-17 shows two graphs. Both graphs have three relatively anomalous points with values of 5. However, these points may appear somewhat more out of place in Panel A where all other data points form a more or less continuous group, and there is a relatively large gap to the anomalous points. Even though there is also a gap for the anomalous points in Panel B, because there are other gaps in the data points, it is not so strange to have a few data points that are separated from the rest—separation seems to be a pattern in Panel B, whereas it is not in Panel A.

```
set.seed(1234)
d <- data.table(
  y1 = rnorm(200, 0, 1),
  y2 = rnorm(200, 0, .2) + rep(c(-3, -1, 1, 3), each = 50))

plot_grid(
  qplot(c(d$y1, rep(5, 3)), geom = "dotplot", binwidth = .1),
  qplot(c(d$y2, rep(5, 3)), geom = "dotplot", binwidth = .1),
  ncol = 1, labels = c("A", "B"))
```

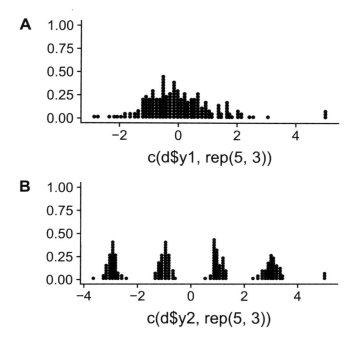

Figure 1-17. *Panel graph showing stacked dot plots with anomalous values*

   Defining anomalous values can also be difficult depending on the shape of the distribution. Figure 1-18 shows two distributions. Panel A is a gamma distribution and shows the characteristic long right tail of a gamma distribution. Even though just a few are fairly extreme, there is no clear "break" in the data, and it is the type of distribution where a continuous, long right tail is typical. Further, it is easy to see from the data that there is a pattern of reducing frequency but fairly extreme positive values. Conversely, the normal distribution in Panel B is more symmetrical and does not evidence such a long tail. One or two extreme positive values added to the graph in Panel B may indeed seem "anomalous."

```
set.seed(1234)
d2 <- data.table(
  y1 = rgamma(200, 1, .1),
  y2 = rnorm(200, 10, 10))

plot_grid(
  qplot(d2$y1, geom = "dotplot", binwidth = 1),
  qplot(d2$y2, geom = "dotplot", binwidth = 1),
  ncol = 1, labels = c("A", "B"))
```

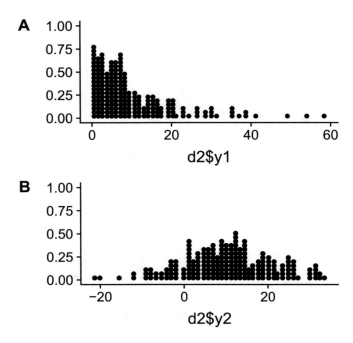

Figure 1-18. *Panel graph showing randomly generated (no added anomalous values) data from a gamma and normal distribution*

These different examples highlight the difficulty in exactly defining what is or is not an anomalous value. Although we cannot offer any single rule to follow, there are some additional tools built into the `testdistr()` function to assist in making these judgments. The `extremevalues` argument can be used to specify whether values that fall below and above a specified percentile of either the empirical data or the theoretical percentiles for the specified distribution should be highlighted. Figure 1-19 shows an example highlighting the bottom 1% and the top 1% of points based on their empirical position. The highlight occurs in the density plot in the rug—the lines underneath the density curve—by coloring them a solid black. In the Q-Q plot, the extreme points are colored solid black as well instead of gray. If no points are colored a solid black, that would indicate that no points fell outside the 1st and 99th empirical percentiles. When using empirical values, unless a large dataset is available, a value such as the top and bottom 1% may be reasonable to examine carefully. However, where possible often a more extreme threshold than the top and bottom 1% is desired, such as the top and bottom 0.10%, to ensure the values are truly very unlikely to occur due to chance alone.

```
testdistr(d$y1, extremevalues = "empirical",
          ev.perc = .01)
```

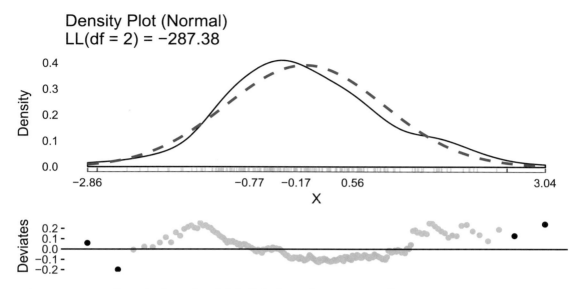

Figure 1-19. *Graph showing highlighting of extreme values*

In addition to defining extreme values empirically, they can be defined based on the specified theoretical distribution. That is, we can look at whether any values are beyond the bottom or top 0.10% for a normal distribution (Figure 1-20) or for the same data whether any points are extreme based on the percentiles from a gamma distribution (Figure 1-21). These graphs show that for the same data, some points may be considered anomalous or outliers if we expected the data to follow a normal distribution, but may not be anomalous if we expected the data to follow a gamma distribution. In practice, as data analysts we must make judgments regarding how extreme or apparently anomalous any data values are incorporating information about the shape or pattern of the distribution and how it will ultimately be analyzed.

```
testdistr(d2$y1, "normal", extremevalues = "theoretical",
          ev.perc = .001)

testdistr(d2$y1, "gamma", extremevalues = "theoretical",
          ev.perc = .001)
```

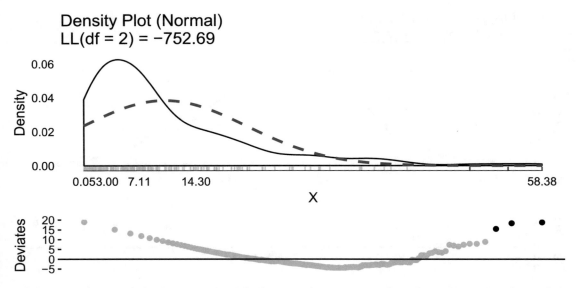

Figure 1-20. *Graph showing highlighting of extreme values based on theoretical percentiles from a normal distribution*

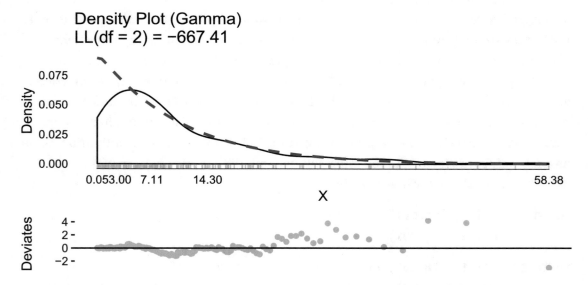

Figure 1-21. *Graph showing highlighting of extreme values based on theoretical percentiles from a gamma distribution*

It is possible to have several anomalous values. However, when using theoretical distributions, if one anomalous value is more extreme than another, the less extreme value may be "masked" because the parameter estimates are influenced by the more extreme value. An example of this is shown in Figure 1-22 where there are two anomalous values: 100 and 1,000. The value of 100 is masked by the value of 1,000 as the mean and variance of the theoretical normal distribution are pulled up so much by the value of 1,000 that the value of 100 is no longer anomalous.

```
testdistr(c(d2$y2, 100, 1000), "normal",
          extremevalues = "theoretical",
          ev.perc = .001)
```

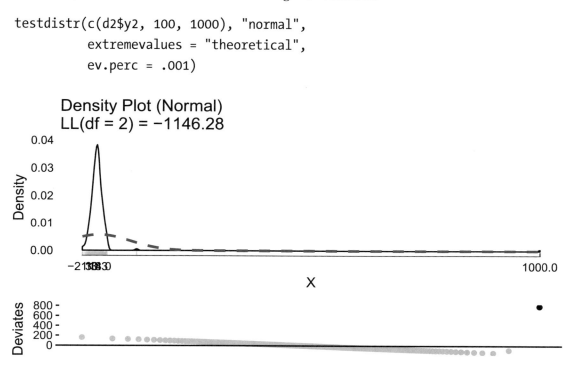

Figure 1-22. *Graph showing an anomalous value of 100 masked by the more extreme anomalous value of 1,000*

If there are multiple anomalous values, an iterative process can be used, addressing the most extreme values and then rechecking until no more anomalous values are apparent. However, the process can be simplified somewhat by using robust methods. When means and (co)variances are the parameters of interest, one such robust method is the minimum covariance determinant (MCD) estimator. Briefly, the MCD estimator finds the subset of the original cases that have the lowest determinant of their sample covariance matrix [82]. In the univariate case, this equates to a subset of the original data cases that have a lower variance. The testdistr() function has an optional robust argument that can be used with the normal distribution. When robust = TRUE, testdistr() uses the covMcd() function from the robustbase package [59], which uses a fast algorithm proposed by [83] to calculate the (approximate) MCD. Results using the robust estimator are shown in Figure 1-23. Using the robust estimator, both outliers are identified, even before removing the more extreme outlier.

```
testdistr(c(d2$y2, 100, 1000), "normal",
          robust = TRUE,
          extremevalues = "theoretical",
          ev.perc = .001)
```

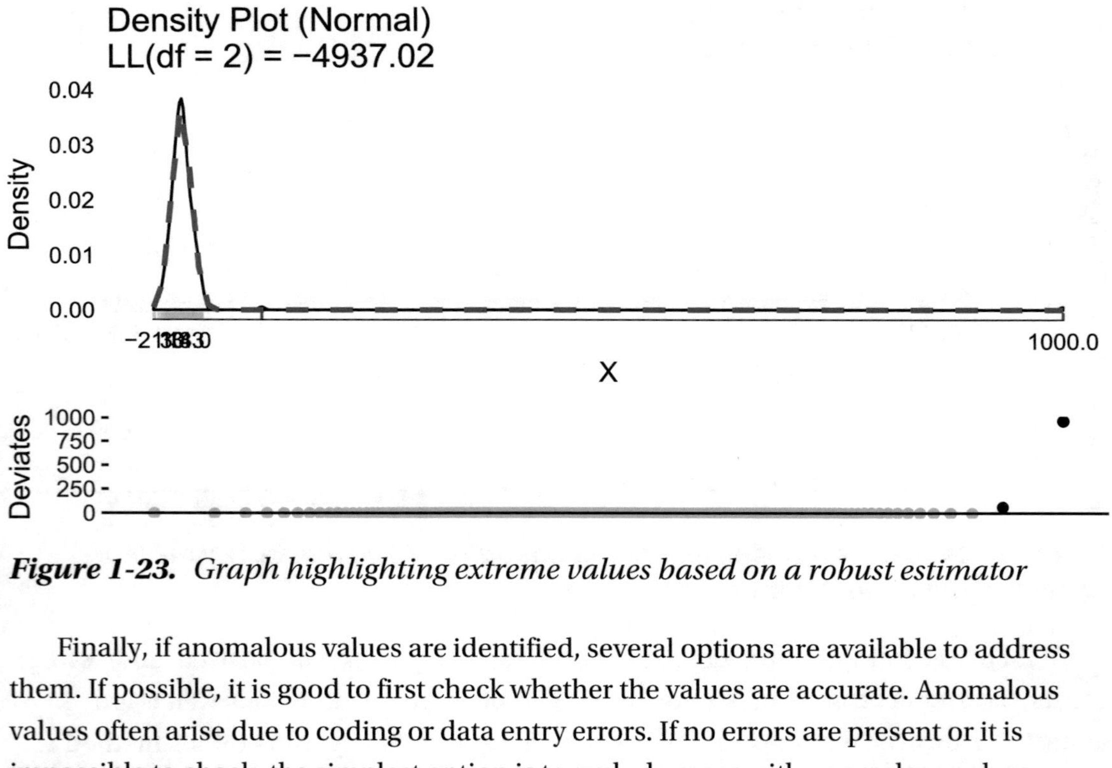

Figure 1-23. *Graph highlighting extreme values based on a robust estimator*

Finally, if anomalous values are identified, several options are available to address them. If possible, it is good to first check whether the values are accurate. Anomalous values often arise due to coding or data entry errors. If no errors are present or it is impossible to check, the simplest option is to exclude cases with anomalous values. Excluding or removing these cases may work particularly well when there are few cases with anomalous values and a large dataset with many cases remaining after excluding those anomalous ones. In smaller datasets where each case counts, excluding anomalous values may result in too much lost data. This can also occur in larger datasets where more cases are anomalous.

An alternative to excluding cases is winsorizing, named after Charles Winsor. Winsorizing takes anomalous values and replaces them with the nearest non-anomalous value [92, pp. 14-20]. One way to accomplish this automatically is to calculate the desired empirical quantiles and set any values outside of those values to the calculated percentiles. Even if anomalous values exist only on one tail of the distribution, the process is applied equally to both the lower and the upper tail. Adjusting both tails helps to ensure that the procedure itself does not tend to move the location of the distribution lower or higher. Winsorizing is easily accomplished in R using the `winsorizor()` function from the `JWileymisc` package. The only required argument is the proportion to winsorize at each tail. The other feature of the `winsorizor()` function is that in addition to returning the winsorized variable, it adds attributes noting the thresholds used for winsorizing and the percentile.

```
winsorizor(1:10, .1)

##   [1] 1.9 2.0 3.0 4.0 5.0 6.0 7.0 8.0 9.0 9.1
## attr(,"winsorizedValues")
##    low high percentile
## 1 1.9  9.1        0.1
```

Figure 1-24 compares the gamma distributed variable we saw earlier before (in Panel A) and after (in Panel B) winsorizing the lower and upper 1%. Panel B shows the characteristic "flattening" of winsorizing, and the raw values now only go to 49.2 instead of 58.4.

```
plot_grid(
  testdistr(d2$y1, "gamma", extremevalues = "theoretical",
            ev.perc = .005, plot=FALSE)$QQPlot,
  testdistr(winsorizor(d2$y1, .01), "gamma", extremevalues = "theoretical",
            ev.perc = .005, plot=FALSE)$QQPlot,
  ncol = 2, labels = c("A", "B"), align = "hv")
```

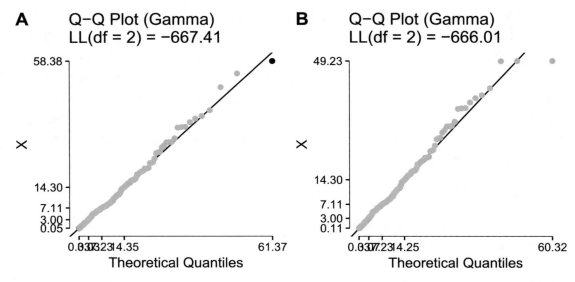

Figure 1-24. *Panel graph comparing data before (A) and after (B) winsorizing the (empirical) bottom and top 1%*

1.3 Summary

In this chapter, we learned various methods for using R to visualize data in raw or amalgamated formats. Additionally, beyond graphical exploratory data analysis, we learned some methods to quantify fits of our data with various distributions. For a summary of key functions used in this chapter, please see Table 1-1.

Table 1-1. *Listing of Key Functions Described in This Chapter and Summary of What They Do*

Function	What It Does
ggplot()	Creates new plots using the ggplot2 package
qplot()	A "quick" plot with simplier (and less nuanced) syntax
geom_dotplot()	Creates a dot plot geometric object—shows all raw data
geom_histogram()	Creates a histogram geometric object
geom_density()	Creates a density distribution—essentially a smoothed histogram

(continued)

Table 1-1. (*continued*)

Function	What It Does
geom_qq()	Q-Q plots graph observed data quantiles against theoretical quantiles
testdistr()	Automatically views a density or Q-Q plot
winsorizor()	Replaces anomalous values with the nearest non-anomalous value
plot_grid()	Places multiple plots into a defined grid
stat_function()	Draws a function on top of the current plot
fitdistr()	Takes data, distribution, and parameters of the distribution
logLik()	LL is how likely the data are to come from a fitdistr( )—bigger is better

CHAPTER 2

Multivariate Data Visualization

The previous chapter covered methods for univariate data visualization. This chapter continues that theme, but moves from visualizing single variables to visualizing multiple variables at a time. In addition to examining distributions and anomalous values as in the previous chapter, we also cover how to visualize the relations between variables. Visualizing relations between variables can help particularly for more traditional statistical models where the data analyst must specify the functional form (e.g., linear, quadratic, etc.). In later chapters we will also cover machine learning models that employ algorithms to learn the functional form in data without the analyst needing to specify it.

```
library(checkpoint)
checkpoint("2018-09-28", R.version = "3.5.1",
  project = book_directory,
  checkpointLocation = checkpoint_directory,
  scanForPackages = FALSE,
  scan.rnw.with.knitr = TRUE, use.knitr = TRUE)

library(knitr)
library(ggplot2)
library(cowplot)
library(MASS)
library(mvtnorm)
library(mgcv)
library(quantreg)
```

© Matt Wiley and Joshua F. Wiley 2019
M. Wiley and J. F. Wiley, *Advanced R Statistical Programming and Data Models*,
https://doi.org/10.1007/978-1-4842-2872-2_2

```
library(JWileymisc)
library(data.table)

options(width = 70, digits = 2)
```

2.1 Distribution

Although it is relatively easy to assess univariate distributions, multivariate distributions are more challenging. The joint distributions of multiple individual variables are made up of their individual distributions. Thus, in addition to all the possible univariate distributions, different combinations of univariate distributions are combined to form the joint distribution being studied. From a visualization perspective, it is also difficult to visualize more than a few dimensions of data in the two dimensions typically used for graphing data.

In this book, the only multivariate distribution we will cover is the multivariate normal (MVN) distribution. In practice, this is not overly restrictive because the majority of analyses focus on one outcome at a time and only require knowing the distribution of the outcome. Factor analyses and structural equation models, two common types of analyses that do model multiple outcomes simultaneously, often assume the data are multivariate normal. If one can assess whether data are multivariate normal, our experience is that covers most analyses that make multivariate distributional assumptions.

The normal distribution is determined by two parameters, the mean, μ, and standard deviation, σ, which control the location and scale of the distribution, respectively. A multivariate normal distribution of p dimensions is governed by two matrices, a px1 matrix of means, μ, and a pxp covariance matrix, Σ.

In the bivariate case, where $p = 2$, it is straightforward to graph a multivariate distribution. We use the rmvnorm() function from the mvtnorm() package to simulate some multivariate normal data. The empirical densities can then be plotted using the geom_density2d() function and ggplot2, shown in Figure 2-1.

```
mu <- c(0, 0)
sigma <- matrix(c(1, .5, .5, 1), 2)

set.seed(1234)
d <- as.data.table(rmvnorm(500, mean = mu, sigma = sigma))
setnames(d, names(d), c("x", "y"))
```

```
ggplot(d, aes(x, y)) +
  geom_point(colour = "grey60") +
  geom_density2d(size = 1, colour = "black") +
  theme_cowplot()
```

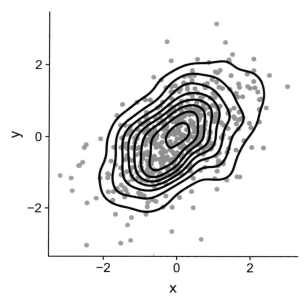

Figure 2-1. *2D empirical density plot for multivariate normal data*

Examining the empirical densities is helpful, but does not exactly compare them against what would be expected under a multivariate normal distribution. To contrast the obtained empirical densities with the expected densities for a multivariate normal distribution, we generate a grid of paired x and y values based on the range of the data, and then using the dmvnorm() function to calculate the density at each x, y pair. The parameters of the multivariate normal distribution, μ, and Σ, are calculated from the observed data. The result is shown in Figure 2-2, this time removing the raw data points for clarity.

```
testd <- as.data.table(expand.grid(
  x = seq(from = min(d$x), to = max(d$x), length.out = 50),
  y = seq(from = min(d$y), to = max(d$y), length.out = 50)))
testd[, Density := dmvnorm(cbind(x, y), mean = colMeans(d), sigma = cov(d))]
```

```
ggplot(d, aes(x, y)) +
  geom_contour(aes(x, y, z = Density), data = testd,
               colour = "blue", size = 1, linetype = 2) +
  geom_density2d(size = 1, colour = "black") +
  theme_cowplot()
```

In Figure 2-2, the empirical and normal densities are quite close, and we might conclude that assuming the data are multivariate normal is reasonable. Next we simulate two normally distributed variables that are not multivariate normal. Figure 2-3 shows the univariate density plots for each variable showing that they appear normally distributed.

```
set.seed(1234)
d2 <- data.table(x = rnorm(500))
d2[, y := ifelse(abs(x) > 1, x, -x)]

plot_grid(
  testdistr(d2$x, plot = FALSE)$Density,
  testdistr(d2$y, plot = FALSE, varlab = "Y")$Density,
  ncol = 2)
```

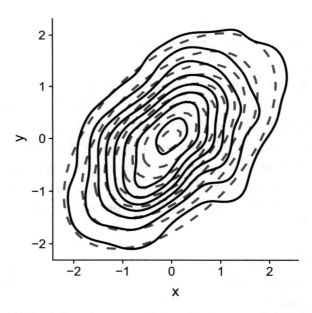

Figure 2-2. *2D empirical density vs. multivariate normal density plot*

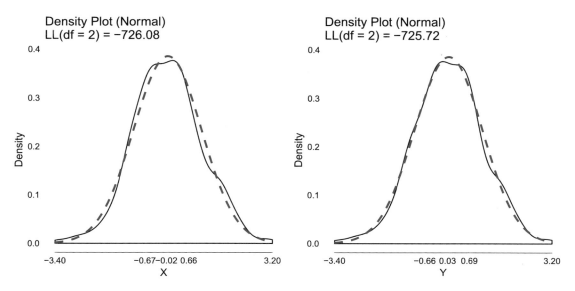

Figure 2-3. *Univariate density plots showing the simulated variables are univariate normal*

Although the variables are individually normal, this does not guarantee that they are multivariate normal. This is a crucial point and highlights the importance of assessing multivariate distributions, if using an analytic technique that makes assumptions about the multivariate distribution (e.g., confirmatory factor analysis). Using the same code we did before, we plot the empirical and multivariate normal densities in Figure 2-4.

```
testd2 <- as.data.table(expand.grid(
  x = seq(from = min(d2$x), to = max(d2$x), length.out = 50),
  y = seq(from = min(d2$y), to = max(d2$y), length.out = 50)))
testd2[, Density := dmvnorm(cbind(x, y), mean = colMeans(d2), sigma = cov(d2))]

ggplot(d2, aes(x, y)) +
  geom_contour(aes(x, y, z = Density), data = testd2,
               colour = "blue", size = 1, linetype = 2) +
  geom_density2d(size = 1, colour = "black") +
  theme_cowplot()
```

Figure 2-4 clearly shows that those two variables are not multivariate normal. Although this is an extreme case, it highlights how misleading univariate assessments may be.

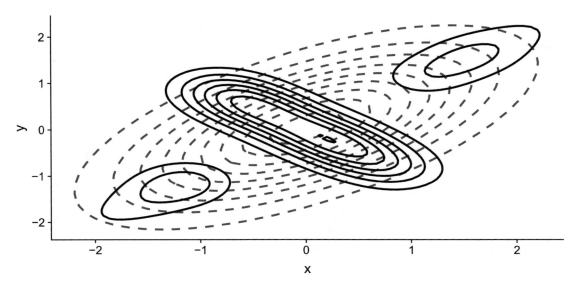

Figure 2-4. *2D density plot showing data that are not multivariate normal*

When $p > 2$, directly visualizing a multivariate normal distribution is difficult. Instead, we can take advantage of the work by Mahalanobis [60] who developed a way to calculate the distance of data from a "center" and the data and center can be multidimensional. The distance measure is named the Mahalanobis distance [60], and using it for each case in our data, we can calculate its distance from the (multivariate) center. Assuming that the data are multivariate normal, these distances will be distributed as a chi-square variable with p degrees of freedom.

The `testdistr()` function we saw earlier also includes an option to graph multivariate normal data, based on the Mahalanobis distances. The results for the bivariate normal data we simulated are shown in Figure 2-5 and for the data we simulated that were univariate normal but not multivariate normal are shown in Figure 2-6.

```
testdistr(d, "mvnorm", ncol = 2)

testdistr(d2, "mvnorm", ncol = 2)
```

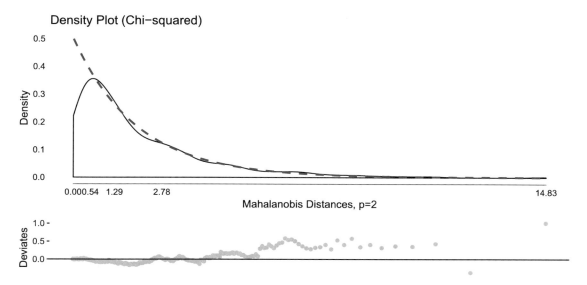

Figure 2-5. *Density plot superimposing multivariate normal distribution and Q-Q plot showing multivariate normal data*

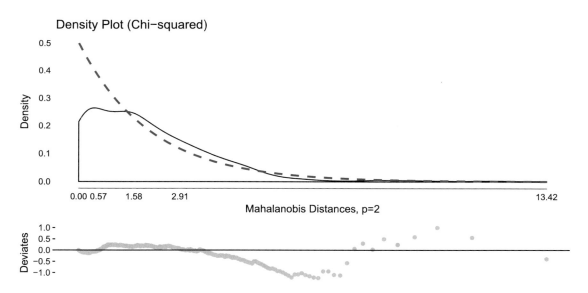

Figure 2-6. *Density plot superimposing multivariate normal distribution and Q-Q plot showing data that are not multivariate normal*

Even when the number of variables grows, the graphs are similar as long as the data are being tested for multivariate normality. For example, we can test the mtcars dataset, which contains 11 variables on different aspects of 32 cars. The results are shown in Figure 2-7 and suggest that the data are approximately multivariate normal, although deviations from the line on the Q-Q plot indicate some non-normality.

```
testdistr(mtcars, "mvnorm", ncol = 2)
```

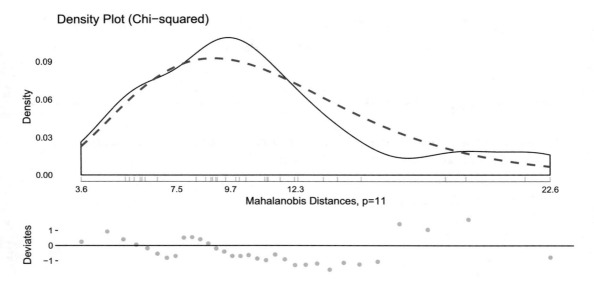

Figure 2-7. *Density plot superimposing multivariate normal distribution and Q-Q plot for mtcars data*

2.2 Anomalous Values

With multivariate data, values may be anomalous for any given variable or anomalous in the multivariate space. If a value is anomalous univariately, it will also be anomalous multivariately. However, just as we saw that univariately normal variables are not necessarily multivariately normal, likewise univariately anomalous values are multivariately anomalous, but univariately non-anomalous values are not guaranteed to be multivariately non-anomalous.

In the code that follows, we simulate strongly positively correlated multivariate normal data and add two anomalous values to V3. Examining V3 on its own (i.e., univariate), one anomalous value is clearly visible, and after removal, no further anomalous values are apparent (Figure 2-8).

```
mu <- c(0, 0, 0)
sigma <- matrix(.7, 3, 3)
diag(sigma) <- 1

set.seed(12345)
d <- as.data.table(rmvnorm(200, mean = mu, sigma = sigma))[order(V1)]
d[c(1, 200), V3 := c(2.2, 50)]

plot_grid(
  testdistr(d$V3, extremevalues = "theoretical", plot=FALSE)$Density,
  testdistr(d[V3 < 40]$V3, extremevalues = "theoretical",
  plot=FALSE)$Density, ncol = 2, labels = c("A", "B"))
```

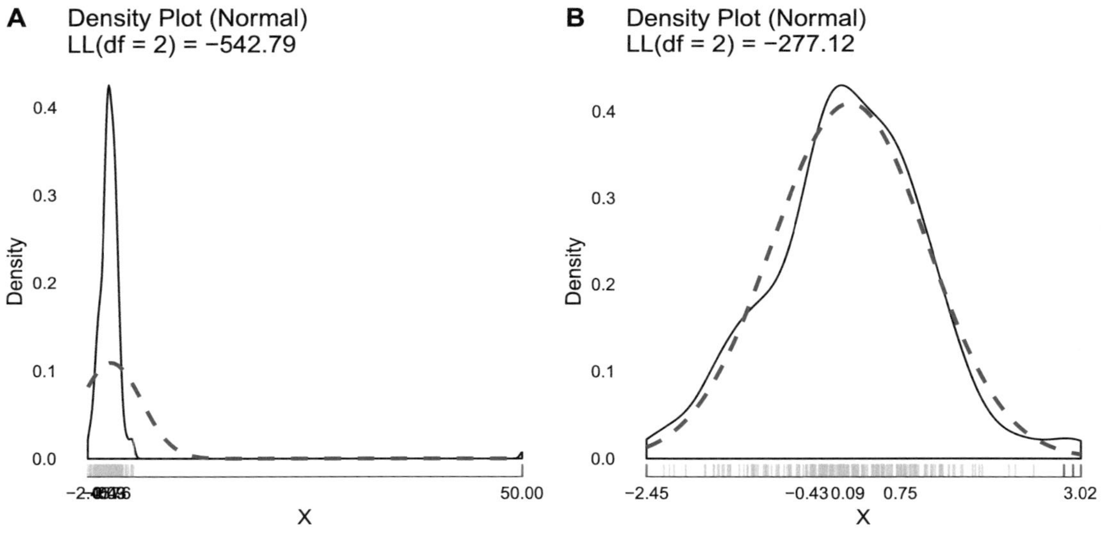

Figure 2-8. *Density plot superimposing normal distribution for data with an anomalous value (Panel A) and with the anomalous value removed (Panel B)*

Next we can again use the testdistr function to look for multivariately anomalous values. The results using all cases are shown in Figure 2-9 and clearly show one anomalous value, with no other points appearing anomalous. However, after excluding the one anomalous case identified, a second multivariately anomalous case emerges (Figure 2-10). This additional anomalous case is unmasked when the more extreme value is removed, resulting in a reduced mean and variance of the variable, V3.

```
testdistr(d, "mvnorm", ncol = 2, extremevalues = "theoretical")

testdistr(d[V3 < 40], "mvnorm", ncol = 2, extremevalues = "theoretical")
```

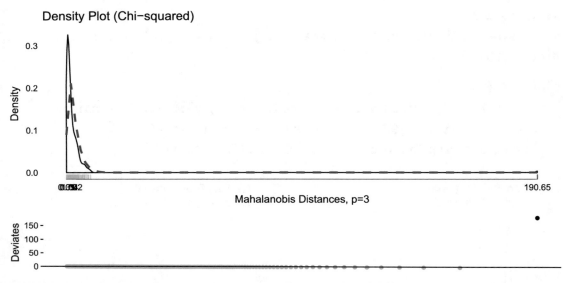

Figure 2-9. *Graph of multivariate normality and (multivariate) anomalous values*

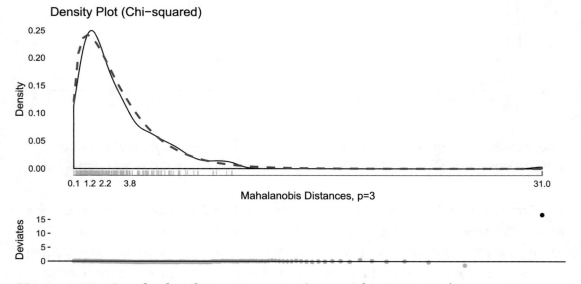

Figure 2-10. *Graph of multivariate normality and (multivariate) anomalous values with one extreme anomalous value removed*

Rather than iteratively removing anomalous cases, multiple anomalous cases can be identified directly by using a robust estimator of the parameters of a multivariate normal distribution: the means and covariance matrix. When the `robust` option is used in conjunction with the `mvnorm` distribution for `testdistr()`, a fast minimum covariance determinant (MCD) estimator [82, 83] from the `robustbase` package is used [59]. The results are shown in Figure 2-11. Using the multivariate normal distribution and robust estimator, both univariately and multivariately anomalous cases are identified in a single pass. After removing them, the data appear approximately multivariate normal (Figure 2-12).

```
testdistr(d, "mvnorm", ncol = 2, robust = TRUE, extremevalues = "theoretical")

testdistr(d[-c(1,200)], "mvnorm", ncol = 2, extremevalues = "theoretical")
```

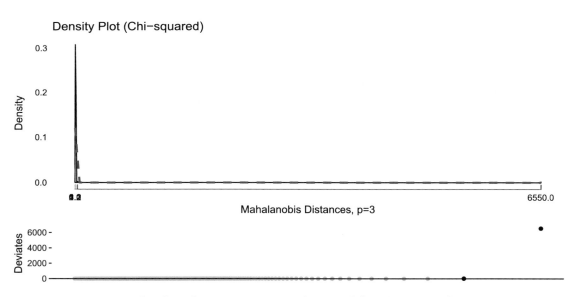

Figure 2-11. *Graph of multivariate normality and (multivariate) anomalous values using the robust estimator*

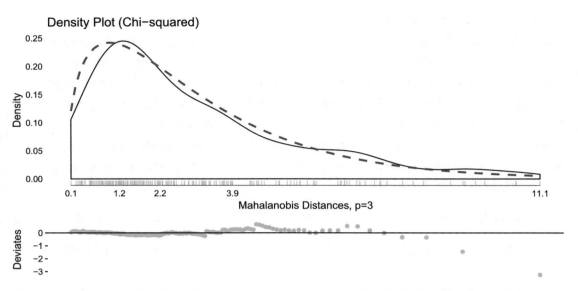

Figure 2-12. *Graph of multivariate normality and (multivariate) anomalous*
values with both anomalous values removed

2.3 Relations Between Variables

For continuous variable, most models assume some functional form, often a linear
relationship between variables. There are many ways to check this, but a quick approach is
to cut the predictor x into k discrete bins. If there are meaningful break points, those may
be used, but it is common to use quintiles, quartiles, or tertiles, depending on the quantity
of data. In large datasets, even fine-grained cuts, such as deciles, may make sense.

After cutting the predictor into discrete groups, boxplots or means can be plotted to
visualize the crude shape of the trend. In the following code, we simulate some data with
a quadratic relationship between the predictor x and outcome y. Using the %+% operator,
we can reuse the same plot and simply update the data to use data that has a different
set of cut points (quartiles, quintiles, deciles). The results are shown in Figure 2-13.
Although the trend is somewhat unclear in Panel A with tertile cuts, in Panel D with
deciles, in this small amount of data, the results become noisier.

```
set.seed(12345)
d2 <- data.table(x = rnorm(100))
d2[, y := rnorm(100, mean = 2 + x + 2 * x^2, sd = 3)]

p.cut3 <- ggplot(
  data = d2[, .(y,
```

```
      xcut = cut(x, quantile(x,
        probs = seq(0, 1, by = 1/3)), include.lowest = TRUE))],
    aes(xcut, y)) +
    geom_boxplot(width=.25) +
    theme(axis.text.x = element_text(
          angle = 45, hjust = 1, vjust = 1)) +
    xlab("")

p.cut4 <- p.cut3 %+% d2[, .(y,
      xcut = cut(x, quantile(x,
        probs = seq(0, 1, by = 1/4)), include.lowest = TRUE))]

p.cut5 <- p.cut3 %+% d2[, .(y,
      xcut = cut(x, quantile(x,
        probs = seq(0, 1, by = 1/5)), include.lowest = TRUE))]

p.cut10 <- p.cut3 %+% d2[, .(y,
      xcut = cut(x, quantile(x,
        probs = seq(0, 1, by= 1/10)), include.lowest = TRUE))]

plot_grid(
  p.cut3, p.cut4,
  p.cut5, p.cut10,
  ncol = 2,
  labels = c("A", "B", "C", "D"),
  align = "hv")
```

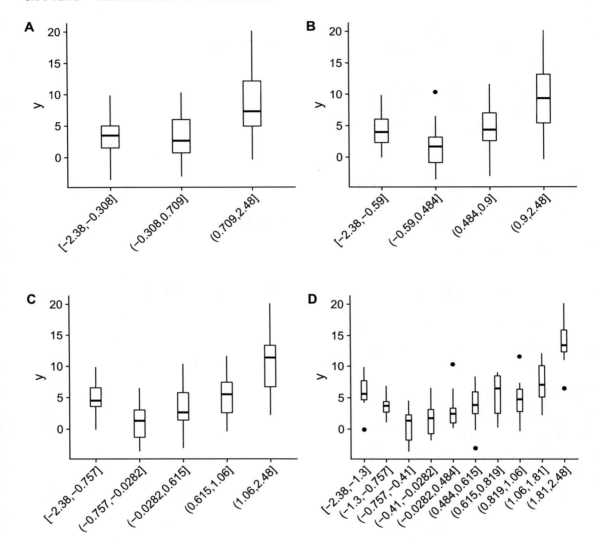

Figure 2-13. *Boxplots from cutting a continuous variable into quartiles showing a non-linear relationship*

Another way to determine the functional form of the relations between two variables is to use a locally weighted regression (loess) line [21]. The idea behind a loess line is similar to fitting a straight line of best fit, but weighted to nearby data points. The following code creates a scatter plot and overlays the loess line and a straight linear regression line (Figure 2-14). The loess line easily identifies the quadratic trend that is missed by the straight line.

```
ggplot(d2, aes(x, y)) +
  geom_point(colour="grey50") +
  stat_smooth(method = "loess", colour = "black") +
  stat_smooth(method = "lm", colour = "blue", linetype = 2)
```

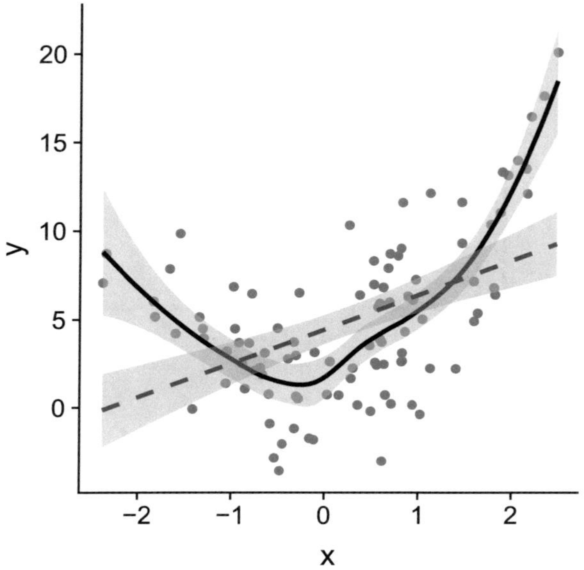

Figure 2-14. *Loess line of best fit showing a non-linear relationship*

Once we know the approximate functional form, we can attempt to approximate it using a parametric function. In the following code, we modify the linear model smooth to include a custom formula indicating that y should be regressed on both x and x^2. The results are much improved and show a close agreement between the loess line and quadratic line (Figure 2-15).

```
ggplot(d2, aes(x, y)) +
  geom_point(colour="grey50") +
  stat_smooth(method = "loess", colour = "black") +
  stat_smooth(method = "lm",
              formula = y ~ x + I(x^2),
              colour = "blue", linetype = 2)
```

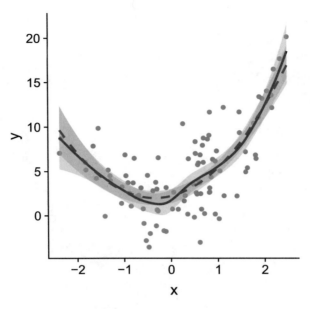

Figure 2-15. *Loess line and quadratic line*

Despite their strengths, loess lines are not infallible. One choice is the degree of smoothing. The smoothing is controlled by the span argument, which is passed to the loess() function that is responsible for estimating the line. In this next example, two loess lines are plotted, one with a low and one with a high span. The higher the span, the smoother the line. Figure 2-16 shows two loess lines with a span of 0.2 and 2.0.

```
ggplot(d2, aes(x, y)) +
  geom_point(colour="grey50") +
  stat_smooth(method = "loess", span = .2,
              colour = "black") +
  stat_smooth(method = "loess", span = 2,
              colour = "black", linetype = 2)
```

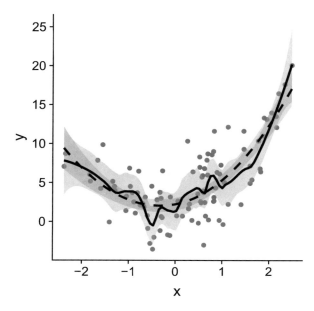

Figure 2-16. *Loess lines with varying degree of smoothing*

Next we will expand beyond two variables and examine ways that many variables can be visualized. The following code simulates an outcome that is a function of three predictor variables, two continuous and one categorical predictor. An initial bivariate only graph of the relation between x and y suggests a relatively weak relationship and perhaps a few outliers (Figure 2-17).

```
set.seed(1234)
d3 <- data.table(
  x = rnorm(500),
  w = rnorm(500),
  z = rbinom(500, 1, .4))
d3[, y := rnorm(500, mean = 3 +
      ifelse(x < 0 & w < 0, -2, 0) * x +
      ifelse(x < 0, 0, 2) * w * x^2 + 4 * z * w,
   sd = 1)]
d3[, z := factor(z)]

ggplot(d3, aes(x, y)) +
  geom_point(colour="grey50") +
  stat_smooth(method = "loess", colour = "black")
```

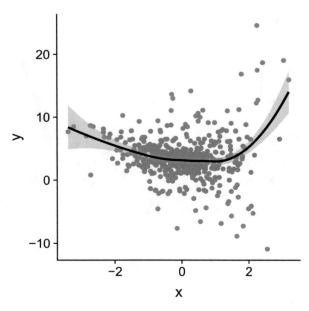

Figure 2-17. *Loess line of bivariate relationship from multivariate data*

Although data visualization is generally limited to two dimensions for fully continuous variables, we can add additional dimensions through shapes, colors, and panels of graphs. The following code again examines predictors of the outcome y, this time bringing all the variables into play. The binary z is used to color the points and lines, and we employ the previous trick to cut w, which is continuous, into quartiles. Although cutting w will not perfectly capture its continuous relationship, it can be enough to hint at what is going on and indicate whether interactions with w are necessary or not. The few relatively extreme values noted from Figure 2-17 are addressed by winsorizing the bottom and top 1% of the data. The final results are shown in Figure 2-18.

```
ggplot(d3, aes(x, winsorizor(y, .01), colour = z)) +
  geom_point() +
  stat_smooth(method = "loess") +
  scale_colour_manual(values = c("1" = "black", "0" = "grey40")) +
  facet_wrap(~ cut(w, quantile(w), include.lowest = TRUE))
```

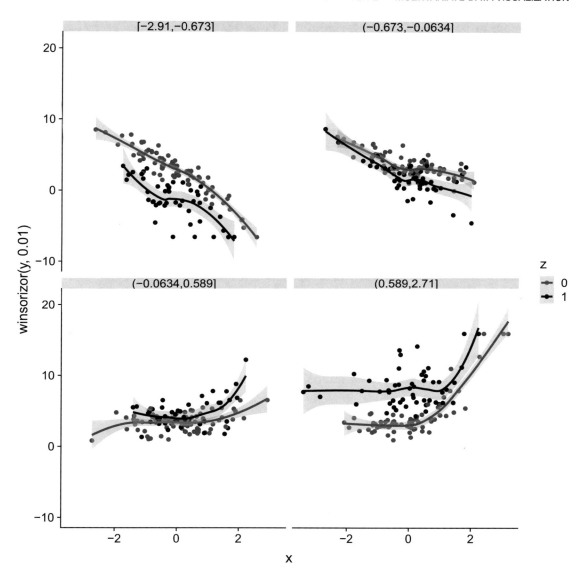

Figure 2-18. *Loess lines for multivariate data*

Figure 2-18 shows that there tends to be a negative, roughly linear, slope between y and x for x values below 0 when w is low, but no relationship when w is high. Depending on the level of w, z appears to shift values down or up.

In addition to loess, generalized additive models (GAMs) are another flexible approach to fitting a line [40]. We will discuss their statistical properties and use later. The advantage now is in fitting flexible, non-linear smooth terms including interactions

between continuous and categorical variables. Specifically we make use of the gam()
function from the mgcv package [122].

In loess, we specified the degree of smoothing using the span argument. In GAMs,
we specify how flexible the line is allowed to be using approximate degrees of freedom
via the k argument. For now, we are not focused on the statistical properties of GAMs
or formal statistical inference, but on using them as a tool to fit data and generate
predictions for graphing and visualizing the (smoothed) patterns in our data. The te()
function is used to allow non-linear interactions between x and w, and we also allow
these to be separate for different levels of z. The fitted model is stored in the object, m.
Next we generate a grid of values to use for prediction. As with densities, we pick equally
spaced points across x and w for all levels of z. After generating predictions, we can plot
the results shown in Figure 2-19. Note that Figure 2-19 looks best in color.

All the predicted y values will be the same for a single contour line in Figure 2-19.
Thus, by tracing one line, you can examine how predictors are related to the outcome.
For instance, when z is 0 (the left panel of Figure 2-19), for x values below 0, whether
w moves between -3 and 0 makes little difference to the predicted y values. In general,
any contour lines that are parallel to a particular dimension indicate little change in
prediction at that point along that dimension. The raster background also helps to show
the predicted y values.

```
m <- gam(winsorizor(y, .01) ~ z + te(x, w, k = 7, by = z), data = d3)

newdat <- expand.grid(
  x = seq(min(d3$x), max(d3$x), length.out = 100),
  w = seq(min(d3$w), max(d3$w), length.out = 100),
  z = factor(0:1, levels = levels(d3$z)))

newdat$yhat <- predict(m, newdata = newdat)

ggplot(newdat, aes(x = x, y = w, z = yhat)) +
  geom_raster(aes(fill = yhat)) +
  geom_contour(colour = "white", binwidth = 1, alpha = .5) +
  facet_wrap(~ z)
```

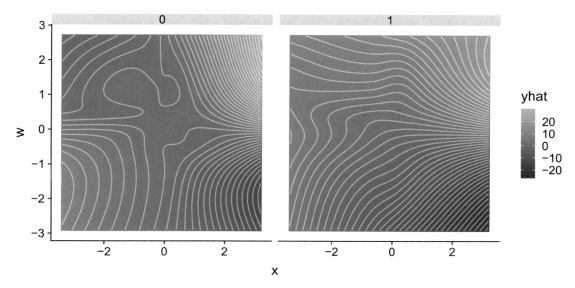

Figure 2-19. *Contour plots showing predicted y values for different combinations of x and w panelled by z*

Assessing Homogeneity of Variance

Homoskedasticity or homogeneity of variance is the idea that there is the same finite variance between groups or for continuous predictors, that the outcome has the same residual variance across levels of the predictor(s). For example, in one-way analysis of variance, the variance of the outcome should be approximately equal for each level of the explanatory variable (often called independent variable).

To assess homogeneity of variance, we make a data table for the iris data and make a first check by simply calculating the variance by species.

```
diris <- as.data.table(iris)
diris[, .(V = var(Sepal.Length)), by = Species]
```

```
##       Species    V
## 1:     setosa 0.12
## 2: versicolor 0.27
## 3:  virginica 0.40
```

It can also be helpful to visualize the data and the distributions. Boxplots or box and whisker diagrams can be a useful way to do this as the "box" part covers the interquartile range, the 25th to 75th percentiles. This is a robust measure of the spread of a distribution. We examine the *spread* of the boxes to get a sense of whether the variability is about equal across species. Boxplots are also more informative than merely computing the variance by group as they can show if there are any anomalous values in the data.

Even if the spread is comparable across species, the medians or locations may be different, which can make it harder to compare the spread. If you do not want to examine the locations and wish to focus solely on the spread, it can be useful to median center prior to graphing. Boxplots without (Panel A) and with (Panel B) median centering are shown in Figure 2-20.

```
plot_grid(
  ggplot(diris, aes(Species, Sepal.Length)) +
    geom_boxplot() +
    xlab(""),
  ggplot(diris[, .(Sepal.Length = Sepal.Length -
                             median(Sepal.Length)), by = Species],
         aes(Species, Sepal.Length)) +
    geom_boxplot() +
    xlab(""),
  ncol = 2, labels = c("A", "B"), align = "hv")
```

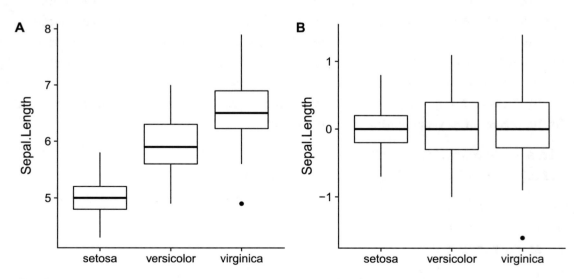

Figure 2-20. *Box and whisker diagrams of sepal length by species. Outliers appear as dots*

We have already seen density plots and used these to assess the distribution of a variable; if density plots are reflected to form a mirror image, it is called a violin plot. We can make it using geom_violin(); it provides similar information as the boxplots. However, violin plots provide more information on the distributions as the boxplot only shows the median (50th percentile), 25th and 75th percentiles, and the range of the data (or somewhat less than the full range if there are outliers). However, as it can be useful to see the median and interquartile range as well, it can be helpful to overlay boxplots with a narrow width onto the violin plots. From the violin plot, we see a steady increase in spread in Figure 2-21 for species with a higher mean (location) of the distribution.

```
ggplot(diris, aes(Species, Sepal.Length)) +
  geom_violin() +
  geom_boxplot(width = .1) +
  xlab("")
```

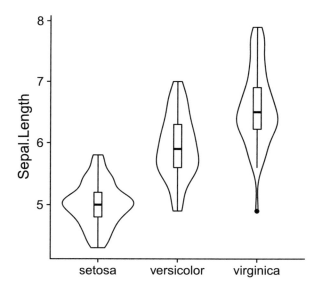

Figure 2-21. *Violin plots with box and whisker diagrams in center*

As we did when examining the relations between variables, violin plots and boxplots can be expanded to handle multiple variables using colors and panels. The following code uses the data we looked at for relationships between variables and cuts all the continuous predictors to facilitate examining the distribution and spread of the outcomes across levels of predictors. The results are in Figure 2-22. Note that we purposefully restricted the range of the y axis to make it easier to see the graphs and place less emphasis on extreme values.

```
## create cuts
d3[, xquartile := cut(x, quantile(x), include.lowest = TRUE)]
d3[, wquartile := cut(w, quantile(w), include.lowest = TRUE)]
d3[, yclean := winsorizor(y, .01)]

## median center y by group to facilitate comparison
d3[, yclean := yclean - median(yclean),
   by = .(xquartile, wquartile, z)]

p <- position_dodge(.5)

ggplot(d3, aes(xquartile, yclean, colour = z)) +
  geom_violin(position = p) +
  geom_boxplot(position = p, width = .1) +
  scale_colour_manual(values = c("1" = "black", "0" = "grey40")) +
  facet_wrap(~ wquartile) +
  theme(axis.text.x = element_text(angle = 45, hjust=1, vjust=1)) +
  coord_cartesian(ylim = c(-5, 5), expand = FALSE)
```

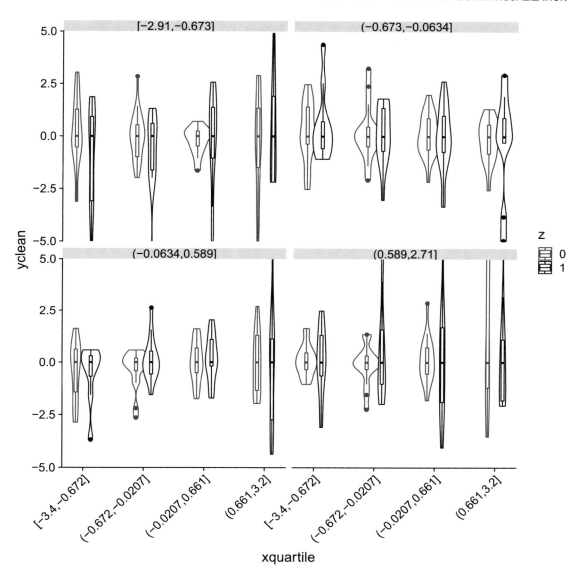

Figure 2-22. *Violin plots with box and whisker diagrams in center by quartiles of x, colored by z*

So far, we have looked at assessing homogeneity of variance for a continuous outcome and discrete explanatory variables or cutting continuous explanatory variables into discrete categories. Now we turn to how this can be accomplished visually for continuous variables. A first start would be to create a scatter plot; however, to see the "spread" we need some continuous estimate of the variance or interquartile range or the like. This can be accomplished using quantile regression. We will not discuss the process at length here, but suffice to say that quantile regression can estimate the quantiles of a distribution as a function of one or more explanatory variables. We will first simulate some homogenous and heterogenous data.

```
set.seed(1234)
d4 <- data.table(x = runif(500, 0, 5))
d4[, y1 := rnorm(500, mean = 2 + x, sd = 1)]
d4[, y2 := rnorm(500, mean = 2 + x, sd = .25 + x)]
```

Panel A is an example of homogenous variability and Panel B is an example of heterogenous variability. We show the visual contrast of these two datasets in Figure 2-23. Quantile regression [51, 50, 107].

```
plot_grid(
  ggplot(d4, aes(x, y1)) +
    geom_point(colour = "grey70") +
    geom_quantile(quantiles = .5, colour = 'black') +
    geom_quantile(quantiles = c(.25, .75),
                  colour = 'blue', linetype = 2) +
    geom_quantile(quantiles = c(.05, .95),
                  colour = 'black', linetype = 3),
  ggplot(d4, aes(x, y2)) +
    geom_point(colour = "grey70") +
    geom_quantile(quantiles = .5, colour = 'black') +
    geom_quantile(quantiles = c(.25, .75),
                  colour = 'blue', linetype = 2) +
    geom_quantile(quantiles = c(.05, .95),
                  colour = 'black', linetype = 3),
  ncol = 2, labels = c("A", "B"))

## Smoothing formula not specified. Using: y ~ x
## Smoothing formula not specified. Using: y ~ x
## Smoothing formula not specified. Using: y ~ x
## Smoothing formula not specified. Using: y ~ x
## Smoothing formula not specified. Using: y ~ x
## Smoothing formula not specified. Using: y ~ x
```

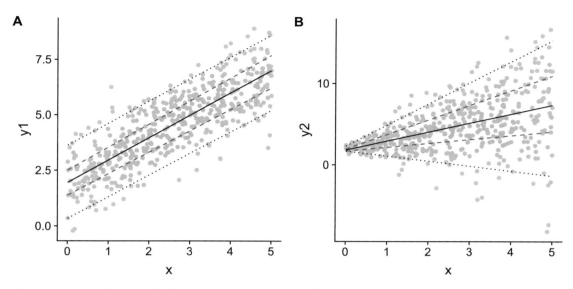

Figure 2-23. *Homoskedasticity vs. heteroskedasticity*

2.4 Summary

This chapter demonstrated techniques for visualizing multivariate data to understand how it compares to the multivariate normal distribution (Table 2-1).

Table 2-1. *Listing of Key Functions Described in This Chapter and Summary of What They Do*

Function	What It Does
geom_quantile()	Fits line(s) at given quantiles (defaults to quartiles)
geom_density2d()	Creates a 2D density contour plot for multivariate normal data points
geom_contour()	Creates a 2D contour plot
geom_violin()	Mirrors the density plot
stat_smooth()	Fits a curve to the data points

CHAPTER 3

GLM 1

Generalized linear models (GLMs) are a broad class of models that encompass both regression analyses and analysis of variance (ANOVA) are other terms or analyses that are often used to refer to GLMs. This chapter uses a number of packages shown as follows. We run our setup code to load these and to make data tables print in a neat fashion.

```
library(checkpoint)
checkpoint("2018-09-28", R.version = "3.5.1",
  project = book_directory,
  checkpointLocation = checkpoint_directory,
  scanForPackages = FALSE,
  scan.rnw.with.knitr = TRUE, use.knitr = TRUE)

library(knitr)
library(data.table)
library(ggplot2)
library(visreg)
library(ez)
library(emmeans)
library(rms)
library(ipw)
library(JWileymisc)
library(RcppEigen)
library(texreg)

options(
  width = 70,
  stringsAsFactors = FALSE,
  datatable.print.nrows = 20,
  datatable.print.topn = 3,
  digits = 2)
```

M. Wiley and J. F. Wiley, *Advanced R Statistical Programming and Data Models*,
https://doi.org/10.1007/978-1-4842-2872-2_3

3.1 Conceptual Background

Generalized linear models provide a general framework and notation that accommodate many specific types of models and analyses. This chapter covers several specific types of GLMs. If you are not familiar with some of the more basic types of GLMs, such as analysis of variance (ANOVA) or linear regression, and you wish to understand the background and conceptual framework for GLMs better, it may be worthwhile to read up on ANOVA and linear regression. An easy to follow and free resource is the Online Statistics Education: A Multimedia Course of Study (`http://onlinestatbook.com/`, Project Leader: David M. Lane, Rice University), which has a section on ANOVA and Regression. For a comprehensive mathematical statistics background on GLMs, we recommend the classic book by McCullagh and Nelder [61], the full reference for which is in the References section.

First, we introduce some commonly used functions in statistics and for GLMs including the following:

- $E(x)$ indicates the expectation of a variable, or its (possibly conditional) mean.

- $Var(x)$ indicates the variance of a variable, or its dispersion.

- $g(x)$ indicates the link function, which takes the raw outcome variable and transforms it to the linear scale that is predicted by the GLM.

- $g^{-1}(x)$ indicates the inverse link function, which takes the linear scale predicted by the GLM and transforms it back to the original scale.

- $exp(x)$ indicates the exponential function.

- $ln(x)$ indicates the natural logarithm function.

GLMs are structured such that there is one outcome or dependent variable, which we call y. One or more predictors or explanatory variables are stored in a matrix with n rows (the number of data points) and k columns (the number of predictor or explanatory variables), which we call X. The regression coefficients, the parameters to be estimated, form a length k (number of predictors) vector called β. The expected value of the outcome is $E(y) = \mu$. The bold Greek lower case μ is shorter way of writing the expected values of the outcome. These will always be on the original or raw data scale. The expected values on the linear scale are another vector called η.

With those conventions in place, we can define the building blocks of a generalized linear model as in the seminal article by Nelder and Wedderburn [71]. We use slightly different notation as Nelder and Wedderburn [71] reflecting more common recent practice in applied statistics. Each GLM has an outcome or dependent variable, *y*. After conditioning on the linear predictor, it is assumed the outcome follows a probability distribution from the exponential family. Further, each GLM has a set of *k* predictor variables, $x_1, \ldots, x_k$, referred to overall as X, and an expected linear outcome:

$$\eta = \mathbf{X}\beta \tag{3.1}$$

Finally, each GLM has a linking function so that

$$\eta = g(\mu) = g\big(E(y)\big) \tag{3.2}$$

and likewise an inverse link function so that

$$E(y) = \mu = g^{-1}(\eta) \tag{3.3}$$

Because the parameters, *β*, are always estimated on the linear predictor scale, *η*, estimation of GLMs is similar regardless of the distribution of the outcome. What changes is the distribution assumed for the outcome and the link and inverse link functions. We cover the distributions and link function (as well as how to analyze them) for three common types of outcome variables: continuous, binary outcomes, and count outcomes. Table 3-1 shows the mostly commonly used (canonical) link functions we cover in this chapter. Note that the distributions listed are not all possible distributions for a given outcome type. For example, for there are many other distributions for continuous outcomes depending on their shape or their bounds (e.g., the Beta distribution for data that are continuous but bounded by 0 and 1).

Table 3-1. *Outcome Types, Distributions, and Corresponding Link Functions*

Outcome Type	Distribution(s)	Link Function	Inverse Link Function
Continuous (real numbers)	Normal (Gaussian)	$\eta = g(\mu) = \mu$	$\mu = g^{-1}(\eta) = \eta$
Binary (0/1)	Bernoulli, multinomial	$\eta = g(\mu) = ln\left(\dfrac{\mu}{1-\mu}\right)$	$\mu = g^{-1}(\eta = \dfrac{1}{1+exp(-\eta)}$
Count (positive integers)	Poisson, negative binomial	$\eta = g(\mu) = \ ln\,(\mu)$	$\mu = g^{-1}(\eta) = \ exp\,(\eta)$

Based on the outcome distribution, it is possible to write a likelihood function. We do not cover the details of likelihood functions in this book as it is not necessary to know what the likelihood function is for applied data analysis. However, the basic idea is that the likelihood function (which depends on the distribution chosen) quantifies how likely the data are to arise from a distribution with a given set of parameters; thus, the generic likelihood function is written as $L(y, \theta)$. Note that here we referred to the parameters as θ because for many (but not all) distributions, additional parameters beyond the regression coefficients must be estimated; specifically, it is often required to estimate a dispersion parameter. For instance, the parameters of a normal distribution are the mean and variance or standard deviation. Another note is that for pragmatic reasons, likelihoods are often reported as log likelihoods. Further details are described elsewhere [71, 61].

Likelihood functions serve several useful purposes. First, they are the basis for maximum likelihood estimation. That is, the parameters of the GLM (and many other models as well) are estimated to maximize the likelihood function for the data. In addition, the overall likelihood from a given model can be compared to the likelihood from another model as a relative comparison which model provides better fit to the data. Next we examine how to apply this general model to specific cases.

3.2 Categorical Predictors and Dummy Coding

Two-Level Categorical Predictors

It is common to examine categorical predictors in GLMs. For instance, one may wish
to test whether there are sex differences in an outcome. However, before doing this,
a system is needed to convert "Women" and "Men" into numeric values. The most
common system for encoding discrete categories into numbers is called dummy coding.
Dummy coding involves coding a series of binary 0/1 variables that represent one
specific category. For sex, we could code two dummy variables, one to represent women
and one to represent men. This is shown in Table 3-2.

Table 3-2. *Example of Dummy Coding Sex*

Sex	D1	D2
Woman	1	0
Man	0	1
Man	0	1
Woman	1	0
Woman	1	0
Man	0	1

In this case, the dummy codes D1 and D2 are perfectly negatively correlated
indicating they capture the same information but reversed. Therefore, we would only
enter one of the two dummy codes into an actual GLM. The general principle is that
for a k-level variable, you can generate k dummy code variables, but only include $k-1$
of the dummy code variables. The omitted variable becomes the reference group. To
understand why this is the case, we can examine a simple design matrix, X, (Table 3-3)
that would correspond to a model testing only sex differences. As is standard for GLMs,
we have a constant column that is the intercept, the expected value of the outcome
when all other predictors are equal to zero. We also have one of our two dummy-coded
variables, D2. In this case, D2 is 1 when participants are men and 0 when they are
women. The corresponding coefficient vector would have two elements, one for the
intercept and one for the dummy code D2.

Table 3-3. *Design Matrix for Sex Differences with an Intercept and One Dummy Variable*

Intercept	D2
1	0
1	1
1	1
1	0
1	0
1	1

The intercept coefficient will be the expected value for women, because all other variables (in this case, just D2) are zero only when participants are women. The D2 coefficient will be the expected change in the outcome for a one unit change in D2. Because we coded D2 using 0/1, a one unit change in D2 is exactly shifting from women (0) to men (1). Thus the coefficient for D2 can be more simply thought of as the expected difference between women and men.

Three- or More Level Categorical Predictors

Dummy coding categorical variables with three or more levels work similarly to the basic two-level approach. Again, we create a set of 0/1 variables encoding each specific level of the variable. Table 3-4 shows an example with a three-level variable of exercise type.

Table 3-4. *Example of Dummy Coding Sex*

Exercise	D1	D2	D3
run	1	0	0
swim	0	1	0
bike	0	0	1
run	1	0	0

(continued)

Table 3-4. (*continued*)

Exercise	D1	D2	D3
bike	0	0	1
bike	0	0	1
run	1	0	0
swim	0	1	0
swim	0	1	0

For analysis, we would include any $k-1 = 3-1 = 2$ of the dummy code variables in the GLM, and again the left out group would be the reference group. The group comparisons, however, become slightly more complicated. Suppose we left out D1, the dummy code for running. The coefficients for the swim and bike dummy codes would capture the expected difference between running and swimming and between running and biking. However, none of the coefficients would directly test the difference between swimming and biking. The general principle is that for a k–level variable, there are $\binom{k}{2}$ possible pairwise comparisons; for a three-level variable, three pairwise comparisons; for a four-level variable, six pairwise comparisons. To test all the pairwise combinations, there are several choices. One choice, albeit not efficient, is to run the GLM as many times as there are dummy code variables, omitting just one dummy code as each iteration. Another approach is to use the coefficient matrix and the parameter covariance matrix estimated from the model to test specific contrasts. In our exercise example, testing whether the coefficient for swim vs. run is different than the coefficient for bike vs. run will provide a test of whether swim is different than bike.

It is also important to consider how to test the overall effect of the variable. For a single continuous variable or a two-level categorical variable, a single coefficient captures the entire effect of the variable. However, with three or more level categorical variables where multiple dummy code variables are entered, the overall test of the significance of that variable is not provided by the test of any specific dummy code variable coefficient. Instead, we need an omnibus test. For GLMs with a normally distributed outcome, this may be an omnibus F test. For other GLMs where degrees of freedom cannot be calculated, the standard tests are a Wald test based on testing that all the dummy variable coefficients are jointly zero, or a likelihood ratio test, which requires refitting the GLM without the variable in question and testing how much the final log likelihood changes, which will be distributed as a χ^2 with degrees of freedom equal to the number of dummy variables excluded.

3.3 Interactions and Moderated Effects

Interactions occur when the relationship between two or more explanatory variables and the outcome depend on the value of each other. For example, Wiley and colleagues [115] found an interaction between how controllable a stressor was and the type of coping response used for how positive people felt. For a stressor over which people had little control, whether people attempted to actively think about how to improve the problem or avoided thinking about it made no difference to their level of positive emotions. However, for controllable stressors, those who avoided thinking about how to fix the problem had lower levels of positive emotions and those who actively thought about how to address it had higher levels of positive emotions.

Interactions between two variables are called two-way interactions. Interactions between three variables are called three-way interactions, etc. Interactions can be tested in GLMs by adding the individual variables to the model on their own as well as adding the product term of the variables for which an interaction is desired. This can be accomplished by creating new variables in the original dataset. For instance, for x_1 and x_2, the "new" interaction variable would be created as $int = x_1 \cdot x_2$. R and other statistical packages also offer the ability to specify that the interaction of two (or more) variables should be tested in the model, in which case the product terms are created automatically without creating additional variables in the dataset. In either case, the end result is the same, with an example design matrix shown in Table 3-5.

For interactions with more than two variables, there are several additional terms. If there are three explanatory variables (x_1, x_2, x_3), there are three two-way interactions ($x_1 \cdot x_2$, $x_1 \cdot x_3$, $x_2 \cdot x_3$) and one three-way interaction ($x_1 \cdot x_2 \cdot x_3$) that can be considered in addition to the three individual variables. Even if the main focus is on the three-way interaction, it is standard practice to include all lower-order terms. So for a three-way interaction, typically all possible two-way interactions and all individual variables would also be included in the analysis.

Because interactions involve more than one variable, it is always possible to interpret them in more than one way. For instance, if there is an interaction between two variables, a and b, the relations between each variable and the outcome, y, depend on the level of the other variable. Thus the relationship between a and y depends on the level of *b*, and also the relationship between *b* and *y* depends on *a*. This fact alters the interpretation of the regression coefficients for the individual variables. Looking at Table 3-5, the coefficient for x1 would be interpreted as the expected change in y for a one unit change in x1 when x2 = 0. Likewise, the coefficient for x2 would be interpreted as the expected change in y for a one unit change in x2 when x1 = 0. Finally, the coefficient for x1x2 would be interpreted as either (1) the expected change in the coefficient for x1 for a one unit change in x2 or (2) the expected change in the coefficient for x2 for a one unit change in x1.

Table 3-5. *Example Design Matrix for Two-Way Interaction*

Intercept	x1	x2	x1x2
1	1	2	2
1	2	2	4
1	3	1	3
1	3	3	9
1	1	1	1
1	2	0	0

The interpretation makes somewhat more sense when written out in standard algebra rather than matrix algebra, as follows:

$$y_i = b_0 + b_1 \cdot x_{1i} + b_2 \cdot x_{2i} + b_3 \cdot (x_{1i} \cdot x_{2i}) \tag{3.4}$$

This can be factored as follows to highlight how the interaction ultimately results in the relations between x1 and x2 with y depending on the other variable in the interaction:

$$\begin{aligned} y_i = b_0 + \\ (b_1 + b_3 \cdot x_{2i}) \cdot x_{1i} + \\ (b_2 + b_3 \cdot x_{1i}) \cdot x_{2i} \end{aligned} \tag{3.5}$$

A similar logic applies to three-way interactions, except that the dependence becomes on two other variables. Because all lower-order two-way interactions are also standard to include, the complexity of the model (the number of parameters) increases dramatically.

$$
\begin{aligned}
y_i = b_0 \\
+ b_1 \cdot x_{1i} + b_2 \cdot x_{2i} + b_3 \cdot x_{3i} \\
+ b_4 \cdot \left(x_{1i} \cdot x_{2i} \right) \\
+ b_5 \cdot \left(x_{1i} \cdot x_{3i} \right) \\
+ b_6 \cdot \left(x_{2i} \cdot x_{3i} \right) \\
+ b_7 \cdot \left(x_{1i} \cdot x_{2i} \cdot x_{3i} \right)
\end{aligned}
\tag{3.6}
$$

This can be factored as follows to highlight the dependence of each variable on both other variables:

$$
\begin{aligned}
y_i = b_0 \\
+ \left(b_1 + b_4 \cdot x_{2i} + b_5 \cdot x_{3i} + b_7 \cdot \left(x_{2i} \cdot x_{3i} \right) \right) \cdot x_{1i} \\
+ \left(b_2 + b_4 \cdot x_{1i} + b_6 \cdot x_{3i} + b_7 \cdot \left(x_{1i} \cdot x_{3i} \right) \right) \cdot x_{2i} \\
+ \left(b_3 + b_5 \cdot x_{1i} + b_6 \cdot x_{3i} + b_7 \cdot \left(x_{1i} \cdot x_{2i} \right) \right) \cdot x_{3i}
\end{aligned}
\tag{3.7}
$$

3.4 Formula Interface

In R, many models and almost all GLMs are specified using a formula interface. Formulae are a flexible way to specify simple to complex models and consist of two parts separated by a tilde ($\sim$). The left hand side (LHS) is on the left of the tilde, and the right hand side (RHS) is on the right of the tilde. The basic form is

```
outcome ~predictor1 + predictor2.
```

The "+" operator adds a variable to the model. R's formula interface is a flexible way to specify models. The main operators are "+" to add terms, "-" to subtract terms, ":" to multiply two terms, and "*" which expands to add both terms and multiply them. Existing formula can be modified using the update() function.

In addition to examining the individual effects of variables from data, GLMs often include the product of two (or more) variables. This is such a common task that the formula interface has a special way to indicate that the product of two terms should be included, the ":" operator. For example, y ~x1 + x2 + x1:x2 includes x1, x2, and their interaction (product term) as predictors of y. One of the nice features of this is that it handles continuous and categorical variables correctly. If x1 and x2 were continuous measures, then x1:x2 will be the regular algebraic product. If one or both are dummy coded, then the products will be appropriately expanded for the dummy codes.

When an interaction term is included, the main effects and each variable's individual effect are almost always included. Because individual effects are almost always included with interactions, the "*" operator can be used to indicate the interaction and individual effects of two variables: y ~x1 * x2 expands to y ~x1 + x2 + x1:x2. Multiple operators can be chained together so that y ~x1 * x2 * x3 expands to y ~x1 + x2 + x3 + x1:x2 + x1:x3 + x2:x3 + x1:x2:x3. Sometimes one variable may moderate an interaction with three or more other predictors, but three- or four-way interactions are not desired. Parentheses can be used to group sets of terms so that operators are distributed to all terms in a group. Thus, y ~x1 * (x2 + x3) expands to y ~x1 + x2 + x3 + x1:x2 + x1:x3.

That covers the most commonly used operators in a formula. Two other details are sometimes helpful, particularly when modifying an existing formula. A dot, ".", can be used as a short way to refer to everything. Finally, terms can be removed using the "-" operator. These are most commonly used when an existing, typically stored, formula is being updated using the update() function. The update() function takes an existing formula object as its first argument and then the desired modifications. The following code shows examples of different types of formula updates that are possible. Note that the other operators, "*" and ":", can also be used with ".". If the dot is completely omitted, then that part of the old formula is not reused at all. We will use the formula interface for most of our model building in R, so it is worth investing the time to learn thoroughly.

```
f1 <- y ~ x1 + x2 + x1:x2

update(f1, . ~ .)

## y ~ x1 + x2 + x1:x2

update(f1, w ~ .)

## w ~ x1 + x2 + x1:x2

update(f1, . ~ . + x3)

## y ~ x1 + x2 + x3 + x1:x2
```

```
update(f1, . ~ . - x1:x2)
## y ~ x1 + x2
```

3.5 Analysis of Variance

Conceptual Background

Analysis of variance (ANOVA) is a statistical technique used to partition variability in the outcome due to different factors. ANOVAs are a special case of the GLM with continuous, normally distributed outcomes and discrete/categorical explanatory variables, such as sex (women, men) or the condition in randomized experimental studies (e.g., Treatment A, Treatment B, or Control). Because of these restrictions, ANOVAs can be conceptualized as testing whether the mean of the outcome is equal in every group. That is, ANOVAs test whether:

$$\mu_{TreatmentA} = \mu_{TreatmentB} = \mu_{Control} \tag{3.8}$$

Traditionally, ANOVAs are used as part of Null Hypothesis Statistical Testing (NHST). Essentially, NHST sets up null hypotheses and asks, what is the probability of obtaining the observed results in this sample of data, given the null hypothesis is true in the population? The flip side of the null hypothesis is the alternative hypothesis. In the case of ANOVA, the alternative hypothesis is that at least one group mean is not equal to the rest. For example, Treatment A may have a lower or higher mean than Treatment B or than the Control group.

To parameterize an ANOVA as a GLM, we write the preceding equality as a series of differences, for example:

$$\mu_{Control} - \mu_{TreatmentA}$$
$$\mu_{Control} - \mu_{TreatmentB} \tag{3.9}$$

These differences are encoded into the design matrix holding the predictor variables, X. Groups are converted into predictors using dummy coding as we covered earlier.

By default, R will create dummy codes using the first level of the factor as the reference group. The reference group is omitted, but the design matrix is augmented with the intercept, a constant column containing ones. The GLM will then estimate the parameters, β one for each column of the design matrix. Because the intercept is in the model, the coefficient for Treatment A (which R labels, ConditionA) will be the difference between the mean for Treatment A and the Control group (the reference level). Likewise, the coefficient for Treatment B will be the difference between the mean for Treatment B and the Control group. The intercept will capture the mean of the Control group.

To see this, we can estimate the regression parameters in R using the function lm() and we extract just the coefficients using the function coef(). We write the model we want R to fit using a formula interface: outcome predictor.

```
set.seed(1234)
example <- data.table(
  y = rnorm(9),
  Condition = factor(rep(c("A", "B", "Control"), each = 3),
                     levels = c("Control", "A", "B")))

coef(lm(y ~ Condition, data = example))

## (Intercept)  ConditionA  ConditionB
##      -0.562       0.614       0.092
```

It is easy enough to check whether these match the differences in group means, by calculating the mean in each group.

```
example[, .(M = mean(y)), by = Condition]

##     Condition      M
## 1:          A  0.052
## 2:          B -0.470
## 3:    Control -0.562
```

We can see immediately that the intercept is the same as the Control group mean. The coefficient for Treatment A is equal to the difference between the Control and Treatment A group means: 0.61 = 0.05 - (-0.56).

If we were to suppress the intercept, then the design matrix would contain dummy codes for each condition, and the regression coefficients would exactly be the group means, which we see in the code that follows. The intercept is suppressed by adding 0 to the formula. Note however that this often only makes sense when there are dummy codes included in a model. In a GLM with continuous explanatory variables, suppressing the intercept would force the intercept to be at exact zero, which is rarely sensible.

```
model.matrix(~ 0 + Condition, data = example)
```

```
##    ConditionControl ConditionA ConditionB
## 1                 0          1          0
## 2                 0          1          0
## 3                 0          1          0
## 4                 0          0          1
## 5                 0          0          1
## 6                 0          0          1
## 7                 1          0          0
## 8                 1          0          0
## 9                 1          0          0
## attr(,"assign")
## [1] 1 1 1
## attr(,"contrasts")
## attr(,"contrasts")$Condition
## [1] "contr.treatment"
```

```
coef(lm(y ~ 0 + Condition, data = example))
```

```
## ConditionControl       ConditionA       ConditionB
##          -0.562            0.052           -0.470
```

The standard ANOVA examines how much variability is in the outcome overall and, of that, how much is between the group means and how much variability remains after accounting for the group means. To test whether there are any differences, a ratio is formed of the amount of variability between group means over the amount of variability within groups (residual variance). This ratio is called an F ratio and can be used to get a p-value, as the proportion of the F distribution is more extreme than the observed F ratio.

The F ratio is based on the ratio of the mean squares, which is the sums of squares (SS) divided by degrees of freedom (DF). There are always two, one for the numerator (the effect of interest) and one for the denominator (error or residual), specifically

$$F = \frac{MS_{model}}{MS_{residual}} = \frac{SS_{model} / DF_{model}}{SS_{residual} / DF_{residual}} \tag{3.10}$$

The degrees of freedom are also used to look up the F ratio against an F distribution using the pf() function.

```
pf(.72, df1 = 1, df2 = 6, lower.tail = FALSE)
```

```
## [1] 0.43
```

For the statistical tests in ANOVAs to be valid, several assumptions must be met. First, observations must be independent (independence). This assumption is violated if, for example, repeated measures are taken on several participants, so that observations are clustered within a person (or within a school, or any other unit of clustering). Second, conditional on the explanatory variables, the outcome must be continuous and normally distributed (normality). Third, the variances within each level of all explanatory variables must be equal (homogeneity of variance). This last assumption is required because a single residual variance is used to estimate uncertainty for all groups.

The previous example had a single grouping factor used in the model. Multiple independent or grouping variables can be tested simultaneously. Including multiple explanatory variables also allows testing of interaction effects—whether the effects of one variable depend on the level of another variable. In one way, this is big advancement over a single explanatory variable as the number of variables doubles, as an additional interaction (moderation) term is added. Viewed from the perspective of the GLM, it is a small change, however. The design matrix gains additional columns, and some of those are the product of dummy codes, rather than the dummy code from a single variable. These are relatively superficial differences. The following code modifies the mtcars dataset we have seen before to convert some of the variables to factor and creates a design matrix for the main effects of two variables as well as their interaction (which is simply the product of the two variables, if both are continuous, or the product of their dummy codes if they are discrete). The interaction (labelled vs1:am1) is 1 only when both of the other dummy codes are 1, and it captures how different the mean is when both vs = 1 and am = 1 beyond what would be expected from their average effects.

```
mtcars <- as.data.table(mtcars)
mtcars[, ID := factor(1:.N)]
mtcars[, vs := factor(vs)]
mtcars[, am := factor(am)]

head(model.matrix(~ vs * am, data = mtcars))
```

```
##    (Intercept) vs1 am1 vs1:am1
## 1            1   0   1       0
## 2            1   0   1       0
## 3            1   1   1       1
## 4            1   1   0       0
## 5            1   0   0       0
## 6            1   1   0       0
```

ANOVA in R

Note   Analysis of variance tests whether group means of a normally distributed outcome are equal or differ. The ezANOVA() function from the ez package can run independent measures, repeated measures, and mixed model ANOVA and provides assumption tests.

To run ANOVA in R, we use the ezANOVA() function from the ez package, which fits various types of ANOVA and provides additional information commonly reported with ANOVAs. To use it, we need to add an ID variable for ezANOVA(). The ezANOVA() function takes a dataset, the outcome variable (dv), the subject ID variable (wid), between subject variables (between). The remaining arguments are optional and control the way variance is calculated when groups are unbalanced and how much output to print. In the code that follows, we use our small example dataset and test the effect of condition overall. It also prints a statistical test whether the homogeneity of variance assumption is met or not, Levene's test. A small F ratio (high p-value) indicates there is little evidence for the variances being significantly different between conditions.

```
example[, ID := factor(1:.N)]

print(ezANOVA(
  data = example,
  dv = y,
  wid = ID,
  between = Condition,
  type = 3,
  detailed = TRUE))
```

```
## Coefficient covariances computed by hccm()

## $ANOVA
##          Effect DFn DFd  SSn SSd    F    p p<.05   ges
## 1 (Intercept)   1   6 0.96   8 0.72 0.43       0.108
## 2    Condition   2   6 0.66   8 0.25 0.79       0.076
##
## $'Levene's Test for Homogeneity of Variance'
##   DFn DFd SSn SSd    F    p p<.05
## 1   2   6 1.5 6.1 0.73 0.52
```

Next we show an ANOVA using the mtcars dataset including an interaction. We add an addition between subjects variable and interaction, but the rest stays the same. The first thing we see is a warning about unequal sample size in each condition. When the sample size is not balanced between groups, the results can differ between different ways of calculating sums of squares, and this is a matter of some debate [55]. These results suggest no significant effect of the interaction term. Both main effects (i.e., on their own) of vs and am are statistically significant and have large effect sizes, suggesting that they are related to miles per gallon. Again there is no evidence of a violation of the homogeneity of variance assumption. Compared with the GLM framework, the F ratios test similar effects to the regression coefficients, but in a slightly different way. When a factor has more than two levels, the F ratios have multiple numerator degrees of freedom and will be equivalent to testing whether multiple regression coefficients are simultaneously zero, rather than testing one coefficient at a time. The last column of the following code's output ges shows the effect size measure of generalized eta-squared. Nevertheless, aside from exactly how the test is constructed, the underlying linear model used by ANOVAs is a subset of those allowed by GLMs.

```
print(ezANOVA(
  data = mtcars,
  dv = mpg,
  wid = ID,
  between = vs * am,
  type = 3,
  detailed = TRUE))
```

```
## Warning: Data is unbalanced (unequal N per group). Make sure you
specified a well-considered value for the type argument to ezANOVA().

## Coefficient covariances computed by hccm()

## $ANOVA
##           Effect DFn DFd   SSn SSd      F       p p<.05   ges
## 1 (Intercept)   1  28 13144 337 1090.6 5.7e-24     * 0.975
## 2              vs   1  28   382 337   31.7 4.9e-06     * 0.531
## 3              am   1  28   284 337   23.5 4.2e-05     * 0.457
## 4           vs:am   1  28    16 337    1.3 2.6e-01       0.045
##
## $'Levene's Test for Homogeneity of Variance'
##   DFn DFd SSn  SSd    F    p p<.05
## 1   3  28  15 156 0.88 0.46
```

We can use Tukey's honestly significant difference (HSD) to test the pairwise difference between cells. The following code creates a new variable that is the combination of vs and am and then creates a graph showing the means and 95% confidence intervals (interpreted as 95% of intervals conducted in the same fashion will include the true population parameter). Any cells that share a letter are not statistically significantly different from each other. Cells that do not share a letter are statistically significantly different. We make the plot using TukeyHSDgg() from the JWileymisc package and adjust the angle of axis labels and remove the x axis title. Figure 3-1 shows the results.

```
mtcars[, Cells := factor(sprintf("vs=%s, am=%s", vs, am))]
TukeyHSDgg("Cells", "hp", mtcars) +
  theme(axis.text.x = element_text(angle=45, hjust=1, vjust=1)) +
  xlab("")
```

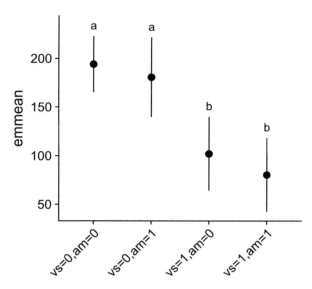

Figure 3-1. *Graph of cell means with confidence intervals. Cells that share letters are not statistically significantly different based on Tukey's honestly significant difference.*

Although brief, hopefully this introduction to ANOVA helps highlight how ANOVA can be used to test for group mean differences in R and how the ANOVA is just a special case of GLMs. Ultimately ANOVAs are a rather limiting special case of the GLM as they do not allow inclusion of continuous explanatory variables. Next we examine linear regression as a more flexible special case of the GLM for continuous normally distributed outcomes that allows both discrete and continuous explanatory variables.

3.6 Linear Regression

Note   Linear regression is a special case of the GLM for continuous, normally distributed outcome variables. Unlike an analysis of variance, linear regression accommodates both discrete and continuous explanatory variables/predictors. The ols() function from the rms package can run linear regression and print comprehensive summary output information.

Conceptual Background

Linear regression is another special case of GLMs where the link and inverse link functions are simply the identity function, that is, $\eta = g(\mu) = \mu$ and $\mu = g^{-1}(\eta) = \eta$ and the outcome is assumed to be distributed as a normal distribution. Specifically, the distribution assumption is written as y ~ $N(\mu, \sigma)$. The critical information here is that y is a vector as is μ but σ is a constant. The mean of a distribution is often referred to as its location and the dispersion parameter or standard deviation as its scale. Another class of models, location scale models, allow both the location and scale parameters to vary as a function of the data, but for now with GLMs, we will assume the scale is constant. Another way the distribution is commonly written is in terms of the distribution of residuals, $\in$ ~ $N(0, \sigma)$ where $\in$ = y$-\mu$. Written that way highlights the fact that the raw data need not follow a normal distribution, they only need to be normally distributed around the expectations. It also highlights that the dispersion parameter, σ, is the dispersion of the residuals. That is, σ captures dispersion around the expected values. In the simplest GLM where the only predictor is a single constant term (the intercept), σ will be the same as the standard deviation of y, but if the regression can explain some or all the variation in y, then σ will go toward zero.

Based on the central limit theorem, the parameter distribution of the ratio of the regression coefficients to their standard errors will converge to a normal distribution as the sample size converges to infinity. In linear regression, rather than test parameters against a normal distribution, we can take into account the fact that we typically work with finite samples and test parameters against the t distribution (for individual regression coefficients). The t distribution converges to a normal distribution when there are infinite degrees of freedom and has somewhat heavier tails when there are finite degrees of freedom. In linear regression, degrees of freedom are calculated based on the sample size and number of parameters ($df = N - k_{parameters}$). In later chapters when we examine other types of GLMs, degrees of freedom cannot be readily calculated and so individual regression coefficients are instead tested against the standard normal distribution instead of the t distribution.

In addition to tests of the individual regression coefficients, the overall model can be tested using a likelihood ratio test, which compares the likelihood ratios for the model we fit against the null model, which only contains an intercept as a predictor. One of the reasons why likelihoods are useful is that they can be readily compared and tested providing accurate tests of multiple variables or other restrictions on a model.

In linear regression, a common effect size is the percentage of variance in the outcome accounted for by the model (or by an individual predictor). The percentage variance accounted for is called the R^2. Before we can calculate the R^2, we need a few pieces. We define the sum of squared deviations (SS) total as the SS for the outcomes deviations from its overall expectation or mean (SSTotal), SSRegression as the SS of our model predicted outcome from the overall expectation of the outcome, and SSResidual as the SS of the differences between the outcome and our model predicted values. This is shown more formally in the following equations:

$$SS_{Total} = \sum_{i=1}^{N} \left(y_i - E(y) \right)^2 \tag{3.11}$$

$$SS_{Regression} = \sum_{i=1}^{N} \left(\mu_i - E(y) \right)^2 \tag{3.12}$$

$$SS_{Residual} = \sum_{i=1}^{N} \left(y_i - \mu_i \right)^2 \tag{3.13}$$

Given these definitions, for linear models with normally distributed outcomes, R^2 can be calculated as follows:

$$R^2 = \frac{SS_{Regression}}{SS_{Total}} = 1 - \frac{SS_{Residual}}{SS_{Total}} = cor(y, \mu)^2 \tag{3.14}$$

Unless there is an infinite sample size, this formulation for calculating R^2 in a sample of data is a biased estimate of the population R^2. Because of this, it is also common to report an adjusted R^2, which takes into account the model degrees of freedom to provide an unbiased estimate of the population R^2. This adjustment for degrees of freedom adjusts the overly optimistic estimate of variance accounted for in the population when we train a model and test it using the same data. We will discuss the concept of overfitting and using separate datasets for model estimation and testing later when we cover machine learning in R. In linear regressions, the bias tends to be minimal as there tend to be relatively few parameters compared to the sample size. As the number of predictors/parameters to observations increases, the problem and bias introduced by overfitting become more problematic. The R^2 can be used to estimate the predictive accuracy of the overall model, but it can also be used to quantify how much an individual predictor adds by comparing how much the model R^2 changes when adding or dropping one predictor at a time.

Although the R^2 is by far the most common fit or discrimination index for linear regression, other indices exist. Another alternative is the g-index, which is based on the Gini index [25]. A key difference between the g-index and R^2 is that the g-index is not standardized, so it depends on the scale of the outcome and predictors. As with R^2, however, higher indicates better discrimination.

Linear Regression in R

As we are moving to applied and practical data analysis, we will start using real data. For this chapter, we will use the Americans' Changing Lives [45] study data. Reading and preparing the data was covered in the Introduction section labelled "Data Setup." Technically, the data have sampling weights, but for simplicity we ignore these weights. The analyses will still be correct without weighting; they just will not reflect the sampled population.

Although not a direct test of the regression assumptions, it is useful to know the approximate distribution of the outcome, satisfaction with life. Figure 3-2 shows a density plot with a normal curve overlayed. The raw density was smoothed more than the default using the `adjust = 2` argument to `testdistr()`. From Figure 3-2 we can see that the satisfaction with life is approximately normally distributed and there are no large outliers.

```
acl <- readRDS("advancedr_acl_data.RDS")

testdistr(acl$SWL_W1, "normal",
        varlab = "Satisfaction with Life", plot = FALSE,
        extremevalues = "theoretical",
        adjust = 2)$DensityPlot
```

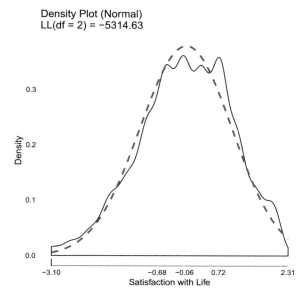

Figure 3-2. *Density plot of satisfaction with life (black line) with normal density overlayed (blue line)*

R has built-in functions to fit linear regression, but we use the `ols()` function from the `rms` package, as it provides convenient features and more comprehensive default output. The name `ols` comes from another name for linear regression: ordinary least squares. This name is based on using the least squared deviations as the criteria to estimate the regression coefficients.

The model output first echoes the formula used to fit the model. It shows the number of observations, the estimate of the residual standard deviation, σ, and the overall model degrees of freedom. The likelihood ratio test provides a test of the statistical significance of all predictors simultaneously, testing the hypothesis that at least one of the coefficients is significantly different than zero. Discrimination indices include R^2 and adjusted R^2 values as well as the g-index. Residuals by construction have a mean of zero, but due to skew or outliers, the median can be quite different. Examining the smallest and largest residual can also be useful for identifying residual outliers. Finally, the regression coefficients along with corresponding standard errors, t-values, and p-values are shown in a table.

```
m.ols <- ols(SWL_W1 ~ Sex + AGE_W1 + SESCategory, data = acl, x = TRUE)
m.ols

## Linear Regression Model
```

```
##
##  ols(formula = SWL_W1 ~ Sex + AGE_W1 + SESCategory, data = acl,
##      x = TRUE)
##
##                     Model Likelihood      Discrimination
##                         Ratio Test              Indexes
##  Obs     3617    LR chi2     118.62     R2         0.032
##  sigma1.0355     d.f.             5     R2 adj     0.031
##  d.f.    3611    Pr(> chi2) 0.0000      g          0.213
##
##  Residuals
##
##       Min       1Q    Median        3Q       Max
##  -3.44270 -0.67206   0.01543   0.75504   2.36635
##
##
##                  Coef    S.E.    t     Pr(>|t|)
##  Intercept       -0.7057 0.0755 -9.35 <0.0001
##  Sex=(2) FEMALE   0.0308 0.0360  0.86 0.3921
##  AGE_W1           0.0103 0.0011  9.75 <0.0001
##  SESCategory=2   -0.0133 0.0447 -0.30 0.7654
##  SESCategory=3    0.2558 0.0482  5.31 <0.0001
##  SESCategory=4    0.2654 0.0635  4.18 <0.0001
##
```

Using the texreg package, we can automatically create nicely formatted tables. We can create tables for screen output, using the screenreg() function, for HTML output using the htmlreg() function, or for LATEX using the texreg() function.

The following example shows how to make a LATEX table, which is given in Table 3-6.

```
texreg(m.ols, single.row = TRUE, label = "tglm1-olstex")
```

Table 3-6. *Statistical Models*

	Model 1
Intercept	−0.71 (0.08)***
Sex=(2) FEMALE	0.03 (0.04)
AGE_W1	0.01 (0.00)***
SESCategory=2	−0.01 (0.04)
SESCategory=3	0.26 (0.05)***
SESCategory=4	0.27 (0.06)***
Num. obs.	3617
R^2	0.03
Adj. R^2	0.03
L.R.	118.62

*** $p < 0.001$, ** $p < 0.01$, * $p < 0.05$

We can also investigate several diagnostics regarding the model. First we can explore the influence of any collinearity (high correlations between explanatory variables) by using the variance inflation factor (VIF). VIF values near one indicate little impact of collinearity. Very high VIF values may indicate that the inclusion of highly correlated explanatory variables inflated the variance of the parameter covariance matrix, which results in very large standard errors and confidence intervals. This occurs most often when two very similar explanatory variables are included.

```
vif(m.ols)

## Sex=(2) FEMALE      AGE_W1 SESCategory=2 SESCategory=3
##             1.0         1.2           1.4           1.5
##  SESCategory=4
##             1.3
```

Next we create a data table with the fitted values and residuals and use this in Figure 3-3 to graph the residuals to check for normality. Figure 3-4 uses quantile regression [51, 50, 107] as we introduced in the chapter on multivariate data visualization to assess heteroskedasticity by graphing the quantile regression lines for the 5th, 25th, 50th, 75th, and 95th percentiles. The lines are relatively flat suggesting little evidence for heteroskedasticity.

```
diagnostic.data <- data.table(
  fitted = fitted(m.ols),
  resid = residuals(m.ols))

testdistr(diagnostic.data$resid,
          "normal",
          varlab = "Satisfaction with Life Residuals", plot = FALSE,
          extremevalues = "theoretical",
          adjust = 2)$DensityPlot

ggplot(diagnostic.data, aes(fitted, resid)) +
  geom_point(alpha = .2, colour = "grey50") +
  geom_quantile(quantiles = .5, colour = 'black', size = 1) +
  geom_quantile(quantiles = c(.25, .75),
                colour = 'blue', linetype = 2, size = 1) +
  geom_quantile(quantiles = c(.05, .95),
                colour = 'black', linetype = 3, size = 1)

## Smoothing formula not specified. Using: y ~ x
## Smoothing formula not specified. Using: y ~ x
## Smoothing formula not specified. Using: y ~ x
```

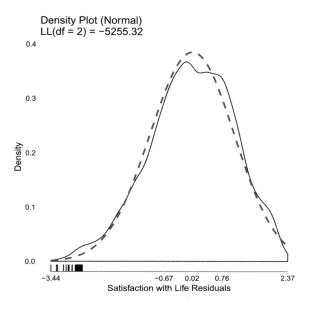

Figure 3-3. *Graphing residuals to assess normality*

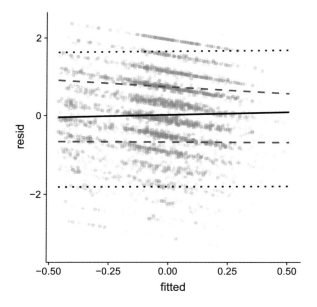

Figure 3-4. *Examining residuals vs. fitted values with quantile regression to explore heteroskedasticity*

We introduced the GLM using matrix notation:

$$\mu = g^{-1}(\eta) = g^{-1}(X\beta)$$

(3.15)

To better understand how to interpret coefficients, it can be helpful to write it out using regular algebra. In the case of linear regression, we can further simplify by removing the inverse link function, as the link and inverse link are (canonically) the identity function. It is also common to write the following equation using subscript i for each variable to indicate the operation is performed for the ith individual. We use bold to indicate the variables are vectors and contain data for multiple individuals.

$$\mu = \beta_0 + \beta_1 \text{Sex} + \beta_2 \text{Age} + \beta_3 \text{SES}_2 + \beta_4 \text{SES}_3 + \beta_5 \text{SES}_4$$

(3.16)

Each coefficient captures the expected change in the outcome for a one-unit change in a predictor. What one unit means depends on the scale of each predictor. For instance, age is coded in years, so one unit means one year. Sex is dummy coded, so one unit means the difference between men and women. Socioeconomic status (SES) is coded into quartiles, with the reference category (omitted) being the lowest quartile of SES. So, for instance, being one year older is associated with 0.01 higher satisfaction with life.

We noted that linear regression is a special case of GLM. R has a built-in function, glm(), which fits GLMs. The benefit of the glm() function is that you can use the same function for many specific types of GLMs. The formula interface is identical to ols(), but glm() allows one to specify different distributions and link functions. If we simply print the stored GLM object, we get minimal output. To get a nice summary, we need to use the summary() function. Although this is slightly more efficient computationally, in our experience most users tend to want this output, so it is convenient that the ols() function provides a good output summary as a default. Another disadvantage of the glm() function is that it does not show the R^2 value. This is because variance accounted for does not make sense for all types of GLMs. There also is no default test of the overall model compared to an independent model and no information regarding the number of missing values on each variable.

```
m.glm <- glm(SWL_W1 ~ Sex + AGE_W1 + SESCategory,
           data=acl, family = gaussian(link="identity"))
m.glm
```

##

```
## Call:  glm(formula = SWL_W1 ~ Sex + AGE_W1 + SESCategory, family =
gaussian(link = "identity"),
##      data = acl)
##
## Coefficients:
##    (Intercept)   Sex(2) FEMALE        AGE_W1    SESCategory2
##        -0.7057          0.0308        0.0103         -0.0133
##   SESCategory3    SESCategory4
##         0.2558          0.2654
##
## Degrees of Freedom: 3616 Total (i.e. Null);  3611 Residual
## Null Deviance:     4000
## Residual Deviance: 3870      AIC: 10500
```

```
summary(m.glm)
```

```
##
## Call:
## glm(formula = SWL_W1 ~ Sex + AGE_W1 + SESCategory, family =
gaussian(link = "identity"),
##      data = acl)
##
## Deviance Residuals:
##    Min      1Q  Median      3Q     Max
## -3.443  -0.672   0.015   0.755   2.366
##
## Coefficients:
##                Estimate Std. Error t value Pr(>|t|)
## (Intercept)    -0.70565    0.07549   -9.35  < 2e-16 ***
## Sex(2) FEMALE   0.03083    0.03602    0.86     0.39
## AGE_W1          0.01030    0.00106    9.75  < 2e-16 ***
## SESCategory2   -0.01333    0.04467   -0.30     0.77
## SESCategory3    0.25579    0.04819    5.31  1.2e-07 ***
## SESCategory4    0.26544    0.06353    4.18  3.0e-05 ***
## ---
## Signif. codes:  0 '***' 0.001 '**' 0.01 '*' 0.05 '.' 0.1 '_' 1
```

```
##
## (Dispersion parameter for gaussian family taken to be 1.1)
##
##      Null deviance: 4000.7  on 3616   degrees of freedom
## Residual deviance: 3871.6  on 3611   degrees of freedom
## AIC: 10525
##
## Number of Fisher Scoring iterations: 2
```

The output is very similar for R's built-in lm() function which is specifically for linear regression models. It does add an estimate of the R^2, but again it requires use of the summary() and does not show missing values per variable.

Oftentimes, multiple related models are estimated. For example, there may be a few focal predictors and the goal is to examine how their effects change when some additional variables (covariates) are in or out of the model. Other times certain predictors may be included for certain and others only included if they are statistically significant or improve the performance of the model. It is also common to try other functional forms. For example, a variable may be entered as a linear effect or added as a linear effect along with the same variable squared (i.e., quadratic). Earlier this chapter, we showed how to update formulae in R. The update() function also has methods for many models. It works similarly as update() on a formula, except that in addition to changing the model formula, it also refits the model. To update and see the results despite being stored in an object, we can wrap the whole call in parentheses, which forces printing. In the code that follows, we update the base model and add employment status to the model.

```
(m.ols2 <- update(m.ols, . ~ . + Employment_W1))

## Linear Regression Model
##
##  ols(formula = SWL_W1 ~ Sex + AGE_W1 + SESCategory + Employment_W1,
##      data = acl, x = TRUE)
##
##                    Model Likelihood    Discrimination
##                      Ratio Test            Indexes
##  Obs    3617     LR chi2    173.43    R2        0.047
##  sigma1.0286    d.f.            12    R2 adj    0.044
```

```
## d.f.   3604    Pr(> chi2) 0.0000    g         0.252
##
## Residuals
##
##      Min        1Q    Median       3Q      Max
## -3.50191 -0.66381   0.03265   0.74125  2.55188
##
##
##                               Coef   S.E.    t      Pr(>|t|)
## Intercept                   -1.1197 0.1244 -9.00 <0.0001
## Sex=(2) FEMALE               0.0109 0.0387  0.28 0.7776
## AGE_W1                       0.0092 0.0013  6.83 <0.0001
## SESCategory=2               -0.0253 0.0451 -0.56 0.5746
## SESCategory=3                0.2179 0.0498  4.37 <0.0001
## SESCategory=4                0.2174 0.0655  3.32 0.0009
## Employment_W1=(2) 2500+HRS   0.5832 0.1098  5.31 <0.0001
## Employment_W1=(3) 15002499   0.4675 0.0985  4.75 <0.0001
## Employment_W1=(4) 500-1499   0.5497 0.1085  5.07 <0.0001
## Employment_W1=(5) 1-499HRS   0.6135 0.1250  4.91 <0.0001
## Employment_W1=(6) RETIRED    0.5345 0.0962  5.55 <0.0001
## Employment_W1=(7) UNEMPLOY   0.2498 0.1233  2.03 0.0428
## Employment_W1=(8) KEEP HS    0.6218 0.0991  6.28 <0.0001
##
```

To test whether SES or Employment is significant overall or not, we need to test all the dummy codes simultaneously. This could be done by comparing the two models or by calling the anova() function on the fitted model. The resulting ANOVA table shows the three degree of freedom test for SES and the seven degree of freedom test for Employment, both of which are statistically significant overall.

```
anova(m.ols2)
```

```
##              Analysis of Variance        Response: SWL_W1
##
## Factor          d.f. Partial SS MS       F       P
## Sex              1 8.4e-02      0.084   0.08 0.78
## AGE_W1           1 4.9e+01     49.364  46.65 <.0001
```

```
## SESCategory        3 3.8e+01    12.631 11.94 <.0001
## Employment_W1      7 5.8e+01     8.318  7.86 <.0001
## REGRESSION        12 1.9e+02    15.608 14.75 <.0001
## ERROR           3604 3.8e+03     1.058
```

There were no sex differences so we may consider dropping sex. We could also explore potential interactions, such as between age and SES. Both of these can be done in one step, updating the model. Again we force printing using parentheses.

```
(m.ols3 <- update(m.ols2, . ~ . + AGE_W1 * SESCategory - Sex))

## Linear Regression Model
##
## ols(formula = SWL_W1 ~ AGE_W1 + SESCategory + Employment_W1 +
##     AGE_W1:SESCategory, data = acl, x = TRUE)
##
##                     Model Likelihood    Discrimination
##                       Ratio Test              Indexes
## Obs      3617    LR chi2     189.72    R2          0.051
## sigma1.0266    d.f.             14    R2 adj      0.047
## d.f.     3602    Pr(> chi2) 0.0000    g           0.256
##
## Residuals
##
##      Min        1Q    Median        3Q       Max
## -3.37389  -0.65254   0.04075   0.72383   2.60671
##
##
##                              Coef    S.E.    t     Pr(>|t|)
## Intercept                   -1.2652 0.1568 -8.07 <0.0001
## AGE_W1                       0.0116 0.0021  5.56 <0.0001
## SESCategory=2               -0.0495 0.1566 -0.32 0.7518
## SESCategory=3                0.6213 0.1678  3.70 0.0002
## SESCategory=4                0.7440 0.2128  3.50 0.0005
## Employment_W1=(2) 2500+HRS   0.5628 0.1092  5.15 <0.0001
## Employment_W1=(3) 15002499   0.4643 0.0984  4.72 <0.0001
## Employment_W1=(4) 500-1499   0.5592 0.1083  5.16 <0.0001
```

```
## Employment_W1=(5) 1-499HRS  0.6284 0.1247  5.04 <0.0001
## Employment_W1=(6) RETIRED   0.5280 0.0961  5.50 <0.0001
## Employment_W1=(7) UNEMPLOY  0.2812 0.1232  2.28 0.0225
## Employment_W1=(8) KEEP HS   0.6293 0.0978  6.43 <0.0001
## AGE_W1 * SESCategory=2       0.0009 0.0026  0.35 0.7248
## AGE_W1 * SESCategory=3      -0.0077 0.0029 -2.62 0.0088
## AGE_W1 * SESCategory=4      -0.0107 0.0041 -2.61 0.0090
##
```

There does seem to be an interaction between age and SES. With the interaction in the model, the coefficient for age is the slope for age and satisfaction with life for the lowest quartile of SES (the reference group of dummy-coded SES and therefore the reference group of the interaction of SES with age). The coefficients for the interaction indicate that the relations between age and satisfaction with life are lower at higher SES categories. To get a better sense of the interaction, we can graph it by showing how the expected value of satisfaction with life changes by age, with results separated by SES categories.

A quick way to plot the results of many regression models in R is to use the `visreg()` function from the `visreg` package [15]. This function allows quick predictions and confidence intervals to be generated and graphed. It has sensible defaults, like holding variables you are not graphing at their median or mode for continuous and categorical variables. The `visreg()` function requires just two arguments at a minimum. First, what is the model, and second, which variable do you want to plot on the x axis (`xvar` argument)? We will use this to ask for age on the x axis and expand it to request different lines by SESCategory, which will make an interaction plot.

By default `visreg()` includes partial residuals, a rug plot, and separates out interaction plots into separate panels. We combine all panels into one plot using `overlay = TRUE`, turn off plotting partial residuals so we only have the predicted lines using `partial = FALSE`, turn off a rug plot showing where data points fall on the x axis using `rug = FALSE`, add some better x and y axis labels using `xlab` and `ylab`, and, finally, change the linetype so that when the graph is printed not in color, it will still be readable by setting `line = list(lty = 1:4)`. The result of all that customization is shown in Figure 3-5.

```
plot(visreg(m.ols3, xvar = "AGE_W1", by = "SESCategory",
            plot = FALSE),
     overlay = TRUE, partial = FALSE, rug = FALSE,
     xlab = "Age (years)", ylab = "Predicted Life Satisfaction",
     line = list(lty = 1:4))
```

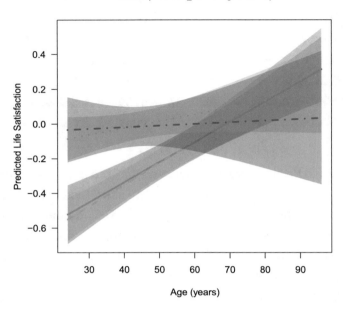

Figure 3-5. *Estimated satisfaction with life across age by SES category. Shaded region indicates 95% confidence intervals for regression estimates.*

The plot is still a bit messy with all the confidence intervals overlapping. For personal understanding, those are likely helpful. For use in say a presentation, it may make viewing the lines difficult. We can turn off confidence intervals using another argument, band = FALSE. We could further alter it for publishing by making the graph grayscale, by passing four colors, used for each of the four lines. The result of this further customization is shown in Figure 3-6.

```
plot(visreg(m.ols3, xvar = "AGE_W1", by = "SESCategory",
        plot = FALSE),
    overlay = TRUE, partial = FALSE, rug = FALSE,
    xlab = "Age (years)", ylab = "Predicted Life Satisfaction",
    line = list(
      lty = 1:4,
      col = c("black", "grey75", "grey50", "grey25")),
    band = FALSE)
```

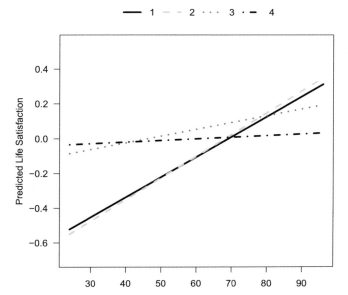

Figure 3-6. *Estimated satisfaction with life across age by SES category. Confidence intervals removed.*

Using the visreg() function is quick and relatively easy, so it makes a good choice in most settings, and certainly just as a way to help understand results yourself. For even more control, we can make the same graph manually. To do this, we need to get the predicted values for a variety of ages and SES categories. We also need to hold the other variables in the model, employment, at some value. It is easy to get predicted values from a model in R, but we first need to create a small dataset containing all the values we want to use as inputs for the prediction. This can be readily accomplished using the expand.grid() function. It is important that factors have the same levels as in the model, which is most easily accomplished by extracting the levels() from the factor in the real data.

```
newdata <- as.data.table(expand.grid(
  AGE_W1=quantile(acl$AGE_W1, .1):quantile(acl$AGE_W1, .9),
  SESCategory = factor(1:4, levels = levels(acl$SESCategory)),
  Employment_W1 = factor("(3) 15002499",
    levels = levels(acl$Employment_W1))))
newdata
```

```
##         AGE_W1 SESCategory Employment_W1
## 1:        30            1  (3) 15002499
## 2:        31            1  (3) 15002499
## 3:        32            1  (3) 15002499
## ---
## 186:      74            4  (3) 15002499
## 187:      75            4  (3) 15002499
## 188:      76            4  (3) 15002499
```

Now we can generate predicted values using the predict() function. We can extract only the predicted values or the predicted values and standard errors for each predicted value. Standard errors are helpful to calculate confidence intervals around each prediction and show the uncertainty in the estimate. To get standard errors, we specify, se.fit = TRUE. The result is a list with the first element containing a vector of the predicted values and the second element containing a vector of the standard errors, which we store back into our data table.

```
newdata[, c("SWL_W1", "SE") :=
        predict(m.ols3, newdata = newdata, se.fit = TRUE)]
newdata
```

```
##         AGE_W1 SESCategory Employment_W1 SWL_W1    SE
## 1:        30            1  (3) 15002499 -0.453 0.076
## 2:        31            1  (3) 15002499 -0.441 0.075
## 3:        32            1  (3) 15002499 -0.430 0.073
## ---
## 186:      74            4  (3) 15002499  0.014 0.121
## 187:      75            4  (3) 15002499  0.015 0.124
## 188:      76            4  (3) 15002499  0.016 0.128
```

Confidence intervals are calculated as

$$Estimate \pm SE x z_{\alpha/2} \tag{3.17}$$

The z refers to quantiles from a unit normal distribution (often referred to as z scores). The $z_{\alpha/2}$ is the quantile based on the desired alpha level (e.g., .05 for 95% confidence intervals). To be more precise, the quantile from a t-distribution with appropriate degrees of freedom could be used, although with such a large sample, the t-distribution will effectively be normal. This can be obtained in R using the qnorm() function.

```
print(qnorm(.05/2), digits = 7)

## [1] -1.959964

print(qnorm(1 - (.05/2)), digits = 7)

## [1] 1.959964
```

Next we create a graph of the predicted values using the ggplot2 package and the theme from the cowplot package. The code is a bit complexqnorm() function, but produces a publication quality graph shown in Figure 3-7. While this is somewhat more tedious than using visreg(), it gives us complete control over what predicted values we are calculating and allows us to potentially do further analysis or work on the predictions, prior to graphing or presentation.

```
ggplot(newdata, aes(AGE_W1, SWL_W1, linetype=SESCategory)) +
  geom_ribbon(aes(ymin = SWL_W1 + SE * qnorm(.025),
                  ymax = SWL_W1 + SE * qnorm(.975)),
              alpha = .2) +
  geom_line(size = 1) +
  scale_x_continuous("Age (years)") +
  ylab("Satisfaction with Life") +
  theme_cowplot() +
  theme(
    legend.position = c(.8, .16),
    legend.key.width = unit(2, "cm"))
```

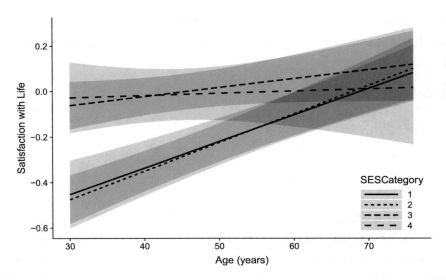

Figure 3-7. *Estimated satisfaction with life across age by SES category. Shaded region indicates 95% confidence intervals for regression estimates.*

Because of their dependence on the regression coefficient and standard error (SE), confidence intervals are closely related to the p-value. However, they are a useful way of showing the uncertainty in the estimate of the true population regression coefficient. In R we can calculate 95% confidence intervals for each regression coefficient using the confint() function.

```
confint(m.ols3)
```

```
##                                   2.5 %    97.5 %
## Intercept                       -1.5726  -0.9579
## AGE_W1                           0.0075   0.0157
## SESCategory=2                   -0.3566   0.2575
## SESCategory=3                    0.2922   0.9504
## SESCategory=4                    0.3267   1.1612
## Employment_W1=(2) 2500+HRS       0.3487   0.7768
## Employment_W1=(3) 15002499       0.2714   0.6572
## Employment_W1=(4) 500-1499       0.3468   0.7715
## Employment_W1=(5) 1-499HRS       0.3839   0.8729
## Employment_W1=(6) RETIRED        0.3396   0.7163
## Employment_W1=(7) UNEMPLOY       0.0396   0.5227
## Employment_W1=(8) KEEP HS        0.4376   0.8211
```

```
## AGE_W1 * SESCategory=2      -0.0041  0.0059
## AGE_W1 * SESCategory=3      -0.0134 -0.0019
## AGE_W1 * SESCategory=4      -0.0186 -0.0027
```

High-Performance Linear Regression

So far we have focused on convenient functions with comprehensive output. This is probably what most users need most of the time. Linear regression is so fast on modern computers that computational time is not an issue most of the time. However, computational speed is a concern in some situations. Bootstrapping is a procedure we will discuss in more depth when we discuss machine learning later, but briefly it entails repeated sampling from a dataset, and estimating some parameters on the resampled data to generate an empirical parameter distribution. For bootstrapping, we may just wish to extract the regression coefficients, which we can do using the coef() function.

People often take hundreds or thousands of bootstrap samples. We will just take 500 for the sake of time and demonstration. First we create a small dataset of just our variables, and because we use Intel's MKL linear algebra library, to get purer timing estimates, we set it to use only a single core. This can be ignored on single core machines or those not using MKL.

```
tmpdat <- na.omit(acl[, .(SWL_W1, AGE_W1, SESCategory, Employment_W1)])
## use if using Microsoft R Open with Intel's MKL linear algebra library
setMKLthreads(1)
```

Our actual code is fairly simple. We use the system.time() function to track how long it takes and then use sapply() to loop through 1 to 500, create an index of the rows to use, and then fit our model and extract the coefficients.

```
set.seed(12345)
t1 <- system.time(ols.boot <- sapply(1:500, function(i) {
  index <- sample(nrow(tmpdat),
                  size = nrow(tmpdat), replace = TRUE)
  coef(ols(SWL_W1 ~ AGE_W1 * SESCategory + Employment_W1,
          data = tmpdat[index]))
}))
```

```
t1
```

```
##    user  system elapsed
##    4.27    0.06    4.33
```

Using the ols() function, it took 4.33 seconds to complete—not so long as to be impossible, but long enough to noticeably slow interactive data analysis. For any real application, we would probably take at least a few thousand bootstrap resamples. The time will increase in a linear fashion with the number of resamples, so 10,000 would take about 86.6 seconds. Next, we use the fastLm() function from the RcppEigen package [4]. It implements linear models using C++ to be faster and more efficient.

```
set.seed(12345)
t2 <- system.time(rcpp.boot1 <- sapply(1:500, function(i) {
  index <- sample(nrow(tmpdat), size = nrow(tmpdat), replace = TRUE)
  coef(fastLm(SWL_W1 ~ AGE_W1 * SESCategory + Employment_W1, data =
tmpdat[index]))
}))
```

```
t2
```

```
##    user  system elapsed
##     2.5     0.0     2.5
```

Now the total time is down to 2.52 seconds, so for 10,000 resamples it would be about 50.4 seconds. Finally, we use the fastLmPure() function, also from the RcppEigen package. The fastLmPure() function is not as smart and requires the user to pass in the outcome as a vector and the model matrix, rather than using a formula interface. We include the explicit calculation of the outcome vector and model matrix inside our system timing and then apply the bootstrap resample indices to these precomputed matrices.

```
set.seed(12345)
t3 <- system.time({
  y <- tmpdat[, SWL_W1]
  X <- model.matrix(~ AGE_W1 * SESCategory + Employment_W1, data = tmpdat)
  N <- nrow(tmpdat)
  rcpp.boot2 <- sapply(1:500, function(i) {
    index <- sample.int(N, size = N, replace = TRUE)
```

```
    fastLmPure(X = X[index, ], y = y[index])$coefficients
  })
})

t3

##     user  system elapsed
##     0.48    0.02    0.50
```

Using this approach, the analysis takes just 0.5 seconds. Since it is so fast, we can easily re-run it using 10,000 resamples.

```
set.seed(12345)
t4 <- system.time({
  y <- tmpdat[, SWL_W1]
  X <- model.matrix(~ AGE_W1 * SESCategory + Employment_W1, data = tmpdat)
  N <- nrow(tmpdat)
  rcpp.boot3 <- sapply(1:10000, function(i) {
    index <- sample.int(N, size = N, replace = TRUE)
    fastLmPure(X = X[index, ], y = y[index])$coefficients
  })
})

t4

##     user  system elapsed
##     9.95    0.21   10.15
```

Using 10,000 resamples takes 10.15 seconds. With parallel processing this number could be brought down even further. We do not use fastLmPure() for interactive data analysis, but for computationally heavy tasks like bootstrapping or if you are trying hundreds of different models, the speed gains can be meaningful. Lastly, we can check that we indeed get the same results from all models using all.equal(). Set check.attributes = FALSE to ignore names as ols() names dummy coefficients slightly differently.

```
all.equal(ols.boot, rcpp.boot1, check.attributes = FALSE)
```

```
## [1] TRUE
```

```
all.equal(ols.boot, rcpp.boot2, check.attributes = FALSE)
```

```
## [1] TRUE
```

3.7 Controlling for Confounds

In a scientific context, GLMs often are used to study the potential effect of one variable
on another. For example, self-efficacy is the tendency for someone to see themselves
as having the ability to make changes or have control in their life. Studies suggest
that people who are high in self-efficacy have an easier time changing behavior (e.g.,
starting an exercise program, quitting smoking, signing up and completing a university
degree). This makes some sense if you imagine someone low in self-efficacy: they tend
to believe they will not succeed at making changes and tend to see their behavior and
circumstances as out of their control (e.g., controlled by the environment, powerful
others, etc.). Regardless of whether someone truly can or cannot influence their own
life, if they believe they cannot, they may give up quicker, be less motivated to try,
and thus be less likely to either begin or sustain any behavior or pursuit of their own
goals. Now suppose we hypothesize that people who are high in self-efficacy are also
less likely to experience symptoms of depression. The ACL data, which we covered
in the introduction, includes a variable capturing self-efficacy and another capturing
symptoms of depression at two waves of data collection. A natural place to begin would
be testing whether self-efficacy at the first wave predicts symptoms of depression at
the second wave. The following code tests this, and we can see that indeed there is a
statistically significant, negative association, with people who are higher in self-efficacy
at wave 1 tending to have lower depression symptoms at wave 2. The results are in
Table 3-7.

```
m0 <- ols(CESD11_W2 ~ SelfEfficacy_W1, data = acl)
```

```
texreg(m0, label = "tglm1-olsunadj")
```

Table 3-7. *Statistical Models*

	Model 1
Intercept	0.02
	(0.02)
SelfEfficacy_W1	−0.36***
	(0.02)
Num. obs.	2867
R²	0.13
Adj. R²	0.13
L.R.	399.71

*** $p < 0.001$, ** $p < 0.01$, * $p < 0.05$

If this were purely a prediction model, we might be content with the results as they stand. However, from a scientific perspective, it is not sufficient to show that two variables are related. An association does not mean that one variable causes another. This is an important distinction. If self-efficacy causes lower depression symptoms, then if we could intervene and increase someone's self-efficacy, we would expect them to have fewer depression symptoms. However, if self-efficacy is not the cause but simply a predictor or associated with depression symptoms, then changing self-efficacy may have no impact on depression symptoms.

Formally, this introduces a concept often called confounding. Two variables may be associated with each other for a number of reasons. One reason for an association is that one variable causes the other. However, in an inaccurate model of reality, you can also find an association between two variables because some third variable causes both. For example, suppose that having chronic health problems both causes lower self-efficacy and higher levels of depression symptoms. If presence of chronic health problems is not somehow accounted for, there would appear to be an association between self-efficacy and depression symptoms. However, once influence of chronic health problems is taken into account, there may be no association of self-efficacy and depression symptoms.

A common way of representing different possible causal configurations is through causal graphs. These can be read as circles representing different variables and directed arrows indicating which variable causes which. For a gentle introduction to such models and a more in-depth overview of causal reasoning, see [81]. An example causal graph is shown in Figure 3-8. In Figure 3-8, Z is the common cause of both X and Y. From our conceptual example, Z would be chronic conditions, X would be self-efficacy, and Y would be depression symptoms at wave 2. Without accounting for Z, an inaccurate, biased estimate of the association of X and Y would be obtained.

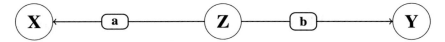

Figure 3-8. *Example diagram where a variable, Z, is a common cause of X and Y. If Z is not accounted for, there will appear to be an association between X and Y.*

There also are other ways in which an inaccurate model can yield biased estimates. Figure 3-9 shows another graph model where Z is now a common outcome of both X and Y, known as a collider variable because the paths from X and Y collide on Z. If the truth is as shown in Figure 3-9, then testing the association between X and Y directly would be appropriate. However, if we tried to control for Z as a confound variable, by adding Z into the model predicting Y, far from reducing confounding by Z, the model would induce a spurious association between X and Y. The lesson from these two examples is that if Z is a common cause (e.g., Figure 3-8), inappropriately *excluding* paths induces bias. Conversely, if Z is a collider (e.g., Figure 3-9), inappropriately *including* paths induces bias.

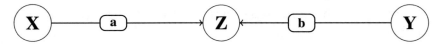

Figure 3-9. *Example diagram where a variable, Z, is a collider (common outcome) of X and Y. If Z is ignored when examining the association between X and Y, that association will be accurately estimated. However, if Z, the collider, is conditioned on, then it will open an association between X and Y.*

A final example is shown in Figure 3-10. Here Z is the mechanism through which X impacts Y. Put differently, Z transmits the effect of X to Y. In this case, we could test an indirect effect of X on Y through Z, but without Z in the model, we should see an association of X with Y, but once Z is added to the model, X should no longer directly be associated with Y because all of the X–Y association is conveyed through Z.

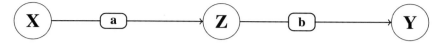

Figure 3-10. *Example diagram where a variable, Z, is a mechanism (mediator) of X and Y. Because Z transmits the effect of X to Y, conditioning on Z will wipe out the association of X and Y. If Z is ignored when examining the association between X and Y, that association will be accurately estimated.*

Many of these ideas, both for cross-sectional data and particularly in a longitudinal context, are covered in more depth in theory developed for marginal structural models (MSMs) [80]. MSMs are generally estimated using inverse probability weighted estimators [80]. In the context of a discrete predictor, X, a basic equation to calculate the inverse probability weight (IPW) for the *ith* person is

$$w_i = \frac{1}{P\left(X = x_i \mid Z = z_i\right)} \tag{3.18}$$

Sometimes these basic weights are extended to calculate so-called stabilized weights, which are just adjusted for the marginal probability that X takes on any specific value, given by the equation

$$sw_i = \frac{P\left(X = x_i\right)}{P\left(X = x_i \mid Z = z_i\right)} \tag{3.19}$$

Although the use of inverse probability weighted (IPW) estimators came from models where the focal predictor was a dichotomous treatment, the IPW idea generalizes readily to continuous predictors as well. In brief, the idea is that suppose you are studying a focal predictor, X, and its association with some outcome, Y, but there is some known confounder variable, Z. One way of adjusting for the effect of Z is to calculate the probability of X given a particular Z. In continuous cases, the same basic idea is used, except that rather than relying on the probability mass function, we must use the probability density function for some assumed distribution. Further, it is standard to use stabilized weights. Assuming we believe that X follows a normal distribution, $X \sim N(\mu, \sigma)$, we estimate stabilized weights for a continuous predictor as

$$sw_i = \frac{f_X(X; \mu_1, \sigma_1)}{f_{X|Z}(X | Z = z_i; \mu_2, \sigma_2)} \tag{3.20}$$

where $f_X(\cdot)$ is the probability density function for a normal distribution. Essentially, by using the probability density function estimating based on an unconditional model for X, we can control for different amounts of variance in X.

Using the IPWs, we can take our model of interest and estimate it using the IPWs, and this is proven to asymptotically yield an unbiased estimate of the association of the predictor X with the outcome Y if both the model for generating weights and the model of the association of X and Y are correctly specified. Misspecification of either the weights model (e.g., by failing to include a required confounding variable, or failing to specify the correct functional form, such as specifying a linear association when in reality it is quadratic) or the focal model will result in biased estimates. One recommendation in the case of a continuous exposure is to winsorize or trim the weights at the bottom and top one percentiles to reduce noise associated with extreme weights [68]. Further information on constructing IPWs is available in the references [22] and [68].

To construct IPWs in R, we can use the excellent `ipw` package [96]. First, we have to decide which variables we wish to adjust for in our IPW-adjusted model. It is common to have a couple of steps. Perhaps as our first step, we include some potential common causes that we believe (based on theory, previous data, hopefully something slightly stronger than just random intuition) to be potential common causes of both self-efficacy and depression symptoms. Sex, race/ethnicity, and age may make good choices, in particular because it is implausible that any of those could be caused by either self-efficacy or depression symptoms. We also may include number of chronic conditions.

To calculate IPWs at a single time point, we use the `ipwpoint()` function from the `ipw` package. The function requires us to specify the variable name of our exposure; the expected distribution, in this case we assume normal; and then a model for the numerator and the denominator probability density functions. Lastly, of course we must tell `ipwpoint()` which dataset to use. We save the results as the variable, "w", in R. The IPWs can be accessed from the results as `ipw.weights`. A quick plot of the distribution of the raw weights and the weights after winsorizing the bottom and top 1 percentiles is shown in Figure 3-11.

```
## weights
w <- ipwpoint(
  exposure = SelfEfficacy_W1,
  family = "gaussian",
  numerator = ~ 1,
  denominator = ~ 1 + Sex + RaceEthnicity + AGE_W1 + NChronic12_W1,
  data = acl)

plot_grid(
  testdistr(w$ipw.weights, plot = FALSE)$DensityPlot,
  testdistr(winsorizor(w$ipw.weights, .01),
          plot = FALSE)$DensityPlot,
  ncol = 1)
```

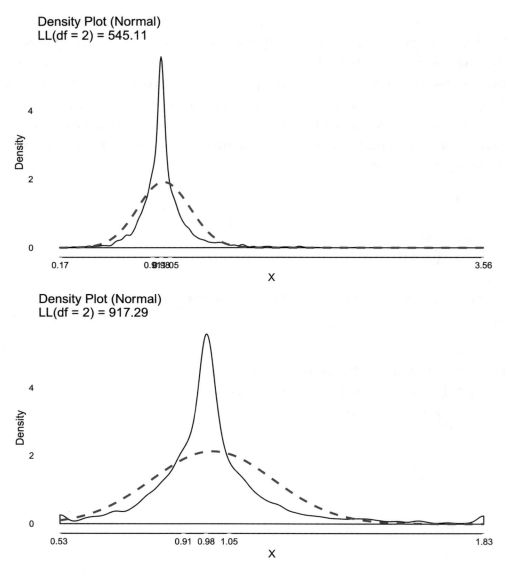

Figure 3-11. *Raw and trimmed inverse probability weights for self-efficacy*

Once we have the weights, we can estimate a weighted model, by passing the weights to the weights argument of ols(). Using IPWs will adjust for sex, race/ethnicity, age, and chronic conditions as potential confounds. For comparison we include the unadjusted model and the results are in Table 3-8.

```
## unweighted, unadjusted
mO <- ols(CESD11_W2 ~ SelfEfficacy_W1, data = acl)
```

```
## weighted, adjusted
m1 <- ols(CESD11_W2 ~ SelfEfficacy_W1, data = acl,
  weights = winsorizor(w$ipw.weights, .01))

texreg(list(m0, m1),
       label = "tglm1-weight1")
```

Table 3-8. *Statistical Models*

	Model 1	Model 2
Intercept	0.02	0.02
	(0.02)	(0.02)
SelfEfficacy_W1	−0.36***	−0.32***
	(0.02)	(0.02)
Num. obs.	2867	2867
R^2	0.13	0.11
Adj. R^2	0.13	0.11
L.R.	399.71	325.23

***p < 0.001, **p < 0.01, *p <0.05*

As a sensitivity analysis, we may also try further adjusting for additional factors that we believe may be confounds or potentially may be mediators or mechanisms conveying the effects of self-efficacy on depression symptoms. In this case we add socioeconomic status category, employment, body mass index, smoking status, and physical activity categories. We estimate the weights and then re-estimate the model using these new weights. A comparison including the unadjusted model and partially and fully adjusted models is given in Table 3-9.

```
# weighted, fully adjusted
w2 <- ipwpoint(
  exposure = SelfEfficacy_W1,
  family = "gaussian",
  numerator = ~ 1,
```

```
denominator = ~ 1 + Sex + RaceEthnicity + AGE_W1 + NChronic12_W1 +
    SESCategory + Employment_W1 + BMI_W1 + Smoke_W1 + PhysActCat_W1,
  data = acl)

m2 <- ols(CESD11_W2 ~ SelfEfficacy_W1, data = acl,
  weights = winsorizor(w2$ipw.weights, .01))

texreg(list(m0, m1, m2),
       label = "tglm1-weight2")
```

Table 3-9. *Statistical Models*

	Model 1	**Model 2**	**Model 3**
Intercept	0.02	0.02	0.02
	(0.02)	(0.02)	(0.02)
SelfEfficacy_W1	−0.36***	−0.32***	−0.29***
	(0.02)	(0.02)	(0.02)
Num. obs.	2867	2867	2867
R^2	0.13	0.11	0.09
Adj. R^2	0.13	0.11	0.09
L.R.	399.71	325.23	261.52

***$p < 0.001$, **$p < 0.01$, *$p < 0.05$

Another approach to adjusting for potential confounding factors is to simply add the potential confounds into the model. Examples of this are shown in the following code, called models m1b and m2b, the "b" indicating that it is an alternative of the IPW "model 1" and "model 2":

```
m1b <- ols(CESD11_W2 ~ Sex + RaceEthnicity + AGE_W1 +
  NChronic12_W1 + SelfEfficacy_W1,
  data = acl)

m2b <- ols(CESD11_W2 ~ Sex + RaceEthnicity + AGE_W1 +
  NChronic12_W1 + SESCategory +
  Employment_W1 + BMI_W1 + Smoke_W1 + PhysActCat_W1 +
  SelfEfficacy_W1, data = acl)
```

Finally, some people have suggested the use of so-called doubly robust estimators. Doubly robust estimators simply include both IPW weights and then include the same confounds used to construct the weights into the model explicitly. Examples of this are shown in the following code and labelled m1c and m2c as these are yet another variant of our two adjusted models.

```
m1c <- ols(CESD11_W2 ~ Sex + RaceEthnicity + AGE_W1 +
  NChronic12_W1 + SelfEfficacy_W1,
  data = acl,
  weights = winsorizor(w$ipw.weights, .01))

m2c <- ols(CESD11_W2 ~ Sex + RaceEthnicity + AGE_W1 +
  NChronic12_W1 + SESCategory +
  Employment_W1 + BMI_W1 + Smoke_W1 + PhysActCat_W1 +
  SelfEfficacy_W1, data = acl,
  weights = winsorizor(w2$ipw.weights, .01))
```

To compare these different methods, we can extract the estimates and confidence intervals from each model and then make a plot so we can easily visualize the differences. This is shown in the code that follows and the result is in Figure 3-12. In this instance, all of the results are quite similar. That often is the case when variables exist at only one point in time. However, the approaches can diverge more in the context of marginal structural models, where IPWs are particularly recommended.

```
## write an extract function
extractor <- function(obj, label) {
  b <- coef(obj)
  ci <- confint(obj)
  data.table(
    Type = label,
    B = b[["SelfEfficacy_W1"]],
    LL = ci["SelfEfficacy_W1", "2.5 %"],
    UL = ci["SelfEfficacy_W1", "97.5 %"])
}

allresults <- rbind(
  extractor(m0,  "M0: Unadjusted"),
  extractor(m1,  "M1: Partial IPW"),
```

```
  extractor(m1b, "M1: Partial Covs"),
  extractor(m1c, "M1: Partial Covs + IPW"),
  extractor(m2,  "M2: Full IPW"),
  extractor(m2b, "M2: Full Covs"),
  extractor(m2c, "M2: Full Covs + IPW"))
allresults[, Type := factor(Type, levels = Type)]

ggplot(allresults, aes(Type, y = B, ymin = LL, ymax = UL)) +
  geom_pointrange() +
  coord_flip() +
  xlab("") + ylab("Estimate + 95% CI")
```

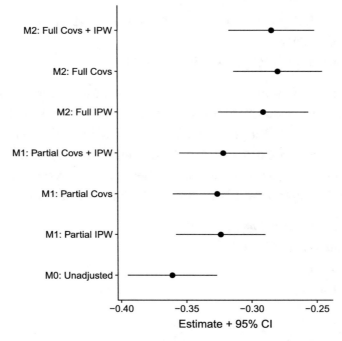

Figure 3-12. *Comparison of the estimate and confidence interval for the association of self-efficacy with depression symptoms from various models. Covs = covariate adjusted models. IPW = inverse probability weight adjusted models. Covs + IPW = models that both include inverse probability weights and the same potential confounds explicitly in the model again.*

3.8 Case Study: Multiple Linear Regression with Interactions

This case study is patterned off a journal article where the researchers were interested in testing a cognitive vulnerability model for sleep and negative mood in adolescents [5]. In this study, approximately 150 adolescents completed measures of negative mood (MOOD, a composite from depression and anxiety symptoms), dysfunctional beliefs about sleep (DBAS), general dysfunctional attitudes (DAS), academic stress (Stress), and subjective sleep quality (SSQ) and wore accelerometers to assess sleep objectively. Specifically, how many minutes it takes to fall asleep at night (sleep onset latency; SOLacti). Measures were collected during school and vacation periods, but here we are focusing on the vacation period. Note that with these data, higher scores on subjective sleep quality indicate poorer sleep quality.

Using the standardized regression coefficients from Table 2 from Bei and colleagues [5] and the means and standard deviations from Table 1 in the same article, we can simulate a dataset that is approximately like the data used in that article.

```
set.seed(12345)
adosleep <- data.table(
  SOLacti = rnorm(150, 4.4, 1.3)^2,
  DBAS = rnorm(150, 72, 26),
  DAS = rnorm(150, 125, 32),
  Female = rbinom(150, 1, .53),
  Stress = rnorm(150, 32, 11))
adosleep[, SSQ := rnorm(150,
            (.36 * 3 / 12.5) * SOLacti +
            (.16 * 3 / 26) * DBAS +
            (.18 * 3 / .5) * Female +
            (.20 * 3 / 11) * Stress, 2.6)]
adosleep[, MOOD := rnorm(150,
            (-.07 / 12.5) * SOLacti +
            (.29  / 3) * SSQ +
            (.14  / 26) * DBAS +
            (.21  / 32) * DAS +
            (.12  / 32) * SSQ * (DAS-50) +
            (.44  / .5) * Female +
```

```
            (.28 / 11) * Stress, 2)]
adosleep[, Female := factor(Female, levels = 0:1,
                            labels = c("Males", "Females"))]
```

As part of a larger model, the researchers hypothesized that subjective sleep quality would be associated with negative mood, but this relationship would be moderated by general dysfunctional attitudes. Specially, dysfunctional attitudes were thought to be a vulnerability, such that adolescents high in dysfunctional attitudes who also perceived poor sleep quality would be vulnerable to more negative mood. Conversely, for those with lower levels of dysfunctional attitudes, even with poor subjective sleep quality, they may be less vulnerable (have a weaker relationship) between subjective sleep quality and negative mood. Sleep onset latency assessed by actigraphy (an objective measure), stress, sex, and dysfunctional beliefs about sleep were also included in the analyses.

To begin with, we check the core variables to examine their distribution and to look for any outliers (Figure 3-13).

```
plot_grid(
  testdistr(adosleep$MOOD, extremevalues = "theoretical",
          plot=FALSE, varlab = "MOOD")$Density,
  testdistr(adosleep$SSQ, extremevalues = "theoretical",
          plot=FALSE, varlab = "SSQ")$Density,
  testdistr(adosleep$SOLacti, extremevalues = "theoretical",
          plot=FALSE, varlab = "SOLacti")$Density,
  testdistr(adosleep$DAS, extremevalues = "theoretical",
          plot=FALSE, varlab = "DAS")$Density,
  ncol = 2)
```

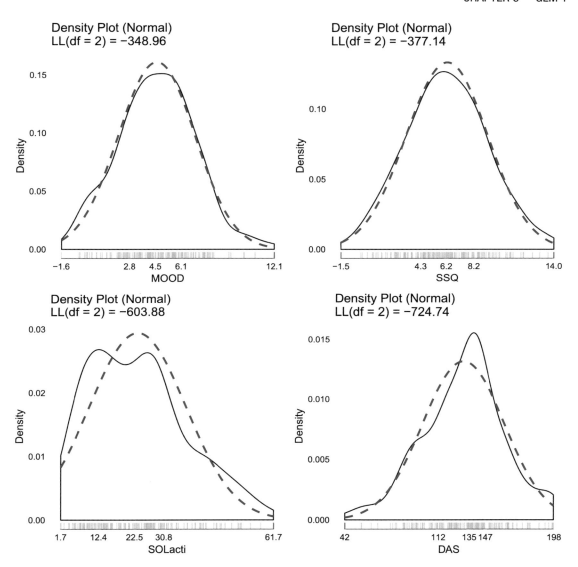

Figure 3-13. *Distributions of case study variables*

Next, we examine the bivariate correlations between (continuous) study variables (Figure 3-14).

```
plot(SEMSummary(
  ~ MOOD + SOLacti + DBAS + DAS + Stress + SSQ,
  data = adosleep), plot = "cor") +
  theme(axis.text.x = element_text(
        angle = 45, hjust = 1, vjust = 1))
```

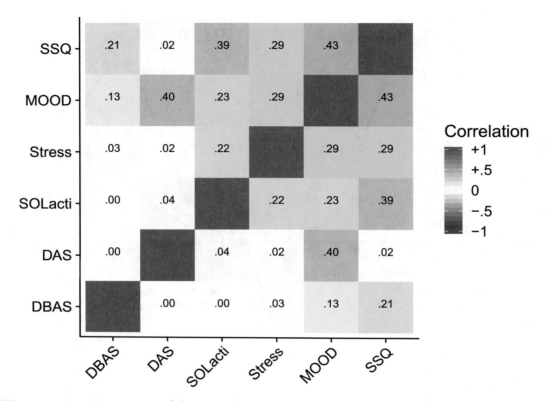

Figure 3-14. *Heatmap of the correlations between study variables*

Next we make a table of descriptive statistics for the study. Although quite basic, a table or graphs of descriptive statistics on the study variables are standard practice in most presentations of results to give readers a better understanding of the distribution and spread of each variable being tested. Here, we make use of the egltable() function from the JWileymisc package to calculate and present the mean and standard deviation for continuous variables and the N and percentage for discrete variables (here just Female).

```
egltable(c("SOLacti", "SSQ", "MOOD", "Stress",
         "DBAS", "DAS", "Female"),
      data = as.data.frame(adosleep))
```

```
##                  M (SD)/N (%)
## 1:   SOLacti   23.33 (13.60)
## 2:       SSQ    6.18 (3.00)
```

```
## 3:       MOOD     4.53 (2.49)
## 4:     Stress  32.84 (10.92)
## 5:       DBAS  72.10 (23.88)
## 6:        DAS 130.57 (30.45)
## 7:     Female
## 8:      Males       67 (44.7)
## 9:    Females       83 (55.3)
```

To get standardized estimates as in Bei and colleagues' paper [5], we can standardize predictors.

```
adosleep[, zMOOD := as.vector(scale(MOOD))]
adosleep[, zDBAS := as.vector(scale(DBAS))]
adosleep[, zDAS := as.vector(scale(DAS))]
adosleep[, zSSQ := as.vector(scale(SSQ))]
adosleep[, zSOLacti := as.vector(scale(SOLacti))]
adosleep[, zStress := as.vector(scale(Stress))]
```

Next we fit three different models for comparison. First we start with all the covariates. Second, we add the main constructs of interest without interactions. Third, we add the hypothesized interaction between subjective sleep quality and global dysfunctional beliefs. Finally, we put all the results together in a table using the screenreg() function. In this case, screenreg() outputs the coefficients and standard errors in parentheses, with the p-value thresholds indicated using asterisks. This layout makes it very easy to compare how coefficients change depending on what other factors are in the model.

```
m.adosleep1 <- ols(zMOOD ~ zSOLacti + zDBAS + Female + zStress,
                 data = adosleep)
m.adosleep2 <- update(m.adosleep1, . ~ . + zSSQ + zDAS)
m.adosleep3 <- update(m.adosleep2, . ~ . + zSSQ:zDAS)

screenreg(list(m.adosleep1, m.adosleep2, m.adosleep3))
```

```
## 
## ========================================================
##                    Model 1      Model 2      Model 3
## --------------------------------------------------------
## Intercept          -0.24 *      -0.28 **     -0.28 **
##                    (0.11)       (0.09)       (0.09)
## zSOLacti            0.17 *       0.04         0.03
##                    (0.08)       (0.07)       (0.07)
## zDBAS               0.14         0.07         0.08
##                    (0.08)       (0.06)       (0.06)
## Female=Females      0.44 **      0.50 ***     0.50 ***
##                    (0.15)       (0.13)       (0.13)
## zStress             0.26 ***     0.19 **      0.20 **
##                    (0.08)       (0.07)       (0.07)
## zSSQ                             0.34 ***     0.34 ***
##                                 (0.07)       (0.07)
## zDAS                             0.41 ***     0.44 ***
##                                 (0.06)       (0.06)
## zSSQ * zDAS                                   0.14 *
##                                              (0.07)
## --------------------------------------------------------
## Num. obs.          150          150          150
## R^2                 0.18         0.43         0.45
## Adj. R^2            0.16         0.41         0.42
## L.R.               29.54        85.59        89.91
## ========================================================
## *** p < 0.001, ** p < 0.01, * p < 0.05
```

Recalling that higher scores on subjective sleep quality indicate worse sleep quality, Model 2 shows that overall worse sleep quality and overall dysfunctional attitudes are significantly associated with more negative mood (both p < .001). Turning to Model 3, the interaction between sleep quality and dysfunctional attitudes is positive and significant, indicating that the more dysfunctional attitudes adolescents' have, the stronger the relationship between poor subjective sleep quality and negative mood.

To make sure that the models seem appropriate, we could check the variance inflation factors and the distribution of residuals. A reasonable starting point is the most complex model as that has the most variables with the highest chance that some may be collinear.

```
vif(m.adosleep3)
```

```
##        zSOLacti           zDBAS Female=Females          zStress
##            1.2             1.1              1.0             1.1
##            zSSQ            zDAS     zSSQ * zDAS
##            1.3             1.1              1.1
```

```
testdistr(resid(m.adosleep3), plot=FALSE, varlab = "Residuals")$QQPlot
```

None of the variance inflation factors are particularly high, and the residuals appear normally distributed (Figure 3-15), so we proceed.

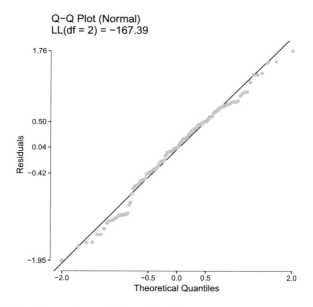

Figure 3-15. *Distribution of model residuals*

To unpack the interaction, it is helpful to create a graph. For graphing purposes, it is common to use the original scale of variables, if the scale of a measure makes sense (e.g., age in years) or the measure is commonly used so that readers are likely to be familiar with the original scale. We also create a new dataset to generate predictions for graphing. All covariates we hold at their mean or mode. As exemplary values of "low" and "high" dysfunctional attitudes, we use the mean - one standard deviation and mean + one standard deviation.

```
## refit model on raw data
m.adosleep.raw <- ols(MOOD ~ SOLacti + DBAS + Female +
                           Stress + SSQ * DAS,
                       data = adosleep)

## create a dataset
adosleep.newdat <- as.data.table(with(adosleep, expand.grid(
  SOLacti = mean(SOLacti),
  DBAS = mean(DBAS),
  Female = factor("Females", levels(Female)),
  Stress = mean(Stress),
  SSQ = seq(from = min(SSQ), to = max(SSQ), length.out = 100),
  DAS = mean(DAS) + c(1, -1) * sd(DAS))))

adosleep.newdat$MOOD <- predict(m.adosleep.raw,
                                newdata = adosleep.newdat,
                                se.fit = FALSE)

adosleep.newdat[, DAS := factor(round(DAS),
  levels = c(100, 161),
  labels = c("M - 1 SD", "M + 1 SD"))]
```

Figure 3-16 shows how the relations between subjective sleep quality and negative mood vary at low and high levels of dysfunctional attitudes, with an exaggerated relation of poor sleep quality with negative mood for vulnerable adolescents with higher levels of dysfunctional attitudes.

```
ggplot(adosleep.newdat, aes(SSQ, MOOD, linetype=DAS)) +
  geom_line(size = 2) +
  scale_x_continuous("Subjective sleep quality\n(higher is worse)") +
  ylab("Negative Mood") +
  theme_cowplot() +
  theme(
    legend.position = c(.85, .15),
    legend.key.width = unit(2, "cm"))
```

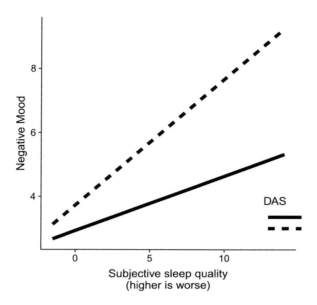

Figure 3-16. *Interaction between subjective sleep quality and overall dysfunctional beliefs predicting negative mood*

Overall, this provides a fairly thorough examination of the question we started with for this case study. For ideas on other ways of presenting results and what more detailed interpretations may look like, we refer readers to the original article [5] which is freely available online: https://doi.org/10.5665/sleep.4508.

3.9 Summary

This chapter introduced the conceptual background to the generalized linear model (GLM) and how that broad framework encompasses many common statistical models as special cases. Two specific special cases of the GLM were also introduced: ANOVA and linear regression, which are appropriate when the focal outcome or dependent variable is continuous and normally distributed. A summary of some of the main modelling functions introduced in this chapter is shown in Table 3-10. The next chapter will continue the GLM theme but focus on GLMs for outcomes that do not follow a normal distribution, such as binary and count data.

Table 3-10. *Listing of Key Functions Described in This Chapter and Summary of What They Do*

Function	What It Does
model.matrix()	Takes a formula describing explanatory predictor variables and generates a design matrix to fit GLMs.
update()	Facilitates updating existing formulae or model objects to either add or remove variable(s).
ezANOVA()	Function from the ez package that provides a framework for fitting many types of ANOVA models as well as fit indices and diagnostics.
anova()	Built-in function for fitting ANOVA models. Also used to get an ANOVA summary table from a GLM model.
ols()	Function from the rms package that fits linear regression models along with a comprehensive default output and diagnostic information.
glm()	Built-in R function for fitting generalized linear models, including linear regressions and more.
summary()	A function that is often called on the results from a fitted GLM to produce additional summary information.
coef()	A function to extract the regression coefficients from linear regression or other fitted GLMs.
vif()	A function to calculate the variance inflation factor for each predictor in a linear regression model to determine how much the parameter covariance matrix is inflated due to having collinear explanatory variables in the model.
predict()	A function to generate predicted values on new data using an existing fitted model. Also useful for generating interaction graphs or graphs of the predictions from a model.
ipwpoint()	A function to calculate inverse probability weights at one time point for either discrete or continuous predictor/exposure variables. Useful for calculating marginal structural models and attempting to account for potential confounding factors.
texreg()	A function to make tables from a variety of models. Also possible to include multiple models in one table for easy comparison.

CHAPTER 4

GLM 2

Generalized linear models (GLMs) also can accommodate outcomes that are not continuous and normally distributed. Indeed, one of the great advantages of GLMs is they provide a unified framework to understand regression models applied to variables assumed to come from a variety of distributions. For this chapter, we will lean heavily on one excellent R package, VGAM, which provides utilities for vector generalized linear models (VGLMs) and vector generalized additive models (VGAMs) [125]. VGLMs and VGAMs are an even more flexible class of models where there may be multiple responses. However, beyond offering flexibility of multiple parameters, the VGAM package implements over 20 link functions, well over 50 different models/assumed distributions. We will only scratch the surface of the VGAM package capabilities in this chapter, but its great flexibility means that we will not need to introduce many different packages nor many different functions. If you would like to learn about VGLMs and VGAMs in far greater depth, we recommend an excellent book by the author of the VGAM package [125].

```
library(checkpoint)
checkpoint("2018-09-28", R.version = "3.5.1",
  project = book_directory,
  checkpointLocation = checkpoint_directory,
  scanForPackages = FALSE,
  scan.rnw.with.knitr = TRUE, use.knitr = TRUE)

library(knitr)
library(data.table)
library(ggplot2)
library(ggthemes)
library(scales)
library(viridis)
library(VGAM)
```

© Matt Wiley and Joshua F. Wiley 2019
M. Wiley and J. F. Wiley, *Advanced R Statistical Programming and Data Models*,
https://doi.org/10.1007/978-1-4842-2872-2_4

```
library(ipw)
library(JWileymisc)
library(xtable)
library(texreg)

options(
  width = 70,
  stringsAsFactors = FALSE,
  datatable.print.nrows = 20,
  datatable.print.topn = 3,
  digits = 2)
```

4.1 Conceptual Background

This chapter covers a few specific types of GLMs that together account for most of the other commonly used GLMs besides the most common linear model.

Logistic Regression

Three types of GLMs for discrete data are binary, ordered, and multinomial logistic regression. They share the canonical link function, the logit, with the general form:

$$\log_e\left(\frac{p}{1-p}\right) \tag{4.1}$$

where p is some probability. For binary and multinomial logistic regression, we could write the probabilities as

$$\log_e\left(\frac{P(Y=j|X)}{P(Y=M+1|X)}\right) \tag{4.2}$$

for j in 1, ..., M where there are $M + 1$ discrete levels of the outcome. In the binary case, there are two levels of the outcome, $M = 1$ and $M + 1 = 2$, because we know that the sum of the probabilities that belong to any given group (level of the outcome) must be 1:

$$\sum_{j=1}^{M+1} P(Y=j|X) = 1 \tag{4.3}$$

Then the logit expression for binary outcomes simplifies to

$$\log_e \left(\frac{P(Y=1|X)}{1-P(Y=1|X)} \right) \tag{4.4}$$

For the ordered logistic regression case, it is standard to use a cumulative logit , which takes the form

$$\log_e \left(\frac{P(Y \le j|X)}{1-P(Y \le j|X)} \right) \tag{4.5}$$

The ratio of probabilities is known as the odds:

$$Odds = \frac{P(Y=1|X)}{1-P(Y=1|X)} \tag{4.6}$$

For example, suppose that given some predictor values and model, $P(Y=1|X) = 0.75$, then

$$Odds = \frac{.75}{1-.75} = \frac{.75}{.25} = 3 \tag{4.7}$$

If $Odds$ = 3 is interpreted as saying that someone with that particular set of predictor values is expected to be three times as likely to have the event ($Y=1$) than not to have the event ($Y=0$). The logit is the log of the odds and ensures that it is at least theoretically possible for it to range from $-\infty$ to $+\infty$.

The regression coefficients are also based on odds. Taking the simplest case of a model with a single predictor, x_1 the model would be

$$\log_e \left(\frac{P(Y_i=1|X=x_{1i})}{1-P(Y_i=1|X=x_{1i})} \right) = \beta_0 + \beta_1 * x_{1i} \tag{4.8}$$

The coefficient, β_1, is then defined as

$$\beta_1 = \log_e \left(\frac{P(Y_i=1|X=x_{1i}+1)}{P(Y_i=1|X=x_{1i})} \right) \tag{4.9}$$

125

This is the natural log of the ratio of odds given x_{i1} vs. $x_{i1} + 1$. It is common practice to report odds ratios in place of log odds ratios, which is accomplished by exponentiating the coefficients.

$$OddsRatio = OR = e^{\beta_1} \tag{4.10}$$

Supposing that $\beta_1 = 0.5$, then $e^{0.5} = 1.65$, and we would interpret this as indicating that a one unit change in x_1 was associated with 1.65 times the odds of having the event. In contrast to Odds which have a fairly natural interpretation, odds ratios are somewhat more difficult to interpret as they represent a multiplicative change in odds, but do not tell us about what the odds were to begin with. To make this clear, an odds ratio of 2 could equally arise from .02/.01 and 2/1. In both cases, the odds ratio is 2, but even though indeed someone has twice the odds, on an absolute basis the odds are still tiny, .02, and in the other case, the new odds indicate that someone is twice as likely to have the event as not to have the event.

Count Regression

The other types of GLMs we will cover in this chapter are models designed for count data. Count data are discrete, similar to outcomes used in logistic regression, but unlike logistic regression, count outcomes may take on many values and are ordered. Formally, the domain for count outcomes is integers in $0 - \infty$. Count outcomes occur in a variety of cases. For example, in medical settings particularly as the population ages, it is increasingly common to consider a count of comorbid conditions. In the insurance industry, it is desirable to build models about how many accidents one person will be in. Many people may have zero accidents, but some will have one, some two, fewer three, four, etc. In production, fault and failure rates are of interest. If one line or factory has a greater count of faulty products, that is valuable information to reduce errors and costs and improve quality control.

The two most common types of GLMs for count data are Poisson regression and negative binomial regression. The Poisson distribution has a single parameter, the rate or mean, conventionally denoted λ. Given that parameter, the probability mass function for a Poisson is

$$P(Y = y; \lambda) = \frac{e^{-\lambda}\lambda^{y}}{y!} \tag{4.11}$$

Both the mean and the variance of the Poisson distribution are λ. For comparison, here are two Poisson distributions, with different rates in Figure 4-1. Note that we use geom_col() instead of geom_bar() when we have precomputed values and want a bar plot.

```
dpoisson <- data.table(X = 0:20)
dpoisson[, Lambda2 := dpois(X, lambda = 2)]
dpoisson[, Lambda6 := dpois(X, lambda = 6)]

ggplot(melt(dpoisson, id.vars = "X"),
       aes(X, value, fill = variable)) +
  geom_col(position = "dodge") +
  scale_fill_viridis(discrete = TRUE) +
  theme(legend.position = c(.7, .8)) +
  xlab("Y Score") + ylab("Poisson Density")
```

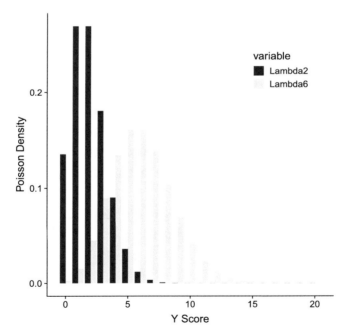

Figure 4-1. *Density for a Poisson distribution with lambda = 2 and lambda = 6*

The canonical link function for both Poisson and negative binomial regression is the natural logarithm. Thus, the standard model for both Poisson and negative binomial regression is

$$\log_e(Y_i) = \beta_0 + \beta_1 * x_1 + \cdots + \beta_k * x_k \tag{4.12}$$

This gives their coefficients a convenient interpretation. For example, β_1 would be how many more times the event is expected to occur for a one unit change in x_1.

Negative binomial regression is very similar to Poisson regression. The only change is that the negative binomial distribution includes an additional parameter that allows the variance to be different from the mean. The probability mass function for the negative binomial distribution is more complex than for the Poisson:

$$P(Y = y; \lambda, v) = \binom{y + v - 1}{y} \left(\frac{\lambda}{\lambda + v}\right)^y \left(\frac{v}{v + \lambda}\right)^v \tag{4.13}$$

The mean of the negative binomial distribution is still λ. However, the variance is now given by $\lambda + \frac{\lambda^2}{v}.v^{-1}$ that is known as the scale or dispersion parameter, sometimes also called the ancillary parameter. As v increases, the negative binomial distribution becomes closer and closer to the Poisson, such that the negative binomial $\lim_{v \to \infty}$ is the Poisson.

In many applied cases, the negative binomial is a better choice than the Poisson, as it is only modestly more complex, and tends to be a more realistic match to data as the assumption that the mean and variance are the same required for the Poisson distribution is frequently violated.

The interpretation of negative binomial regression is more or less identical to Poisson regression, so no extra effort is needed there.

4.2 R Examples

For the examples in this chapter, we will again use the Americans' Changing Lives [45] study data. Reading and preparing the data was covered in the "Introduction" section labelled "Data Setup." Technically, the data have sampling weights, but for simplicity we ignore these weights. The analyses will still be correct without weighting; they just will not reflect the sampled population.

```
acl <- readRDS("advancedr_acl_data.RDS")
```

Binary Logistic Regression

To try a binary logistic regression model, we need a binary outcome variable. We achieve this by converting smoking status into a binary current smoker vs. not (former or never smoker). Then we use the `vglm()` function with `family = binomialff()` and a logit link to run our logistic regression model. As with other models, the `summary()` function provides a summary of the model and coefficients.

```
acl$CurSmoke <- as.integer(acl$Smoke_W1 == "(1) Cur Smok")

m.lr <- vglm(CurSmoke ~ Sex,
             family = binomialff(link = "logit"),
             data = acl, model = TRUE)
summary(m.lr)

##
## Call:
## vglm(formula = CurSmoke ~ Sex, family = binomialff(link = "logit"),
##     data = acl, model = TRUE)
##
##
## Pearson residuals:
##                 Min   1Q Median   3Q Max
## logit(prob) -0.712 -0.603 -0.603 1.4 1.66
##
## Coefficients:
##               Estimate Std. Error z value Pr(>|z|)
## (Intercept)    -0.6788     0.0574  -11.82  < 2e-16 ***
## Sex(2) FEMALE  -0.3314     0.0746   -4.44  8.8e-06 ***
## ---
## Signif. codes:  0 '***' 0.001 '**' 0.01 '*' 0.05 '.' 0.1 '_' 1
##
## Number of linear predictors: 1
##
## Name of linear predictor: logit(prob)
##
## Residual deviance: 4356 on 3615 degrees of freedom
```

```
##
## Log-likelihood: -2178 on 3615 degrees of freedom
##
## Number of iterations: 4
##
## No Hauck-Donner effect found in **any** of the estimates
```

In this simple example with a binary outcome and a single binary predictor, we can readily calculate the odds ratio directly using the frequencies. In a 2 x 2 frequency table, the odds ratio is the ratio of the odds, and the coefficient reported from the logistic regression is the natural logarithm of the odds ratio. Specifically, using the table shown in Table 4-1, the odds ratio is defined as

$$\frac{\dfrac{a}{c}}{\dfrac{b}{d}}$$

Table 4-1. *Hypothetical Frequency Table*

Predictor	No Smoke	Smoke
Male	A	B
Female	C	D

For our data, the actual frequency table is in Table 4-2:

```
or.tab <- xtabs(~ Sex + CurSmoke, data = acl)
or.tab.res <- (or.tab[1,1]/or.tab[2,1])/(or.tab[1,2]/or.tab[2,2])
xtable(or.tab, caption = "Observed frequency table",
       label = "tglm2-obsfreq")
```

Table 4-2. *Observed Frequency Table*

	0	1
(1) MALE	901	457
(2) FEMALE	1656	603

The resulting odds ratio can be calculated as

$$\frac{\dfrac{901}{1656}}{\dfrac{457}{603}} = 0.72$$

The natural logarithm of the odds is -0.33 which is the same as the coefficient from our logistic regression model, -0.33.

Understanding how odds ratios are calculated is helpful for understanding how to interpret them. The simple equation for odds ratios from a 2 x 2 frequency table

$$\frac{\dfrac{a}{c}}{\dfrac{b}{d}}$$

also explains the requirement of logistic regression that there cannot be any zero cells. If any of the cells are zero, the result is either undefined (division by zero) if c, b, or d is zero or the result is exactly zero if a is zero, and the logarithm of zero is negative infinity, meaning the coefficient from the logistic regression would be negative infinity, which is also problematic.

Despite being able to read the equation to calculate an odds ratio or the log odds ratio, for most people, it is not an intuitive value to interpret. Many people find the probability scale easier to interpret. For example, we could find the probability (proportion or percentage) of men and women who smoke. Referring back to 4-1, the equation for the probability of smoking for men is

$$\frac{b}{a+b}$$

and the equation for the probability of smoking for women is

$$\frac{d}{c+d}$$

We can also report the difference in the two probabilities, to quantify how different the probability of smoking is for men and women. The easiest and most general way to get the probabilities is to predict them based on data and our model. We can use the `predict()` function. To get predictions on the probability scale rather than the log odds scale, we use the optional argument `type = "response"`, so that results are backtransformed to the original scale, shown in Figure 4-2.

```
preddat <- data.table(Sex = levels(acl$Sex))
preddat$yhat <- predict(m.lr, newdata = preddat,
        type = "response")
```

```
ggplot(preddat, aes(Sex, yhat)) +
  geom_bar(stat = "identity") +
  scale_y_continuous("Smoking Probability", labels = percent) +
  theme_tufte()
```

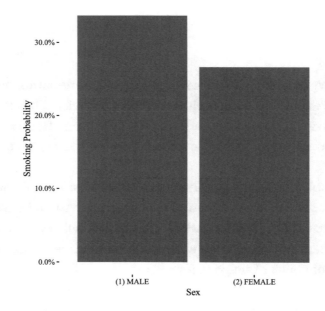

Figure 4-2. *Graph showing the probability of smoking by sex*

We can make a nice table of results using the `xtable` function, with the results shown in Table 4-3.

```
xtable(coef(summary(m.lr)), digits = 2,
       caption = paste(
  "Summary of logistic regression model",
  "including coefficients, standard errors",
  "and p-values."), label = "tglm2-orsimple")
```

Table 4-3. *Summary of Logistic Regression Model Including Coefficients, Standard Errors, and p-Values*

	Estimate	Std. Error	z value	Pr(>lzl)
(Intercept)	−0.68	0.06	−11.82	0.00
Sex(2) FEMALE	−0.33	0.07	−4.44	0.00

We can interpret the results as indicating that women have -0.33 lower log odds of smoking than do men. We can also interpret the results on the odds ratio scale. In that case, we would say that women have 0.72 times the odds of smoking as do men.

As another example, we go back to examining self-efficacy as we did in the previous chapter. Some potential confounding factors are sex, race/ethnicity, and age. We can calculate inverse probability weights and use these to adjust our model of self-efficacy and smoking.

```
## unadjusted model
m0.lr <- vglm(CurSmoke ~ SelfEfficacy_W1,
          family = binomialff(link = "logit"),
          data = acl, model = TRUE)

## estimate IPWs
w <- ipwpoint(
  exposure = SelfEfficacy_W1,
  family = "gaussian",
  numerator = ~ 1,
  denominator = ~ 1 + Sex + RaceEthnicity + AGE_W1,
  data = acl)

## adjusted logistic regression model
m1.lr <- vglm(CurSmoke ~ SelfEfficacy_W1,
          family = binomialff(link = "logit"),
          data = acl, model = TRUE,
          weights = winsorizor(w$ipw.weights, .01))
```

We can then make a table comparing estimates from the raw and adjusted models again using the `xtable()` function, shown in Table 4-4. In this case, the results are actually slightly stronger after adjusting for sex, race/ethnicity, and age.

```
xtable(rbind(
  data.table(Type = "Raw", coef(summary(m0.lr))),
  data.table(Type = "Adj", coef(summary(m1.lr)))),
  digits = 2,
  caption = paste("Comparison of unadjusted (raw)",
    "and adjusted regression models"),
  label = "tglm2-lrcompare")
```

Table 4-4. *Comparison of Unadjusted (Raw) and Adjusted Regression Models*

	Type	Estimate	Std. Error	z value	Pr(>|z|)
1	Raw	−0.88	0.04	−24.13	0.00
2	Raw	−0.06	0.03	−1.71	0.09
3	Adj	−0.88	0.04	−24.12	0.00
4	Adj	−0.08	0.03	−2.36	0.02

We can interpret the results as indicating that a one unit increase in self-efficacy is associated with a -0.08 lower log odds of smoking. We can also interpret the results on the odds ratio scale. In that case, we would say that a one unit increase in self-efficacy is associated with 0.92 times the odds of smoking.

Often, because odds ratios do not have an absolute interpretation, it can be helpful to present results in terms of absolute probabilities. We can again rely on the `predict()` function to generate specific predicted probabilities and graph them. To do this, we might generate predictions for a range of self-efficacy values. The results in Figure 4-3 show a decline in probability of being a current smoker as self-efficacy increases. In this instance, it is approximately linear, even on the probability scale, because no probabilities approach 0 or 1.

```
preddat2 <- data.table(SelfEfficacy_W1 =
  seq(from = min(acl$SelfEfficacy_W1, na.rm = TRUE),
      to = max(acl$SelfEfficacy_W1, na.rm = TRUE),
```

```
    length.out = 1000))
preddat2$yhat <- predict(m1.lr, newdata = preddat2,
                    type = "response")

ggplot(preddat2, aes(SelfEfficacy_W1, yhat)) +
  geom_line() +
  scale_x_continuous("Self-Efficacy") +
  scale_y_continuous("Smoking Probability", label = percent) +
  theme_tufte() + coord_cartesian(ylim = c(.25, .40))
```

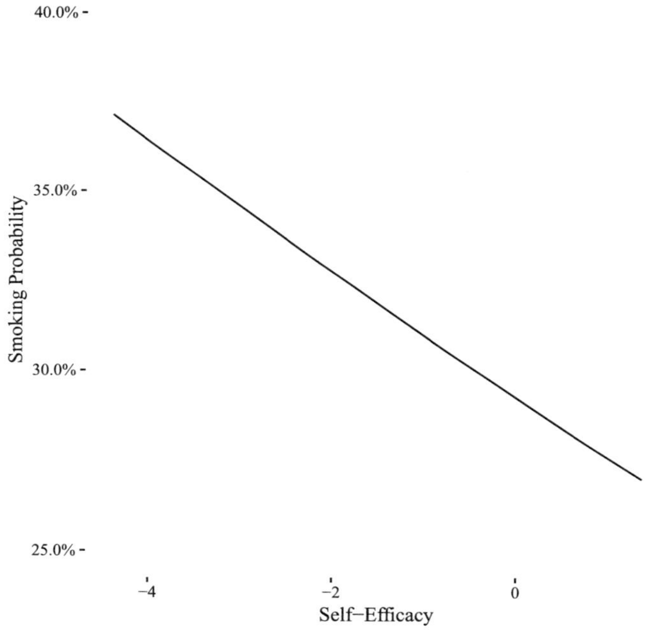

Figure 4-3. *Graph showing the probability of smoking by self-efficacy*

Finally, sometimes people calculate the average change in probability based
on the dataset. Because on the probability scale the results are not linear (although in
Figure 4-3 they are approximately linear), the change in probability associated with a
change in self-efficacy depends on what someone's initial level of self-efficacy is. Further,
if there were other variables in the model, the change also would depend on those other
variables. One approach to deal with this is to generate predicted probabilities using
the actual dataset and then using the actual dataset but slightly increasing everyone's
self-efficacy. In this was, we do not have to make any unrealistic assumptions about what

people's self-efficacy or other predictor scores should be, we use their actual scores. Then we can find the predicted change in probability of smoking for each person and finally average across all of these, to get the average, marginal change in probability.

```
## delta value for change in self efficacy
delta <- .01

## create a copy of the dataset
## where we increase everyone's self-efficacy by delta
aclalt <- copy(acl)
aclalt$SelfEfficacy_W1 <- aclalt$SelfEfficacy_W1 + delta

## calculate predicted probabilities
p1 <- predict(m1.lr, newdata = acl, type = "response")
p2 <- predict(m1.lr, newdata = aclalt, type = "response")

## calculate the average, marginal change in probabilities
## per unit change in self efficacy
## in percents and rounded
round(mean((p2 - p1) / delta) * 100, 1)

## [1] -1.7
```

The code shows that in this sample, the average, marginal effect is such that a one unit increase in self-efficacy is predicted to result in 1.7% lower chance of currently smoking.

Ordered Logistic Regression

Ordered logistic regression is useful when an outcome is discrete and categorical but also where the categories have a natural ordering. In the ACL dataset, physical activity is measured in five categories from least active to most active.

```
acl$PhysActCat_W2 <- factor(acl$PhysActCat_W2, ordered = TRUE)

## adjusted ordered logistic regression model
m0.or <- vglm(PhysActCat_W2 ~ SelfEfficacy_W1,
              family = propodds(),
              data = acl)
```

```
## estimate IPWs
w <- ipwpoint(
  exposure = SelfEfficacy_W1,
  family = "gaussian",
  numerator = ~ 1,
  denominator = ~ 1 + Sex + RaceEthnicity + AGE_W1,
  data = acl)

## adjusted ordered logistic regression model
m1.or <- vglm(PhysActCat_W2 ~ SelfEfficacy_W1,
              family = propodds(),
              data = acl, model = TRUE,
              weights = winsorizor(w$ipw.weights, .01))
```

In the ordered logistic regression model, there are multiple intercepts, one less than the number of unique levels in the outcome. Because we used a proportional odds model, it is assumed that the association of self-efficacy with the outcome is proportional across all levels, so only a single coefficient is estimated for self-efficacy. Comparisons of the raw, unadjusted model and adjusted model are shown in Table 4-5.

```
xtable(rbind(
  data.table(Type = "Raw",
             Labels = rownames(coef(summary(m0.or))),
             coef(summary(m0.or))),
  data.table(Type = "Adj",
             Labels = rownames(coef(summary(m1.or))),
             coef(summary(m1.or)))),
  digits = 2,
  caption = paste("Comparison of unadjusted (raw) and",
    "adjusted ordered logistic regression models"),
  label = "tglm2-orcompare")
```

Table 4-5. *Comparison of Unadjusted (Raw) and Adjusted Ordered Logistic Regression Models*

	Type	Labels	Estimate	Std. Error	z value	Pr(>|z|)
1	Raw	(Intercept):1	0.80	0.04	19.70	0.00
2	Raw	(Intercept):2	0.08	0.04	2.03	0.04
3	Raw	(Intercept):3	−1.14	0.04	−26.11	0.00
4	Raw	(Intercept):4	−1.88	0.06	−34.20	0.00
5	Raw	SelfEfficacy_W1	0.22	0.03	6.60	0.00
6	Adj	(Intercept):1	0.79	0.04	19.56	0.00
7	Adj	(Intercept):2	0.07	0.04	1.93	0.05
8	Adj	(Intercept):3	−1.14	0.04	−26.15	0.00
9	Adj	(Intercept):4	−1.89	0.06	−34.22	0.00
10	Adj	SelfEfficacy_W1	0.19	0.03	5.87	0.00

As with binary logistic regression, it can be helpful to plot the predicted probabilities to provide a more directly interpretable view of the data. Because there are multiple categories, we end up with multiple probabilities and then use the `melt()` function to make these into one long dataset. The resulting plot is shown in Figure 4-4. There we can see that with increasing self-efficacy, there are modest increases in probability of membership in the top four categories, offset by a dramatic decline in probability of membership in the lowest category.

```
preddat3 <- data.table(SelfEfficacy_W1 =
  seq(from = min(acl$SelfEfficacy_W1, na.rm = TRUE),
      to = max(acl$SelfEfficacy_W1, na.rm = TRUE),
    length.out = 1000))
preddat3 <- cbind(preddat3,
  predict(m1.or, newdata = preddat3,
        type = "response"))
preddat3 <- melt(preddat3, id.vars = "SelfEfficacy_W1")

ggplot(preddat3, aes(SelfEfficacy_W1, value,
                colour = variable, linetype = variable)) +
```

```
geom_line(size = 2) +
scale_color_viridis(discrete = TRUE) +
scale_x_continuous("Self-Efficacy") +
scale_y_continuous("Activity Probability", label = percent) +
coord_cartesian(ylim = c(0, .6), expand = FALSE) +
theme_tufte() +
theme(legend.position = c(.7, .8),
      legend.key.width = unit(2, "cm"))
```

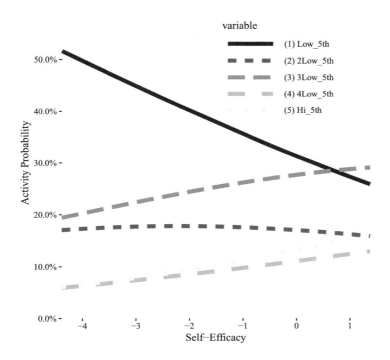

Figure 4-4. *Graph showing the probability of different physical activity categories by self-efficacy*

As with the binary logistic regression, we can calculate the average marginal change in predicted probabilities for a unit change in self-efficacy. With multiple categories, we get the average marginal change for each category.

```
## delta value for change in self efficacy
delta <- .01

## create a copy of the dataset
## where we increase everyone's self-efficacy by delta
```

```
aclalt <- copy(acl)
aclalt$SelfEfficacy_W1 <- aclalt$SelfEfficacy_W1 + delta

## calculate predicted probabilities
p1 <- predict(m1.or, newdata = acl, type = "response")
p2 <- predict(m1.or, newdata = aclalt, type = "response")

## average marginal change in probability of
## membership in each category
round(colMeans((p2 - p1) / delta) * 100, 1)

## (1) Low_5th (2) 2Low_5th (3) 3Low_5th (4) 4Low_5th (5) Hi_5th
##         -4.2           -0.6          1.3           1.3         2.2
```

Multinomial Logistic Regression

Multinomial logistic regression is similar to ordered logistic regression in that it applies when the outcome variable has more than two levels. However, unlike ordered logistic regression that assumes proportional odds, the multinomial logistic model does not make assumptions about any proportional effects or ordering of the levels. This flexibility comes at a cost, however, of a greater number of parameters and increased complexity in terms of interpreting the results.

We will look at the employment information in the ACL data for this. ACL codes people who are employed based on the number of hours they work. For simplicity, we will collapse this into a single employed category.

```
acl[, EmployG_W2 := as.character(Employment_W2)]
acl[EmployG_W2 %in% c(
  "(2) 2500+HRS", "(3) 15002499",
  "(4) 500-1499", "(5) 1-499HRS"),
  EmployG_W2 := "(2) EMPLOYED"]
acl[, EmployG_W2 := factor(EmployG_W2)]
```

After recoding employment, the resulting frequency table is shown in Table 4-6.

```
xtable(as.data.frame(table(acl$EmployG_W2)),
      caption = "Frequency table of employment",
      label = "tglm2-freqtab")
```

Table 4-6. *Frequency Table of Employment*

	Var1	Freq
1	(1) DISABLED	122
2	(2) EMPLOYED	1476
3	(6) RETIRED	724
4	(7) UNEMPLOY	86
5	(8) KEEP HS	459

Next we can estimate a multinomial logistic regression model using the vglm() function as before. The only change is that we specify the argument, family = multinomial(). As in our previous examples, we can estimate both an unadjusted and an adjusted model, where the adjusted model uses IPWs to account for confounding by sex, race/ethnicity, and age.

```
## unadjusted multinomial logistic regression model
m0.mr <- vglm(EmployG_W2 ~ SelfEfficacy_W1,
            family = multinomial(),
            data = acl, model = TRUE)

## estimate IPWs
w <- ipwpoint(
  exposure = SelfEfficacy_W1,
  family = "gaussian",
  numerator = ~ 1,
  denominator = ~ 1 + Sex + RaceEthnicity + AGE_W1,
  data = acl)

## adjusted multinomial logistic regression model
m1.mr <- vglm(EmployG_W2 ~ SelfEfficacy_W1,
```

```
        family = multinomial(),
        data = acl, model = TRUE,
        weights = winsorizor(w$ipw.weights, .01))
```

Next we can make a table comparing the estimates and coefficients from the unadjusted (raw) and adjusted models. In multinomial logistic regression, no assumptions are made that the effect of a predictor is equivalent across levels of the outcome. Instead, k–1 separate parameters are estimated for each predictor, where k is the number of unique levels of the outcome. Another way of thinking about multinomial logistic regression is that if one level of the outcome is chosen as the reference group, then effectively a series of k–1 binary logistic regressions are run. The only real change is that there also is a constraint that the probability of belonging to any group must sum to 1, capturing the reality that people can only belong to one group and everyone must belong to some group. The coefficients from vglm() are labelled numerically, and these are based on the orders of the factor levels. The results are shown in Table 4-7.

```
xtable(rbind(
  data.table(Type = "Raw",
             Labels = rownames(coef(summary(m0.mr))),
             coef(summary(m0.mr))),
  data.table(Type = "Adj",
             Labels = rownames(coef(summary(m1.mr))),
             coef(summary(m1.mr)))),
  digits = 2,
  caption = paste("Comparison of unadjusted (raw) and",
    "adjusted multinomial logistic regression models"),
  label = "tglm2-mrcompare")
```

Table 4-7. *Comparison of Unadjusted (Raw) and Adjusted Multinomial Logistic Regression Models*

	Type	Labels	Estimate	Std. Error	z value	Pr(> z)
1	Raw	(Intercept):1	−1.44	0.11	−12.98	0.00
2	Raw	(Intercept):2	1.18	0.05	21.83	0.00
3	Raw	(Intercept):3	0.46	0.06	7.73	0.00
4	Raw	(Intercept):4	−1.72	0.12	−14.00	0.00
5	Raw	SelfEfficacy_W1:1	−0.33	0.09	−3.70	0.00
6	Raw	SelfEfficacy_W1:2	0.22	0.05	4.28	0.00
7	Raw	SelfEfficacy_W1:3	0.23	0.06	3.91	0.00
8	Raw	SelfEfficacy_W1:4	−0.17	0.11	−1.66	0.10
9	Adj	(Intercept):1	−1.44	0.11	−13.01	0.00
10	Adj	(Intercept):2	1.17	0.05	21.80	0.00
11	Adj	(Intercept):3	0.46	0.06	7.65	0.00
12	Adj	(Intercept):4	−1.73	0.12	−14.07	0.00
13	Adj	SelfEfficacy_W1:1	−0.40	0.09	−4.50	0.00
14	Adj	SelfEfficacy_W1:2	0.15	0.05	2.96	0.00
15	Adj	SelfEfficacy_W1:3	0.17	0.06	2.82	0.00
16	Adj	SelfEfficacy_W1:4	−0.23	0.10	−2.21	0.03

We can make a plot to show how the predicted probability of being in any particular employment category varies as a function of self-efficacy. The results are graphed in Figure 4-5. This figure shows us that as self-efficacy increases, people are less likely to be disabled and more likely to be employed or retired. It also highlights how in these models there are not always linear changes over time. Probability of being disabled declines rapidly from a self-efficacy of -4 to -2 and then more slowly at higher levels of self-efficacy.

```
preddat4 <- data.table(SelfEfficacy_W1 =
  seq(from = min(acl$SelfEfficacy_W1, na.rm = TRUE),
      to = max(acl$SelfEfficacy_W1, na.rm = TRUE),
```

```
      length.out = 1000))
preddat4 <- cbind(preddat4,
  predict(m1.mr, newdata = preddat4,
          type = "response"))
preddat4 <- melt(preddat4, id.vars = "SelfEfficacy_W1")

ggplot(preddat4, aes(
  SelfEfficacy_W1, value,
  colour = variable, linetype = variable)) +
  geom_line(size = 2) +
  scale_color_viridis(discrete = TRUE) +
  scale_x_continuous("Self-Efficacy") +
  scale_y_continuous("Probability", label = percent) +
  coord_cartesian(ylim = c(0, .65), expand = FALSE) +
  theme_tufte() +
  theme(legend.position = c(.18, .82),
        legend.key.width = unit(2, "cm"))
```

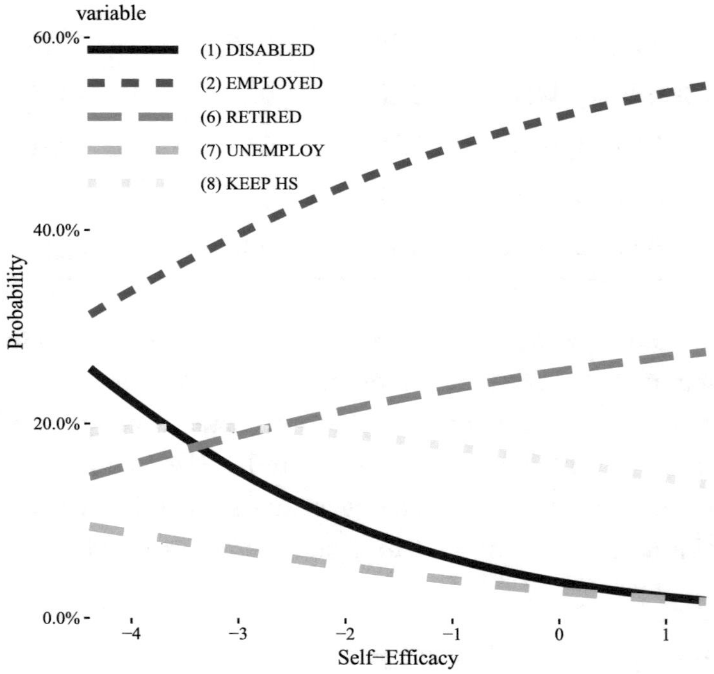

Figure 4-5. *Graph showing the probability of different employment categories by self-efficacy*

Finally, we can calculate the average marginal change in predicted probabilities for a unit change in self-efficacy. With multiple categories, we get the average marginal change for each category. These results show us that on average, a one unit increase in self-efficacy is associated with the largest change in employed (2.9% increase, on average), followed by disabled (2.1% decrease, on average), with somewhat smaller amounts of change for the other categories.

```
## delta value for change in self efficacy
delta <- .01

## create a copy of the dataset
## where we increase everyone's self-efficacy by delta
aclalt <- copy(acl)

aclalt$SelfEfficacy_W1 <- aclalt$SelfEfficacy_W1 + delta

## calculate predicted probabilities
p1 <- predict(m1.mr, newdata = acl, type = "response")
p2 <- predict(m1.mr, newdata = aclalt, type = "response")

## average marginal change in probability of
## membership in each category
round(colMeans((p2 - p1) / delta) * 100, 1)

## (1) DISABLED (2) EMPLOYED (6) RETIRED (7) UNEMPLOY (8) KEEP HS
##          -2.1           2.9          1.7          -1.0          -1.5
```

Poisson and Negative Binomial Regression

For count outcomes, we can use Poisson regression. In the ACL data, one variable is a count of chronic conditions experienced in the past 12 months. This sort of variable potentially lends itself well to a Poisson regression.

The first thing we might do is look at the distribution and get some basic descriptive statistics. For count outcomes, a mean and standard deviation may not be meaningful, so something like the median and interquartile range are often a better summary.

We can get a quick summary of the median and interquartile range using the `egltable()` function, shown as follows.

```
egltable(c("NChronic12_W1", "NChronic12_W2"),
         data = acl, parametric = FALSE)
```

```
##                      Mdn (IQR)
## 1: NChronic12_W1 1.00 (2.00)
## 2: NChronic12_W2 1.00 (2.00)
```

We can use the `ggplot()` function to make a bar plot of the frequencies of each number of chronic conditions, to get a broader sense of the distribution at each wave. The results are in Figure 4-6.

```
plot_grid(
  ggplot(acl, aes(NChronic12_W1)) +
  geom_bar() + theme_tufte(),
  ggplot(acl, aes(NChronic12_W2)) +
  geom_bar() + theme_tufte(),
  ncol = 1,
  labels = c("Wave 1", "Wave 2"),
  label_x = .8)
```

```
## Warning: Removed 750 rows containing non-finite values (stat_count).
```

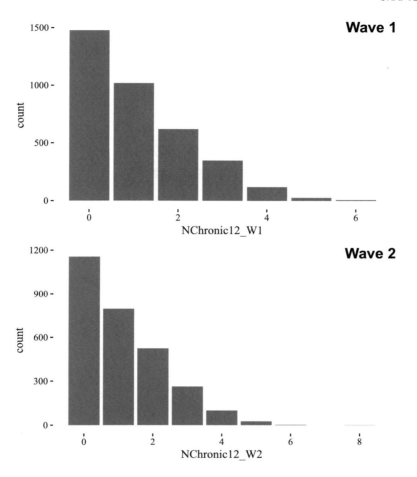

Figure 4-6. *Graph showing the frequency of each number of chronic conditions at each wave in the ACL data*

Next we can estimate a Poisson regression model using the vglm() function. For Poisson regression, we specify the family argument as family = poissonff(). A call to summary() provides a quick summary of the model results and estimates.

```
## unadjusted poisson regression model
m0.pr <- vglm(NChronic12_W2 ~ SelfEfficacy_W1,
          family = poissonff(),
          data = acl, model = TRUE)

summary(m0.pr)

##
```

```
## Call:
## vglm(formula = NChronic12_W2 ~ SelfEfficacy_W1, family = poissonff(),
##     data = acl, model = TRUE)
##
##
## Pearson residuals:
##                   Min    1Q Median    3Q  Max
## loge(lambda) -1.34 -1.01 -0.117 0.811 5.67
##
## Coefficients:
##                  Estimate Std. Error z value Pr(>|z|)
## (Intercept)        0.0954     0.0179    5.33  9.8e-08 ***
## SelfEfficacy_W1   -0.1347     0.0165   -8.16  3.3e-16 ***
## ---
## Signif. codes:  0 '***' 0.001 '**' 0.01 '*' 0.05 '.' 0.1 '_' 1
##
## Number of linear predictors: 1
##
## Name of linear predictor: loge(lambda)
##
## Residual deviance: 4075 on 2865 degrees of freedom
##
## Log-likelihood: -4126 on 2865 degrees of freedom
##
## Number of iterations: 5
##
## No Hauck-Donner effect found in any of the estimates
```

However, before proceeding, it is a good idea to check whether the assumptions for the Poisson regression model are reasonable. Namely, it often occurs there is overdispersion and the assumption that the variance is the same as the mean is violated. To examine this assumption, the easiest approach is to also fit a negative binomial regression model and then compare the relative fit of the two models to determine whether the negative binomial improves fit.

To compare the Poisson and negative binomial results, we first need to fit a negative binomial regression model. The only change required for this is to change `family = poissonff()` to `family = negbinomial()`. Then we can use the `AIC()` and `BIC` to return the Akaike information criterion (AIC) and Bayesian information criterion (BIC) which are based on the model likelihoods, penalized for number of parameters. Lower AIC and BIC scores indicate better fit, even after accounting for model complexity. Use of AIC and BIC is preferable over use of simple model fit measures because typically more complex models provide a better fit. What we want to know is whether the improvement in fit is worth the added complexity; hence, some penalty term for number of parameters is needed, which both the AIC and BIC include.

Comparing the AIC and BIC reveals that the negative binomial regression model has both a lower AIC and lower BIC, pointing to the negative binomial model as superior for these data compared to the Poisson.

```
## unadjusted negative binomial regression model
m0.nbr <- vglm(NChronic12_W2 ~ SelfEfficacy_W1,
            family = negbinomial(),
            data = acl, model = TRUE)

AIC(m0.nbr) - AIC(m0.pr)

## [1] -97

BIC(m0.nbr) - BIC(m0.pr)

## [1] -91
```

Another helpful sanity check is to examine whether simulated values from a model seem consistent with the true observed data. We can accomplish this easily using the `simulate()` function built into the VGAM package. All it requires is a model, but we can also specify the number of simulations to generate, we will just ask for one, and set the random seed so results are reproducible. Next we build a single dataset with the true outcome scores, the simulations from the Poisson, and the simulations from the negative binomial model. Finally, we plot all of these shown in Figure 4-7. The plot shows us that neither of our models perfectly reproduce the true distribution. However, we can also see that the simulations from the negative binomial model are closer to the truth than are the simulations from the Poisson. Plots such as Figure 4-7 are very useful both for comparing models and for evaluating whether a model is a remotely reasonable approximation of the observed data. Sometimes the "best" model may still be a terrible model, and we would want to know that in advance.

```
test.pr <- simulate(m0.pr, nsim = 1, seed = 1234)$sim_1
test.nbr <- simulate(m0.nbr, nsim = 1, seed = 1234)$sim_1
test.all <- data.table(
  Type = rep(c("Truth", "Poisson", "Negative\nBinomial"),
             times = c(
               nrow(model.frame(m0.pr)),
               length(test.pr),
               length(test.nbr))),
  Score = c(
    model.frame(m0.pr)$NChronic12_W2,
    test.pr,
    test.nbr))

ggplot(test.all, aes(Score, fill = Type)) +
  geom_bar(position = "dodge") +
  scale_fill_viridis(discrete = TRUE) +
  theme_tufte() +
  theme(legend.position = c(.8, .8))
```

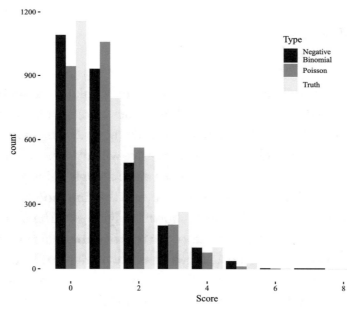

Figure 4-7. *Graph showing the frequency of each number of chronic conditions based on the true data, simulations from the negative binomial model, and simulations from the Poisson regression model.*

At this point, both from comparing the AIC and BIC scores and from visualizing the simulated values from the Poisson and negative binomial regression model, it is clear that we should proceed with the negative binomial model. If we want to compare unadjusted and adjusted results, we can calculate IPWs and use these to estimate an adjusted model, accounting for the effects of sex, race/ethnicity, and age.

```
## estimate IPWs
w <- ipwpoint(
  exposure = SelfEfficacy_W1,
  family = "gaussian",
  numerator = ~ 1,
  denominator = ~ 1 + Sex + RaceEthnicity + AGE_W1,
  data = acl)

## adjusted negative binomial regression model
m1.nbr <- vglm(NChronic12_W2 ~ SelfEfficacy_W1,
               family = negbinomial(),
               data = acl, model = TRUE,
               weights = winsorizor(w$ipw.weights, .01))
```

Next we can make a table comparing the estimates and coefficients from the unadjusted (raw) and adjusted models. The negative binomial regression models include two intercepts, one for the location, called μ, and the other for the overdispersion parameter, called the size parameter. For the model of the mean number of chronic conditions, only the first intercept is relevant. The results are shown in Table 4-8.

```
xtable(rbind(
  data.table(Type = "Raw",
             Labels = rownames(coef(summary(m0.nbr))),
             coef(summary(m0.nbr))),
  data.table(Type = "Adj",
             Labels = rownames(coef(summary(m1.nbr))),
             coef(summary(m1.nbr)))),
  digits = 2,
  caption = paste("Comparison of unadjusted (raw) and",
   "adjusted negative binomial regression models"),
  label = "tglm2-nbrcompare")
```

Table 4-8. *Comparison of Unadjusted (Raw) and Adjusted Negative Binomial Regression Models*

	Type	Labels	Estimate	Std. Error	z value	Pr(>|z|)
1	Raw	(Intercept):1	0.10	0.02	4.65	0.00
2	Raw	(Intercept):2	1.23	0.12	10.41	0.00
3	Raw	SelfEfficacy_W1	−0.13	0.02	−6.98	0.00
4	Adj	(Intercept):1	0.10	0.02	4.73	0.00
5	Adj	(Intercept):2	1.23	0.12	10.41	0.00
6	Adj	SelfEfficacy_W1	−0.13	0.02	−6.69	0.00

Because both Poisson and negative binomial regression models use a natural log link function, the coefficients are on the log scale. If we looked at the unadjusted results in Table 4-8, the coefficient for self-efficacy indicates that a one unit increase in self-efficacy is associated with a -0.13 log unit change in number of chronic conditions. Alternately, we could exponentiate the coefficient, in which case the interpretation is that a one unit increase in self-efficacy is associated with 0.87 times as many chronic conditions.

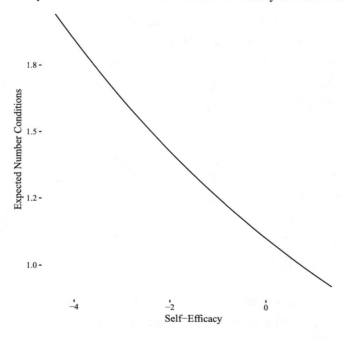

Figure 4-8. *Graph showing the predicted number of chronic conditions as a function of self-efficacy*

If we wanted, we also can plot the predicted average number of conditions as a function of self-efficacy. When generating predictions, as with logistic regression, we specify `type = "response"` to indicate that we want predictions on the original scale, not on the link scale, here the log transformed scale. The results are graphed in Figure 4-8. This figure shows us that as self-efficacy increases, people are expected to have fewer chronic conditions, on average.

```
preddat5 <- data.table(SelfEfficacy_W1 =
  seq(from = min(acl$SelfEfficacy_W1, na.rm = TRUE),
      to = max(acl$SelfEfficacy_W1, na.rm = TRUE),
    length.out = 1000))
preddat5$yhat <- predict(m1.nbr, newdata = preddat5,
        type = "response")

ggplot(preddat5, aes(SelfEfficacy_W1, yhat)) +
  geom_line() +
  scale_x_continuous("Self-Efficacy") +
  scale_y_continuous("Expected Number Conditions") +
  theme_tufte()
```

4.3 Case Study: Multinomial Logistic Regression

By default in multinomial logistic regression, parameters (e.g., odds ratios) are calculated with respect to the reference group. Although that is sufficient to specify the model, it is common when multinomial logistic regression is applied in practice to consider all (or at least key) pairwise comparisons between groups [14, 87]. For example, simply knowing that Group B and Group C are significantly different from Group A does not tell you whether Group B and Group C are different from each other.

It also is common to simultaneously evaluate the effects of several predictors (e.g., [14, 87]), which requires slightly different handling than when evaluating a single predictor. In this case study, we will build a complete example from question through to final presentation of results and interpretation.

The ACL dataset includes smoking status at wave 1 and wave 2. Beyond looking at smoking status at one wave, an interesting question is who changes (either starting or stopping) smoking over time, and what factors may predict change. To begin with, we need to create a new variable for smoking that indicates change or stability over time. We do the recoding as follows. The resulting frequency table is shown in Table 4-9.

```
acl[, Smoke_W2W1 := NA_character_]
acl[Smoke_W1 == "(3) Nevr Smo" &
    Smoke_W2 == "(3) W2 Never Smoker",
    Smoke_W2W1 := "Stable Never Smoker"]
acl[Smoke_W1 == "(2) Past Smo" &
    Smoke_W2 == "(2) W2 Former Smoker",
    Smoke_W2W1 := "Stable Former Smoker"]
acl[Smoke_W1 == "(1) Cur Smok" &
    Smoke_W2 == "(1) W2 Current Smoker",
    Smoke_W2W1 := "Stable Current Smoker"]
acl[Smoke_W1 %in% c("(2) Past Smo", "(3) Nevr Smo") &
    Smoke_W2 == "(1) W2 Current Smoker",
    Smoke_W2W1 := "New Smoker"]
acl[Smoke_W1 == "(1) Cur Smok" &
    Smoke_W2 == "(2) W2 Former Smoker",
    Smoke_W2W1 := "Recently Quit Smoker"]

acl[, Smoke_W2W1 := factor(Smoke_W2W1,
  levels = c("Stable Never Smoker", "Stable Former Smoker",
             "Stable Current Smoker", "Recently Quit Smoker",
             "New Smoker"))]

xtable(as.data.frame(table(acl$Smoke_W2W1)),
       caption = "Frequency table of smoking over time",
       label = "tglm2-freqtab-smoke")
```

Table 4-9. *Frequency Table of Smoking Over Time*

	Var1	Freq
1	Stable Never Smoker	1292
2	Stable Former Smoker	705
3	Stable Current Smoker	641
4	Recently Quit Smoker	167
5	New Smoker	62

Earlier in the chapter, we focused on just one predictor. In real settings, we may be interested in several potential predictors. One interesting question is whether sociodemographic, psychosocial, or health type variables are better predictors of smoking over time. We will estimate models using the vglm() function as before with family = multinomial() for a multinomial outcome.

```
acl[, SES := as.numeric(SESCategory)]

mr.ses <- vglm(Smoke_W2W1 ~ Sex + SES + AGE_W1,
  family = multinomial(),
  data = acl, model = TRUE)

mr.psych <- vglm(Smoke_W2W1 ~ SWL_W1 + InformalSI_W1 +
  FormalSI_W1 + SelfEfficacy_W1 + CESD11_W1,
  family = multinomial(),
  data = acl, model = TRUE)

mr.health <- vglm(Smoke_W2W1 ~ PhysActCat_W1 +
  BMI_W1 + NChronic12_W1,
  family = multinomial(),
  data = acl, model = TRUE)
```

We can compare the relative performance of each model, penalized for complexity using the AIC and BIC, shown in Table 4-10. The results suggest that the sociodemographic factors are the best predictors of smoking status and change over time.

```
xtable(
  data.table(
  Model = c("Sociodemographics", "Psychosocial", "Health"),
  AIC = c(AIC(mr.ses), AIC(mr.psych), AIC(mr.health)),
  BIC = c(BIC(mr.ses), BIC(mr.psych), BIC(mr.health))),
  caption = "Model Comparisons",
  label = "tglm2-modelcomparisons")
```

155

Table 4-10. *Model Comparisons*

	Model	AIC	BIC
1	Sociodemographics	7056.34	7151.72
2	Psychosocial	7203.73	7346.79
3	Health	7340.54	7507.45

We can examine the individual coefficients from the sociodemographics model using the summary() function. However, by default these are only the comparisons to the reference level, which by default in the VGAM package is the last level, for us, "New Smoker."

```
summary(mr.ses)

## 
## Call:
## vglm(formula = Smoke_W2W1 ~ Sex + SES + AGE_W1, family = multinomial(),
##     data = acl, model = TRUE)
## 
## 
## Pearson residuals:
##                         Min     1Q Median     3Q  Max
## log(mu[,1]/mu[,5]) -7.39 -0.744 -0.412  0.811 1.97
## log(mu[,2]/mu[,5]) -6.95 -0.441 -0.306 -0.186 2.98
## log(mu[,3]/mu[,5]) -6.33 -0.420 -0.289 -0.184 3.58
## log(mu[,4]/mu[,5]) -5.88 -0.202 -0.155 -0.118 4.82
## 
## Coefficients:
##                 Estimate Std. Error z value Pr(>|z|)
## (Intercept):1    -0.79722    0.61458   -1.30  0.19457
## (Intercept):2    -1.16638    0.62965   -1.85  0.06397 .
## (Intercept):3     1.41356    0.61771    2.29  0.02212 *
## (Intercept):4    -1.18639    0.70648   -1.68  0.09309 .
## Sex(2) FEMALE:1   0.76073    0.27200    2.80  0.00516 **
## Sex(2) FEMALE:2  -0.46076    0.27545   -1.67  0.09437 .
## Sex(2) FEMALE:3  -0.04184    0.27459   -0.15  0.87888
## Sex(2) FEMALE:4   0.02589    0.30782    0.08  0.93297
```

```
## SES:1              0.51292    0.14821    3.46  0.00054 ***
## SES:2              0.50412    0.15079    3.34  0.00083 ***
## SES:3              0.20550    0.15032    1.37  0.17159
## SES:4              0.39726    0.16691    2.38  0.01731 *
## AGE_W1:1           0.04439    0.00858    5.18  2.3e-07 ***
## AGE_W1:2           0.05430    0.00877    6.19  5.9e-10 ***
## AGE_W1:3           0.01111    0.00871    1.28  0.20181
## AGE_W1:4           0.02732    0.00971    2.82  0.00487 **
## ---
## Signif. codes:  0 '***' 0.001 '**' 0.01 '*' 0.05 '.' 0.1 '_' 1
##
## Number of linear predictors:   4
##
## Names of linear predictors:
## log(mu[,1]/mu[,5]), log(mu[,2]/mu[,5]), log(mu[,3]/mu[,5]), log(mu[,4]/mu[,5])
##
## Residual deviance: 7024 on 11452 degrees of freedom
##
## Log-likelihood: -3512 on 11452 degrees of freedom
##
## Number of iterations: 6
##
## No Hauck-Donner effect found in any of the estimates
##
## Reference group is level 5 of the response
```

The other thing we can notice in looking at these results is that although it is statistically significant, the coefficient for age is quite small, because a 1-year change in age is a relatively small change. We might consider converting age to decades so a one unit difference is more meaningful.

```
acl[, AGE_W1 := AGE_W1 / 10]
```

If we wanted to explicitly export the contrasts between other categories, we can change the reference level. This can be accomplished as an option argument to the multinomial() function. For example, setting refLevel = 1 will make the first category the reference, in our case that is "Stable Never Smoker." We could also re-run for "Stable Former Smoker," level 2, and "Stable Current Smoker," level 3. Note that mathematically, all of these models are the same, what differs is the default comparisons in the results. When re-running, we do not need to re-specify the entire model, we can use the update() function to simply update an existing model.

```
mr.ses1 <- vglm(Smoke_W2W1 ~ Sex + SES + AGE_W1,
            family = multinomial(refLevel = 1),
            data = acl, model = TRUE)
mr.ses2 <- update(mr.ses1,
                family = multinomial(refLevel = 2))
mr.ses3 <- update(mr.ses1,
                family = multinomial(refLevel = 3))
```

Next, it is common to report odds ratios instead of the log odds, which are the default output. It also is common to report confidence intervals, which we can calculate using the confint() function. We can create a table of results by combining the coefficients and confidence intervals, and after exponentiating them, we have odds ratios and confidence intervals for the odds ratios.

For example, if we look at the odds ratio for AGE_W1:1 when the reference group is "Stable Never Smoker," it tells us that a one decade increase in age is associated with 1.1 times the odds of being a "Stable Former Smoker" as being a "Stable Never Smoker."

If we look at the odds ratio for AGE_W1:2 when the reference group is "Stable Never Smoker," it tells us that a one decade increase in age is associated with 0.72 times the odds of being a "Stable Current Smokerx" as being a "Stable Never Smoker."

In contrast, if we look at the odds ratio for AGE_W1:2 when the reference group is "Stable Current Smoker," it tells us that a one decade increase in age is associated with 1.54 times the odds of being a "Stable Former Smoker" as being a "Stable Current Smoker."

Finally, if we look at the odds ratio for AGE_W1:3 when the reference group is "Stable Current Smoker," it tells us that a one decade increase in age is associated with 1.18 times the odds of being a "Recently Quit Smoker" as being a "Stable Current Smoker."

```
data.table(
  Ref = "Stable Never Smoker",
  Term = names(coef(mr.ses1)),
  OR = exp(coef(mr.ses1)),
  exp(confint(mr.ses1)))
```

```
##                          Ref            Term   OR 2.5 % 97.5 %
## 1: Stable Never Smoker      (Intercept):1 0.69  0.42   1.15
## 2: Stable Never Smoker      (Intercept):2 9.12  5.55  14.98
## 3: Stable Never Smoker      (Intercept):3 0.68  0.30   1.54
## 4: Stable Never Smoker      (Intercept):4 2.22  0.67   7.40
## 5: Stable Never Smoker Sex(2) FEMALE:1 0.29  0.24   0.36
## 6: Stable Never Smoker Sex(2) FEMALE:2 0.45  0.36   0.55
## 7: Stable Never Smoker Sex(2) FEMALE:3 0.48  0.34   0.67
## 8: Stable Never Smoker Sex(2) FEMALE:4 0.47  0.27   0.80
## 9: Stable Never Smoker            SES:1 0.99  0.90   1.10
## 10: Stable Never Smoker           SES:2 0.74  0.66   0.82
## 11: Stable Never Smoker           SES:3 0.89  0.75   1.06
## 12: Stable Never Smoker           SES:4 0.60  0.45   0.80
## 13: Stable Never Smoker        AGE_W1:1 1.10  1.04   1.17
## 14: Stable Never Smoker        AGE_W1:2 0.72  0.67   0.76
## 15: Stable Never Smoker        AGE_W1:3 0.84  0.76   0.94
## 16: Stable Never Smoker        AGE_W1:4 0.64  0.54   0.76
```

```
data.table(
  Ref = "Stable Current Smoker",
  Term = names(coef(mr.ses3)),
  OR = exp(coef(mr.ses3)),
  exp(confint(mr.ses3)))
```

```
##                          Ref            Term    OR   2.5 % 97.5 %
## 1: Stable Current Smoker     (Intercept):1 0.110  0.067   0.18
## 2: Stable Current Smoker     (Intercept):2 0.076  0.043   0.13
## 3: Stable Current Smoker     (Intercept):3 0.074  0.032   0.17
## 4: Stable Current Smoker     (Intercept):4 0.243  0.072   0.82
## 5: Stable Current Smoker Sex(2) FEMALE:1 2.231  1.811   2.75
## 6: Stable Current Smoker Sex(2) FEMALE:2 0.658  0.525   0.82
## 7: Stable Current Smoker Sex(2) FEMALE:3 1.070  0.752   1.52
## 8: Stable Current Smoker Sex(2) FEMALE:4 1.043  0.609   1.79
## 9: Stable Current Smoker           SES:1 1.360  1.221   1.51
## 10: Stable Current Smoker          SES:2 1.348  1.195   1.52
## 11: Stable Current Smoker          SES:3 1.211  1.006   1.46
```

```
## 12: Stable Current Smoker          SES:4 0.814   0.606   1.09
## 13: Stable Current Smoker        AGE_W1:1 1.395   1.309   1.49
## 14: Stable Current Smoker        AGE_W1:2 1.540   1.432   1.66
## 15: Stable Current Smoker        AGE_W1:3 1.176   1.054   1.31
## 16: Stable Current Smoker        AGE_W1:4 0.895   0.754   1.06
```

Another way to present results would be to calculate predicted probabilities. However, this becomes more complicated when there are multiple predictors, because where we hold the other predictors will influence the results. Instead, with multiple predictors, calculating the average marginal probabilities, which hold predictors at their observed values and vary one predictor at a time, may be most sensible.

```
## delta value for change in age and SES
delta <- .01

## create a copy of the dataset
## where we increase everyone's age by delta
aclage <- copy(acl)
aclage[, AGE_W1 := AGE_W1 + delta]

## create a copy of the dataset
## where we increase everyone's SES by delta
aclses <- copy(acl)
aclses[, SES := SES + delta]

## create two copies of the data
## one where we set everyone to "female" and another to "male"
aclfemale <- copy(acl)
aclfemale[, Sex := factor("(2) FEMALE",
                          levels = levels(acl$Sex))]

aclmale <- copy(acl)
aclmale[, Sex := factor("(1) MALE",
                        levels = levels(acl$Sex))]

## calculate predicted probabilities
p.ref <- predict(mr.ses1, newdata = acl,
              type = "response")
```

```
p.age <- predict(mr.ses1, newdata = aclage,
                 type = "response")
p.ses <- predict(mr.ses1, newdata = aclses,
                 type = "response")
p.female <- predict(mr.ses1, newdata = aclfemale,
                    type = "response")
p.male <- predict(mr.ses1, newdata = aclmale,
                  type = "response")
```

Finally, we can calculate all the average marginal changes in predicted probabilities and put these together into a table for easy presentation. The end results are in Table 4-11. This highlights the robust impact of sex, with women being much more likely to be stable never smokers. We can also see how older age and higher socioeconomic status are associated with about a 5% lower probability of being a stable current smoker.

```
xtable(
data.table(
  Level = colnames(p.ref),
  Age = colMeans((p.age - p.ref) / delta) * 100,
  SES = colMeans((p.ses - p.ref) / delta) * 100,
  Female = colMeans(p.female - p.male) * 100),
  digits = 2,
  caption = "Average marginal change in predicted probability",
  label = "tglm2-margprobs")
```

Table 4-11. *Average Marginal Change in Predicted Probability*

	Level	Age	SES	Female
1	Stable Never Smoker	2.83	3.66	23.34
2	Stable Former Smoker	3.94	1.85	−16.96
3	Stable Current Smoker	−5.46	−4.52	−5.06
4	Recently Quit Smoker	−0.55	−0.12	−0.93
5	New Smoker	−0.76	−0.87	−0.39

Although it takes some additional work, creating such a table showing the average marginal change in predicted probability for key predictors is a very helpful way to present results in a more intuitive format than odds ratios. Combined with odds ratios and confidence intervals to give estimates of uncertainty, this provides a relatively comprehensive presentation of results.

4.4 Summary

This chapter showed how generalized linear models (GLMs) can be used to build regression models for discrete outcomes including dichotomous outcomes, ordered and unordered categorical outcomes, as well as outcomes that are counts or number of events. Although these outcomes are some of the most common uses of GLMs, GLMs can accommodate many other types of outcome variables and many other distributions than the ones introduced here. The excellent VGAM package has support for both common and uncommon types of distributions, so if you have data that do not appear to readily be normally distributed nor any of the distributions introduced in this chapter, it is very likely that it can still be modelled using the vglm() with a different distribution. For far more details and coverage of the features of the VGAM package, see the book by its author [125].

This chapter also introduced tools and functions to help us interpret GLMs by generating predictions or getting effects on more directly interpretable scales, such as probabilities. Although such code is not strictly part of GLMs, it often can make the results clearer and help both analysts and readers appreciate the implications of a model. A summary of the main functions introduced and what they do is presented in Table 4-12.

Table 4-12. *Listing of Key Functions Described in This Chapter and Summary of What They Do*

Function	What It Does
vglm()	Flexible function fitting numerous types of specific generalized linear models via a "family" function with hundreds of combinations of distributions and link functions available.
binomialff()	VGAM family function fitting GLMs to binomial, discrete/categorical outcomes (e.g., smoking or non-smoking).
propodds()	VGAM family function fitting GLMs to ordered, discrete/categorical outcomes (e.g., physical activity levels).
multinomial()	VGAM family function fitting GLMs to discrete/categorical outcomes with no assumption as to any inherent order (e.g., over time, different predictors may be more or less influential toward a particular outcome).
poissonff()	VGAM family function fitting GLMs to Poisson outcomes (e.g., chronic conditions).
negbinomial()	VGAM family function fitting GLMs to count data where variance exceeds mean (generalized Poisson).
summary()	Generic function to print a summary of an object, including vglm models.
coef()	Generic function that extracts coefficients from a model, including vglm models.
confint()	Generic function to compute confidence intervals from a model, including vglm models.
update()	Updates an existing model, without requiring unchanged parts to be re-written (e.g., changing default reference for smoking levels).
ipwpoint()	Estimates inverse probability weights.
xtable()	Exports tables nicely to LATEXor HTML.
winsorizor()	Clips outliers at specified percentile.
predict()	Takes new data predictors, applies the model to them to estimate the most likely response outcome(s), and, when used with type = "response", converts back to the original data's scale.

(*continued*)

Table 4-12. (*continued*)

Function	What It Does
`simulate()`	Generates simulated data from a model which can be used to compare model and original data distributions for `vglm` models.
`AIC()`	Returns model likelihood based on Akaike information criterion (penalized by parameter count).
`BIC()`	Returns model likelihood based on Bayesian information criterion (penalized by parameter count).

CHAPTER 5

GAMs

Generalized additive models (GAMs) are extensions of the generalized linear models (GLMs) we discussed in earlier chapters. Like GLMs, GAMs accommodate outcomes that are both continuous and discrete. However, unlike GLMs that are fully parametric models, GAMs are semi-parametric models. GAMs allow a mix of a parametric and a nonparametric association between outcome and predictors. For this chapter, we will lean heavily on one excellent R package, VGAM, which provides utilities for vector generalized linear models (VGLMs) and vector generalized additive models (VGAMs) [125]. VGAMs are an even more flexible class of models than GAMs where there may be multiple responses. However, beyond offering flexibility of multiple parameters, the VGAM package implements over 20 link functions, well over 50 different models/assumed distributions. We will only scratch the surface of the VGAM package capabilities in this chapter, but its great flexibility means that we will not need to introduce many different packages nor many different functions. If you would like to learn about VGAMs in far greater depth, we recommend an excellent book by the author of the VGAM package [125].

```
library(checkpoint)
checkpoint("2018-09-28", R.version = "3.5.1",
  project = book_directory,
  checkpointLocation = checkpoint_directory,
  scanForPackages = FALSE,
  scan.rnw.with.knitr = TRUE, use.knitr = TRUE)

library(knitr)
library(data.table)
library(ggplot2)
library(ggthemes)
library(scales)
library(viridis)
```

© Matt Wiley and Joshua F. Wiley 2019
M. Wiley and J. F. Wiley, *Advanced R Statistical Programming and Data Models*,
https://doi.org/10.1007/978-1-4842-2872-2_5

```
library(car)
library(mgcv)
library(VGAM)
library(ipw)
library(JWileymisc)
library(xtable)

options(
  width = 70,
  stringsAsFactors = FALSE,
  datatable.print.nrows = 20,
  datatable.print.topn = 3,
  digits = 2)
```

5.1 Conceptual Overview

Generalized additive models (GAMs) are semi-parametric, additive models that relax the linearity assumption of generalized linear models (GLMs) using nonparametric smoothing functions [40]. The use of nonparametric smoothing functions provides GAMs great flexibility by allowing them to model the association between predictor(s) and an outcome, even when the functional form is not known and without requiring the (correct) specification of the functional form. This flexibility has made GAMs adopted in multiple disciplines including psychology [53] and medicine [75].

The next section introduces the concept of smoothing splines, which are foundational to GAMs. If you are interested in an excellent, comprehensive introduction to modern GAMs in R, see [122]. To read about foundational details on estimation and inference for GAMs, see [57, 121, 123]. Beyond the scope of this chapter, GAMs also have been extended to model not only the location but also the scale and shape of distributions. For further information, see the excellent papers [78] and [88] describing the theoretical aspects, and for practical aspects in R, see the GAMLSS package [88].

Smoothing Splines

A key concept behind GAMs that allows them to model unknown functional forms is smoothing splines. Smoothing splines rely in part on polynomials. It can be shown that a sufficiently high-order polynomial can approximate any function. However, often an adequate approximation can take very high-order polynomials. Although in theory true, in practice it often is difficult to build a sufficiently high-order polynomial to approximate observed data. In particular, it often is the case that polynomials provide very bad approximations at either extreme. An example of an intercept-only, linear, and second-, third-, fourth-, and tenth-degree polynomial for the association of age with depression symptoms in the ACL data is shown in Figure 5-1. The lower-order polynomials generate more moderate predictions at their extremes than do the high-order polynomials. Even the quadratic polynomial generates relatively extreme predictions for the youngest and oldest participants (Figure 5-1).

```
acl <- readRDS("advancedr_acl_data.RDS")

ggplot(acl, aes(AGE_W1, CESD11_W1)) +
  stat_smooth(method = "lm", formula = y ~ 1,
    colour = viridis(6)[1], linetype = 1, se = FALSE) +
  stat_smooth(method = "lm", formula = y ~ x,
    colour = viridis(6)[2], linetype = 4, se = FALSE) +
  stat_smooth(method = "lm", formula = y ~ poly(x, 2),
    colour = viridis(6)[3], linetype = 2, se = FALSE) +
  stat_smooth(method = "lm", formula = y ~ poly(x, 3),
    colour = viridis(6)[4], linetype = 3, se = FALSE) +
  stat_smooth(method = "lm", formula = y ~ poly(x, 4),
    colour = viridis(6)[5], linetype = 1, se = FALSE) +
  stat_smooth(method = "lm", formula = y ~ poly(x, 10),
    colour = viridis(6)[6], linetype = 5, se = FALSE)
```

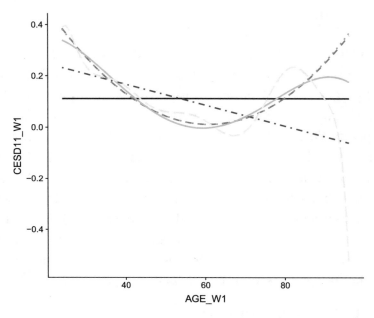

Figure 5-1. *Graph showing an intercept-only (flat line) and progressively higher-order polynomials*

Regression splines attempt to solve the problem of requiring very high-order polynomials to accommodate different functional forms and the issue of extreme predictions at the lower and upper ends of the distribution. Splines were not initially used in statistics. Splines initially referred to thin pieces of wood, bent to form smooth curves between knots. Regression splines use this idea as they are essentially piecewise models, where each piece is a polynomial model and they are smoothly connected at the edge of each piece, the knots. The simplest spline model is a step function. To create this step, we define logical statements with the boundary points, in this case: $x > 42$ *and* $x \leq 65$ and $x > 65$ *and* $x \leq 96$. To achieve these bounds in R, we utilize an extended logical operator: %gle%. The left side should be a number of vector. The right side should be a vector of length two, and it will return TRUE or FALSE if the number or vector on the left hand side is greater than the lowest value and less than or equal to the highest value on the right hand side. These are demonstrated in the following simple examples.

```
## > and <
1:5 %gl% c(2, 4)

## [1] FALSE FALSE  TRUE FALSE FALSE
```

```
## > and <=
1:5 %gle% c(2, 4)

## [1] FALSE FALSE   TRUE   TRUE FALSE

## >= and <
1:5 %gel% c(2, 4)

## [1] FALSE   TRUE   TRUE FALSE FALSE

## >= and <=
1:5 %gele% c(2, 4)

## [1] FALSE   TRUE   TRUE   TRUE FALSE
```

As the polynomial degree increases, linear and quadratic trends emerge, between each knot. Results for two inner knots and step, linear, quadratic, and cubic polynomials are shown in Figure 5-2.

```
ggplot(acl, aes(AGE_W1, CESD11_W1)) +
  stat_smooth(method = "lm",
    formula = y ~ 1 +
      ifelse(x %gle% c(42, 65), 1, 0) +
      ifelse(x %gle% c(65, 96), 1, 0),
    colour = viridis(6)[1], linetype = 1, se = FALSE) +
  stat_smooth(method = "lm",
    formula = y ~ bs(x, df = 3, degree = 1L),
    colour = viridis(6)[2], linetype = 2, se = FALSE) +
  stat_smooth(method = "lm",
    formula = y ~ bs(x, df = 4, degree = 2L),
    colour = viridis(6)[3], linetype = 3, se = FALSE) +
  stat_smooth(method = "lm",
    formula = y ~ bs(x, df = 5, degree = 3L),
    colour = viridis(6)[4], linetype = 4, se = FALSE)
```

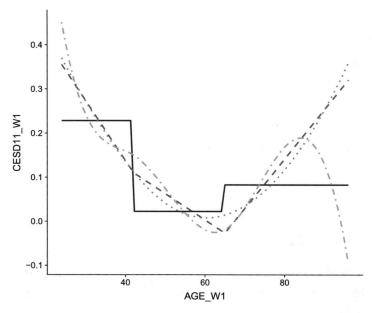

Figure 5-2. *Graph showing step function splines, linear splines, and quadratic splines, all with two inner knots*

B-splines for basis or basic splines attempt work by having relatively little overlap in the basis functions for a given region, which helps them to be computationally more stable and has contributed to their being a very popular type of spline. The following code helps visualize the overlap by showing the basis functions for B-splines with fixed knots. Because the plots are the same except that they are based on different datasets and have a different title, instead of repeating the code to make the plots (which is fairly long), we store the results in an R object, p1. Then we replace the original data with a new dataset using the %+% operator. The results are shown in Figure 5-3.

```
knots <- c(33, 42, 57, 65, 72)
x <- seq(from = min(acl$AGE_W1),
         to = max(acl$AGE_W1), by = .01)

p1 <- ggplot(melt(bs(x, degree = 1,
          knots = knots, intercept = TRUE)),
          aes(Var1, value, colour = factor(Var2))) +
  geom_line() +
  scale_color_viridis("Basis", discrete = TRUE) +
  theme_tufte()
```

```
plot_grid(
  p1 +
    ggtitle("5 Knots, Degree = 1"),
  p1 %+% melt(bs(x, degree = 2,
          knots = knots, intercept = TRUE)) +
    ggtitle("5 Knots, Degree = 2"),
  p1 %+% melt(bs(x, degree = 3,
          knots = knots, intercept = TRUE)) +
    ggtitle("5 Knots, Degree = 3"),
  p1 %+% melt(bs(x, degree = 4,
          knots = knots, intercept = TRUE)) +
    ggtitle("5 Knots, Degree = 4"),
  ncol = 2)
```

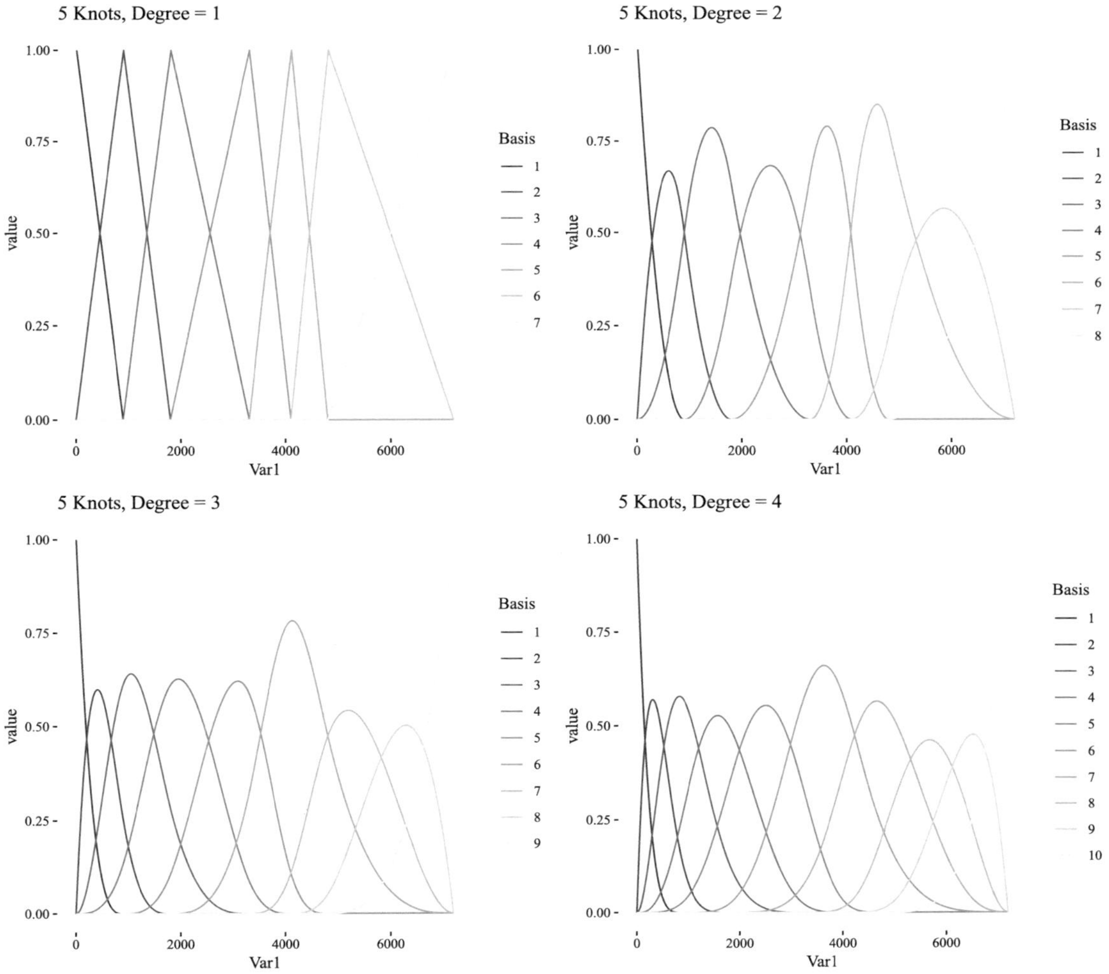

Figure 5-3. *Graph showing B-splines (basis splines)*

An extension of splines is known as smooth splines. The basic idea of smoothing splines is that rather than directly specifying the knots and polynomial degree, which would require knowing a priori what is required, ideally we could automatically learn the appropriate degree of smoothing.

The process of automatically learning the appropriate degree of smoothing generally works by allowing many knots and a high degree of flexibility and using penalties to reduce flexibility (impose smoothness) based on some criteria. A common approach to this for smoothing splines is the generalized cross validation (GCV) criteria or, if the scale is known (which often it is not), what is essentially a variant of the Akaike information criterion (AIC), called the unbiased risk estimator (UBRE). For details on GCV, see [104]. A final option is using restricted maximum likelihood (REML) where the smooth components are viewed as random effects, with one "variance component" per smooth.

A final note on smooth splines is that it often is desirable to have some method of quantifying how smooth or flexible they are. Such a value is useful descriptively, and also plays a role in calculating values such as model AIC, where the likelihood is penalized by complexity. The general solution is to use the "effective degrees of freedom" or EDFs. Depending whether EDFs are calculated including the intercept (constant) term, an EDF of 1 or 2 may correspond to a linear function. Sometimes EDFs are reported counting the intercept, in which case an EDF = 2 corresponds to a linear trend. However, if the intercept is not counted, sometimes referred to as effective non-linear degrees of freedom (ENDF), then an ENDF = 1 corresponds to a linear trend. As the EDF/ENDF increases, the flexibility of the fit likewise increases. For more details on splines and smoothing in regression, see [39].

For the purposes of this chapter, a rough understanding of splines provides the foundation for generalized additive models (GAMs). GAMs extend GLMs by allowing a mix of parametric (e.g., assuming a linear association for one predictor) and nonparametric (e.g., using a smooth spline for another). GAMs are still additive as each term is added, and they take the general form:

$$g(y) = \eta = b_0 + f_1(x_1) + f_2(x_2) + \cdots + f_k(x_k) \tag{5.1}$$

In this parameterization, we have the familiar intercept; however, in place of the regression coefficients for each predictor, we have functions, f_1, etc. These functions may be specified in advance, for instance, if $f_1(x_1) = b_1 * x_1$, or these functions may be smoothing splines where the degree of smoothness is learned from the data based on some criteria, such as the GCV or UBRE. To be a GAM, there generally must be at least one smoothing term. However, like GLMs, GAMs in theory accommodate many smoothing terms and many regular parametric terms. The ability to mix smooth and non-smooth terms makes GAMs a highly flexible class of models that can be applied in many circumstances. For example, it may be that the smoothing term is of substantive interest, such as weight growth charts for children by age, where the growth often is non-linear but the functional form is unknown. However, GAMs also can be applied in other cases. For example, a substantive question may concern the effect of one variable with a known or assumed parametric form. However, there may be concerns about confounding and the effect of the confounders has an unknown functional form. In this case, any parameters for the confounding variables are essentially nuisance parameters as the only goal is to adequately model their (unknown) functional form. There is no interest in actually understanding their functional form. In such cases the "hypothesis test" may be on a variable with a prespecified parametric form, thus avoiding risks of overfitting, but confounding variables can be robustly captured using smoothing terms, as statistical inference regarding them is of little interest.

Beyond the use of smoothing splines, GAMs essentially function as other GLMs. Thus distributional assumptions and the different distribution families that can be used are comparable. Admittedly, computation of degrees of freedom and standard errors tend to be different and likely more approximate in GAMs due to the data-driven nature of determining the appropriate smoothness for the splines. The next section introduces GAMs in R including how to estimate them, plot results, and present or use the estimated models.

5.2 GAMs in R
Gaussian Outcomes
Basic GAMs

GAMs can be fit to Gaussian outcomes using the `vgam()` function, which is set up almost the same as the `vglm()` function we are familiar with from previous chapters. The main difference is the use of smoothing splines, which are added to the model using another function, `s()`. The `s()` function takes an argument, `df,` which controls the maximum flexibility of the smoothing spline for a variable. The example that follows fits a GAM with two predictors: sex and a smoothing spline for age. The `summary()` function provides a summary primarily of the smooth terms and the p-values test whether the parameters differ significantly from a linear trend. Note that the `uninormal()` family can model both the location and the scale of the normal distribution, although by default all predictors are for the scale only. Thus there are two intercepts, one for the location, the "usual" intercept common to GLMs and GAMs, and one for the *scale* of the distribution, which is based on the natural logarithm of the variance. The reason for this is that the VGAM package is preparing to allow predictors both of the location and scale of a distribution, although the "classic" focus has been only on predicting the location of a distribution.

```
mgam <- vgam(CESD11_W1 ~ Sex + s(AGE_W1, df = 3), data = acl,
        family = uninormal(), model = TRUE)

summary(mgam)

## 
## Call:
## vgam(formula = CESD11_W1 ~ Sex + s(AGE_W1, df = 3), family = uninormal(),
##     data = acl, model = TRUE)
## 
## 
## Number of linear predictors:     2
## 
## Names of linear predictors: mean, loge(sd)
## 
## Dispersion Parameter for uninormal family:    1
## 
## Log-likelihood: -5290 on 7228 degrees of freedom
## 
## Number of iterations:   4
## 
## DF for Terms and Approximate Chi-squares for Nonparametric Effects
## 
##                     Df Npar Df Npar Chisq P(Chi)
## (Intercept):1        1
## (Intercept):2        1
## Sex                  1
## s(AGE_W1, df = 3)    1       2          20      0
```

Coefficients from the GAM can be viewed using the coef() function. However, coefficients for smooth terms are not readily interpretable. But coefficients for the parametric terms, in this model only sex, are interpretable just as from regular GLMs, with the exception that the effect of sex is controlling for a smooth spline function of age, rather than simply a linear trend of age.

```
coef(mgam)
```

```
##      (Intercept):1    (Intercept):2    Sex(2) FEMALE
##             0.2158           0.0435           0.2393
## s(AGE_W1, df = 3)
##            -0.0047
```

We can get hypothesis test of the parametric coefficients using the linearHypothesis() function from the car package.

```
## test parametric coefficient for sex
linearHypothesis(mgam, "Sex(2) FEMALE",
  coef. = coef(mgam), vcov = vcov(mgam))
```

```
## Linear hypothesis test
##
## Hypothesis:
## Sex(2) FEMALE = 0
##
## Model 1: restricted model
## Model 2: CESD11_W1 ~ Sex + s(AGE_W1, df = 3)
##
## Note: Coefficient covariance matrix supplied.
##
##    Res.Df Df Chisq Pr(>Chisq)
## 1    7229
## 2    7228  1  44.1    3.2e-11 ***
## ---
## Signif. codes:  0 '***' 0.001 '**' 0.01 '*' 0.05 '.' 0.1 '_' 1
```

It is possible to use linearHypothesis() to test more complex linear hypotheses of the parametric coefficients. For example, we could test whether the intercept and sex coefficient are simultaneously equal to zero.

```
## test parametric coefficient for
## intercept and sex simultaneously
linearHypothesis(mgam,
  c("(Intercept):1", "Sex(2) FEMALE"),
  coef. = coef(mgam), vcov = vcov(mgam))

## Linear hypothesis test
##
## Hypothesis:
## (Intercept):1 = 0
## Sex(2) FEMALE = 0
##
## Model 1: restricted model
## Model 2: CESD11_W1 ~ Sex + s(AGE_W1, df = 3)
##
## Note: Coefficient covariance matrix supplied.
##
##   Res.Df Df Chisq Pr(>Chisq)
## 1   7230
## 2   7228  2 79.1      <2e-16 ***
## ---
## Signif. codes:  0 '***' 0.001 '**' 0.01 '*' 0.05 '.' 0.1 '_' 1
```

We can visualize the results with 95% confidence intervals based on ±2 *se*. These should be treated as approximate 95% confidence intervals as the degrees of freedom are not directly estimated and thus rely on the central limit theorem and because the estimate of the standard error (se) is approximate due to the smoothing splines. Plots are easily made using the plot() function which includes methods for VGAM objects. The results are presentedin Figure 5-4, which shows the parametric effect of sex and the smooth term for age. Rug plots are added by default, which is not so useful for sex but helps to show the spread of data for age. Colors come from the viridis() palette.

```
par(mfrow = c(1, 2))
plot(mgam, se = TRUE,
     lcol = viridis(4)[1], scol = viridis(4)[2])
```

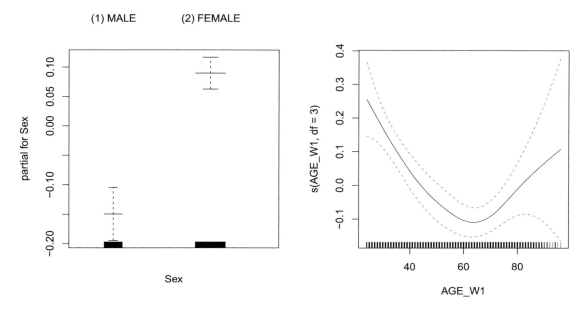

Figure 5-4. *Plot of model results for a generalized additive model with sex as a parametric term and age as a smooth spline*

It can be helpful to compare the results of our GAM to more familiar alternatives. To accomplish this we will fit two regular GLMs. The first includes a linear term for age and the second includes a quadratic term for age using the poly() function to make a degree 2 polynomial.

```
mlin <- vglm(CESD11_W1 ~ Sex + AGE_W1, data = acl,
        family = uninormal(), model = TRUE)
mquad <- vglm(CESD11_W1 ~ Sex + poly(AGE_W1, 2), data = acl,
        family = uninormal(), model = TRUE)
```

To compare the fits of these GLMs with our GAM, we can make a panel of two plots, shown in Figure 5-5. The dark purple is the GAM fit, and on the left we have the linear fit superimposed, while on the right we have the quadratic polynomial fit imposed. Obviously the linear fit differs dramatically from the GAM fit. The quadratic fit (on the right panel of Figure 5-5) is relatively close to the GAM except at the right most tail. The GAM begins to flatten at the tail, whereas the quadratic trend continues to increase rapidly.

```
par(mfrow = c(1, 2))
plot(mgam, se = TRUE, which.term = 2,
     lcol = viridis(4)[1], scol = viridis(4)[1])
plot(as(mlin, "vgam"), se = TRUE, which.term = 2,
     lcol = viridis(4)[2], scol = viridis(4)[2],
     overlay = TRUE, add = TRUE)

plot(mgam, se = TRUE, which.term = 2,
     lcol = viridis(4)[1], scol = viridis(4)[1])
plot(as(mquad, "vgam"), se = TRUE, which.term = 2,
     lcol = viridis(4)[3], scol = viridis(4)[3],
     overlay = TRUE, add = TRUE)
```

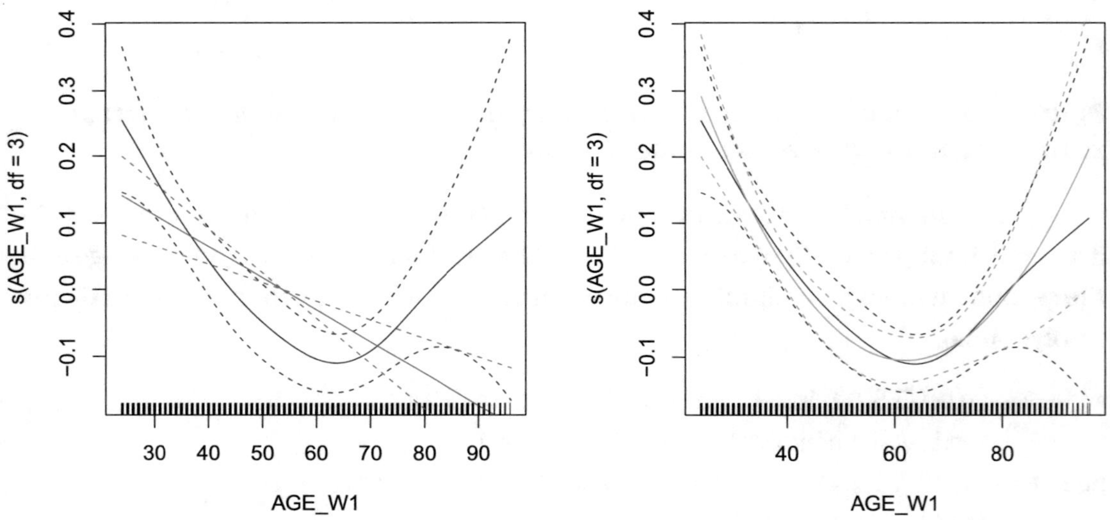

Figure 5-5. *Two-panel plot showing the predicted depression symptom level by age from a generalized additive model vs. a linear fit on the left and a quadratic fit on the right.*

Sometimes, GAMs may be used to choose a simpler trend function. For example, we may decide based on the graph that a quadratic polynomial for age is sufficient and switch to a polynomial model. However, if the results are not clearly well approximated by any simple polynomial, we may wish to remain with the GAM as our final model. In such cases, it can be helpful to derive some inference from the GAMs.

First, let us look at another example of a GAM. Here we predict depression symptoms at the second wave from sex, a smooth spline of depression symptoms at the first wave and a smooth spline of age at the first wave. In this new model, we see that the summary() suggests that the non-linearity in age is not statistically significant.

```
mgam2 <- vgam(CESD11_W2 ~ Sex +
                s(CESD11_W1, df = 3) +
                s(AGE_W1, df = 3), data = acl,
        family = uninormal(), model = TRUE)

summary(mgam2)

##
## Call:
## vgam(formula = CESD11_W2 ~ Sex + s(CESD11_W1, df = 3) + s(AGE_W1,
##     df = 3), family = uninormal(), data = acl, model = TRUE)
##
##
## Number of linear predictors:    2
##
## Names of linear predictors: mean, loge(sd)
##
## Dispersion Parameter for uninormal family:   1
##
## Log-likelihood: -3657 on 5725 degrees of freedom
##
## Number of iterations:   5
##
## DF for Terms and Approximate Chi-squares for Nonparametric Effects
##
##                       Df Npar Df Npar Chisq P(Chi)
## (Intercept):1          1
## (Intercept):2          1
## Sex                    1
## s(CESD11_W1, df = 3)   1       2          31    0.0
## s(AGE_W1, df = 3)      1       2           4    0.1
```

If one of the smoothing splines does not significantly differ from a linear term, we might consider falling back to a linear fit for that term, shown in the following code.

```
mgam3 <- vgam(CESD11_W2 ~ Sex +
              s(CESD11_W1, df = 3) +
              AGE_W1, data = acl,
       family = uninormal(), model = TRUE)
```

```
summary(mgam3)
```

```
##
## Call:
## vgam(formula = CESD11_W2 ~ Sex + s(CESD11_W1, df = 3) + AGE_W1,
##      family = uninormal(), data = acl, model = TRUE)
##
##
## Number of linear predictors:    2
##
## Names of linear predictors: mean, loge(sd)
##
## Dispersion Parameter for uninormal family:   1
##
## Log-likelihood: -3659 on 5727 degrees of freedom
##
## Number of iterations:   5
##
## DF for Terms and Approximate Chi-squares for Nonparametric Effects
##
##                     Df Npar Df Npar Chisq P(Chi)
## (Intercept):1        1
## (Intercept):2        1
## Sex                  1
## s(CESD11_W1, df = 3) 1       2          31 2e-07
## AGE_W1               1
```

We already saw how to test the statistical significance of the parametric terms using the linearHypothesis() function. We can use that to test both age and sex, now. We can use the names() function and the coef() function to get the names of each parameter to pass for testing.

```
names(coef(mgam3))
```

```
## [1] "(Intercept):1"       "(Intercept):2"
## [3] "Sex(2) FEMALE"       "s(CESD11_W1, df = 3)"
## [5] "AGE_W1"
```

```
linearHypothesis(mgam3,
  "Sex(2) FEMALE",
  coef. = coef(mgam3), vcov = vcov(mgam3))
```

```
## Linear hypothesis test
##
## Hypothesis:
## Sex(2) FEMALE = 0
##
## Model 1: restricted model
## Model 2: CESD11_W2 ~ Sex + s(CESD11_W1, df = 3) + AGE_W1
##
## Note: Coefficient covariance matrix supplied.
##
##   Res.Df Df Chisq Pr(>Chisq)
## 1   5728
## 2   5727  1  4.09       0.043 *
## ---
## Signif. codes:  0 '***' 0.001 '**' 0.01 '*' 0.05 '.' 0.1 '_' 1
```

```
linearHypothesis(mgam3,
  "AGE_W1",
  coef. = coef(mgam3), vcov = vcov(mgam3))
```

```
## Linear hypothesis test
##
## Hypothesis:
## AGE_W1 = 0
##
## Model 1: restricted model
## Model 2: CESD11_W2 ~ Sex + s(CESD11_W1, df = 3) + AGE_W1
##
## Note: Coefficient covariance matrix supplied.
##
##   Res.Df Df Chisq Pr(>Chisq)
## 1   5728
## 2   5727  1  3.56      0.059 .
## ---
## Signif. codes:  0 '***' 0.001 '**' 0.01 '*' 0.05 '.' 0.1 '_' 1
```

We visualize the results of this final model in Figure 5-6. Here we can see that for levels of depression symptoms below about 2 at wave 1, higher depression symptoms predict higher depression symptoms at wave 2. However, for depression symptoms above about a score of 2, there is not much association, evidenced by the flattening of the predicted values.

```
par(mfrow = c(2, 2))
plot(mgam3, se = TRUE,
     lcol = viridis(4)[1],
     scol = viridis(4)[2])
```

As in other cases, we can generate predictions for the GAMs. Because it is not really possible to describe smoothing splines verbally, it is standard to present GAMs graphically if the results from the smoothing splines are of interest. First we set up a new dataset for predictions. We get all levels of the factor variable, Sex, a sequence of depression symptoms at wave 1 from the minimum score to the maximum score, with an evenly spaced grid of 1,000 points, and a five number summary of age at wave 1. Prediction works the same as usual in R utilizing the predict() function which has methods in place for models from the VGAM package.

```r
## generate new data for prediction
## use the whole range of sex and depression symptoms
## and a five number summary of age
## (min, 25th 50th 75th percentiles and max)
newdat <- as.data.table(expand.grid(
  Sex = levels(acl$Sex),
  CESD11_W1 = seq(
    from = min(acl$CESD11_W1, na.rm=TRUE),
    to = max(acl$CESD11_W1, na.rm=TRUE),
    length.out = 1000),
  AGE_W1 = fivenum(acl$AGE_W1)))

newdat$yhat <- predict(mgam3, newdata = newdat)

## Warning in `[<-.data.table`(x, j = name, value = value): 2 column matrix
RHS of := will be treated

## Warning in `[<-.data.table`(x, j = name, value = value): Supplied 20000
items to be assigned to 10000
```

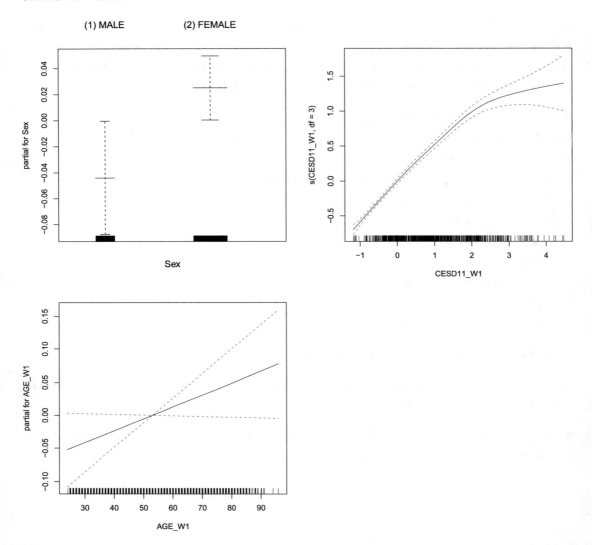

Figure 5-6. *Plot of model results for a generalized additive model with sex and age as parametric terms and wave 1 depression symptoms as a smooth spline*

Once we have a dataset of predictions, we can make our final plot using ggplot(). The results are shown in Figure 5-7. The figure clearly highlights that the strongest predictor of subsequent depression symptoms is previous depression symptoms. Although there are some minor differences between different ages and sexes, these are dwarfed by comparison.

```
ggplot(newdat,
       aes(CESD11_W1, yhat,
           colour = factor(AGE_W1),
           linetype = factor(AGE_W1))) +
  geom_line() +
  scale_color_viridis("Age", discrete = TRUE) +
  scale_linetype_discrete("Age") +
  facet_wrap(~ Sex) +
  theme(legend.position = c(.75, .2),
        legend.key.width = unit(1.5, "cm")) +
  xlab("Depression Symptoms (Wave 1)") +
  ylab("Depression Symptoms (Wave 2)")
```

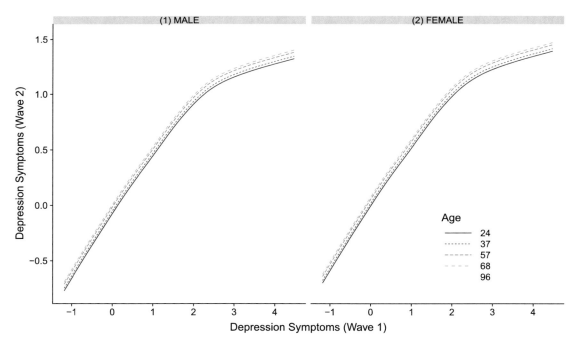

Figure 5-7. *Predicted depression symptoms at wave 2 across levels of wave 1 depression symptoms at varying ages and sexes*

185

Finally, although the VGAM package has a wide breadth of functionality and is one of the only choices for vector GAMs, for single outcome GAMs as we have shown here, there are some features of GAMs that have yet to be implemented. The mgcv package written by one of the leaders of theory and implementation of GAMs, Simon Wood, has some additional features that can be helpful. We will briefly overview the use of the mgcv package here, but far more details are available in the book on GAMs by Simon Wood [122].

Before we can use the mgcv package, we have to deal with some conflicts. Specifically, both the VGAM and mgcv packages implement a function called s() for smoothing splines, but they are not the same function. With both packages loaded, the package loaded last "masks" the previous packages' function, meaning that when we type s() in our R console, we get the results from the last package loaded, not necessarily the package we wanted. In our case the easiest way to resolve this is to detach the unwanted package and ensure that the desired package is loaded. We do that using the following code. Note that after this is run, you will not be able to use the vgam() function until you reload the VGAM package.

```
detach("package:VGAM")
library(mgcv)
```

Now we can fit a GAM using the gam() function from the mgcv package. We indicate smoothing splines again using the s() function, but note that for mgcv the argument to control the maximum flexibility is k instead of df. Also the appropriate family function is gaussian(). Although the rest of the code may appear similar, here are some differences under the hood in the types of smoothing splines and estimation used by default. Specifically, the mgcv package defaults to thin plate regression splines and the GCV criterion for learning the appropriate degree of flexibility.

```
mgam4 <- gam(CESD11_W2 ~ Sex +
               s(CESD11_W1, k = 3) +
               s(AGE_W1, k = 3), data = acl,
         family = gaussian())
```

One of the benefits of using the gam() function is that the default summary includes a variety of information that can be useful. Specifically, it automatically calculates statistical inference for parametric terms, and it provides approximate significance tests for the smooth terms. Note that unlike vgam() which tested whether the smooth terms differed significantly from a linear trend, gam() tests the overall significance of the smooth terms (i.e., it includes the linear trend as well as any non-linearity).

```
summary(mgam4)
```

```
##
## Family: gaussian
## Link function: identity
##
## Formula:
## CESD11_W2 ~ Sex + s(CESD11_W1, k = 3) + s(AGE_W1, k = 3)
##
## Parametric coefficients:
##                 Estimate Std. Error t value Pr(>|t|)
## (Intercept)      -0.0202     0.0272   -0.74    0.457
## Sex(2) FEMALE     0.0681     0.0342    1.99    0.046 *
## ---
## Signif. codes:  0 '***' 0.001 '**' 0.01 '*' 0.05 '.' 0.1 '_' 1
##
## Approximate significance of smooth terms:
##               edf Ref.df      F p-value
## s(CESD11_W1) 1.95   2.00 514.03  <2e-16 ***
## s(AGE_W1)    1.63   1.86   2.04   0.085 .
## ---
## Signif. codes:  0 '***' 0.001 '**' 0.01 '*' 0.05 '.' 0.1 '_' 1
##
## R-sq.(adj) =  0.271   Deviance explained = 27.2%
## GCV = 0.75609  Scale est. = 0.75461   n = 2867
```

Also because the mgcv provides ready access to the estimated degrees of freedom, it is easier to examine whether we should allow greater flexibility. For example, if we increase k = 3 to k = 4 and refit the model, we can see that the estimated degrees of freedom changed very little for age, but increased for depression symptoms.

```
mgam5 <- gam(CESD11_W2 ~ Sex +
                s(CESD11_W1, k = 4) +
                s(AGE_W1, k = 4), data = acl,
        family = gaussian())
```

```
summary(mgam5)
```

```
##
## Family: gaussian
## Link function: identity
##
## Formula:
## CESD11_W2 ~ Sex + s(CESD11_W1, k = 4) + s(AGE_W1, k = 4)
##
## Parametric coefficients:
##                Estimate Std. Error t value Pr(>|t|)
## (Intercept)     -0.0207     0.0272   -0.76    0.447
## Sex(2) FEMALE    0.0688     0.0342    2.01    0.044 *
## ---
## Signif. codes:  0 '***' 0.001 '**' 0.01 '*' 0.05 '.' 0.1 '_' 1
##
## Approximate significance of smooth terms:
##                edf Ref.df      F p-value
## s(CESD11_W1) 2.86   2.99  344.7  <2e-16 ***
## s(AGE_W1)    1.78   2.16    2.4   0.076 .
## ---
## Signif. codes:  0 '***' 0.001 '**' 0.01 '*' 0.05 '.' 0.1 '_' 1
##
## R-sq.(adj) =  0.272   Deviance explained = 27.3%
## GCV = 0.75508  Scale est. = 0.75333   n = 2867
```

We plot the two models side by side in Figure 5-8. The results show a subtle difference in the trend of depression symptoms at wave 1, with a more pronounced plateau in the GAM where we set $k = 4$.

```
par(mfrow = c(2, 2))
plot(mgam4, se = TRUE, scale = 0, main = "k = 3")
plot(mgam5, se = TRUE, scale = 0, main = "k = 4")
```

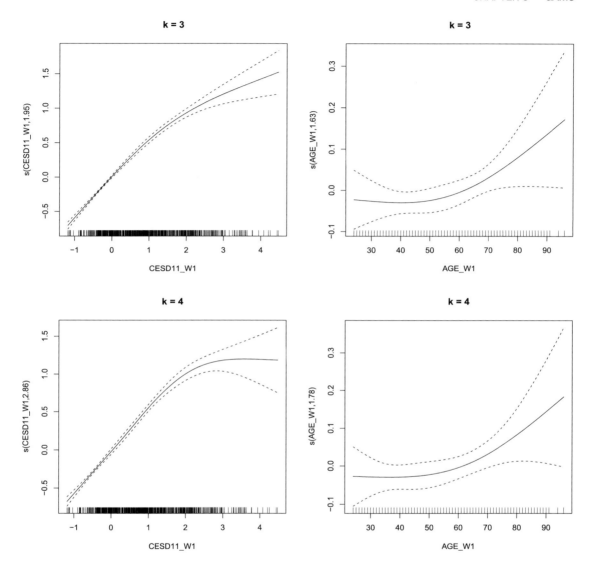

Figure 5-8. *Plot of model results for two generalized additive models varying the maximum flexibility of the smooth splines*

GAMs with Interactions

Another feature that is available in the mgcv package but is not yet available in the VGAM package is the ability to include smoothing splines for interactions. For example, suppose that we believe it is possible that the effect of either depression symptoms or

age varies by sex. We can accomplish this with relative ease by adding the argument by =
Sex to the s() functions we believe may vary by sex. The summary does not suggest large
differences, which is borne out by the plots which appear similar although there are
some differences by sex (Figure 5-9).

```
mgam6 <- gam(CESD11_W2 ~ Sex +
             s(CESD11_W1, k = 4, by = Sex) +
             s(AGE_W1, k = 4, by = Sex),
          data = acl,
       family = gaussian())
```

```
summary(mgam6)
```

```
##
## Family: gaussian
## Link function: identity
##
## Formula:
## CESD11_W2 ~ Sex + s(CESD11_W1, k = 4, by = Sex) + s(AGE_W1, k = 4,
##     by = Sex)
##
## Parametric coefficients:
##               Estimate Std. Error t value Pr(>|t|)
## (Intercept)    -0.0218     0.0276   -0.79    0.428
## Sex(2) FEMALE   0.0693     0.0343    2.02    0.044 *
## ---
## Signif. codes:  0 '***' 0.001 '**' 0.01 '*' 0.05 '.' 0.1 '_' 1
##
## Approximate significance of smooth terms:
##                            edf Ref.df      F p-value
## s(CESD11_W1):Sex(1) MALE   2.52   2.83 119.68  <2e-16 ***
## s(CESD11_W1):Sex(2) FEMALE 2.76   2.96 234.18  <2e-16 ***
## s(AGE_W1):Sex(1) MALE      1.80   2.17   1.58     0.2
## s(AGE_W1):Sex(2) FEMALE    1.00   1.00   2.68     0.1
## ---
## Signif. codes:  0 '***' 0.001 '**' 0.01 '*' 0.05 '.' 0.1 '_' 1
##
```

```
## R-sq.(adj) =  0.272    Deviance explained = 27.4%
## GCV = 0.75643  Scale est. = 0.75377    n = 2867

par(mfrow = c(2, 2))
plot(mgam6, ask = FALSE, scale = 0)
```

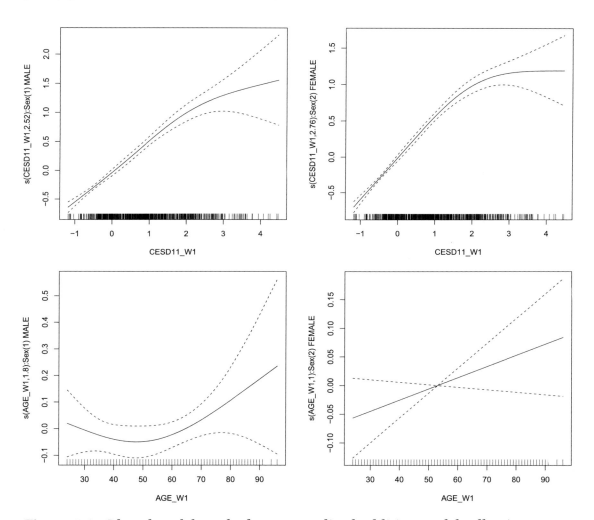

Figure 5-9. *Plot of model results for a generalized additive models allowing splines to vary by sex*

We could use the Akaike information criterion or Bayesian information criterion as a quick way to compare results between the two models. In this case, both indices point to the model without interactions by sex as the superior model balancing fit and parsimony.

```
AIC(mgam5, mgam6)
```

```
##             df    AIC
## mgam5  7.6   7333
## mgam6 11.1   7338
```

```
BIC(mgam5, mgam6)
```

```
##             df    BIC
## mgam5  7.6   7378
## mgam6 11.1   7404
```

Smoothing splines for interactions of two continuous variables becomes even more complex. However, the `mgcv` package allows this through tensor product smooths. The details behind tensor product smooths are not easy to understand, but roughly they can be thought of as taking each variable, separating it by knots and fitting polynomials (as with regular splines). These are assumed to be product terms, and although this may be unrealistic across the whole range of data, it is hoped that it provides a reasonable approximation as the possible space is broken down by knots on both variables. From a practical perspective, imagine an unknown three-dimensional surface where the depth and width are defined by two predictor variables and the height is the level of the outcome. Then imagine draping a heavy sheet over the top. The heavy material will provide a degree of "smoothness," but the shape can vary flexibly as you move in any direction, and you cannot speak of the "effect" of one variable without specifying the level of the other variable. The other practical note is that tensor product smooths are more computationally demanding and thus can be slower to fit.

In the code that follows, we fit a GAM with sex as a parametric term and a tensor product smooth between depression symptoms and self-esteem at wave 1 predicting depression symptoms at wave 2.

```
mgam7 <- gam(CESD11_W2 ~ Sex +
                te(CESD11_W1, SelfEsteem_W1, k = 4^2),
            data = acl,
        family = gaussian())

summary(mgam7)

##
## Family: gaussian
## Link function: identity
##
## Formula:
## CESD11_W2 ~ Sex + te(CESD11_W1, SelfEsteem_W1, k = 4^2)
##
## Parametric coefficients:
##              Estimate Std. Error t value Pr(>|t|)
## (Intercept)   -0.0226     0.0268   -0.84    0.400
## Sex(2) FEMALE  0.0718     0.0337    2.13    0.033 *
## ---
## Signif. codes:  0 '***' 0.001 '**' 0.01 '*' 0.05 '.' 0.1 '_' 1
##
## Approximate significance of smooth terms:
##                            edf Ref.df    F p-value
## te(CESD11_W1,SelfEsteem_W1) 12.1   14.5 77.4  <2e-16 ***
## ---
## Signif. codes:  0 '***' 0.001 '**' 0.01 '*' 0.05 '.' 0.1 '_' 1
##
## R-sq.(adj) =  0.286   Deviance explained = 28.9%
## GCV = 0.74276  Scale est. = 0.7391    n = 2867
```

The summary indicates that overall the tensor product smooth is statistically significant, although it does not break out which variable contributes most. Another challenge with such interactions is that they are even more difficult to visualize. Now we use the `vis.gam()` function which can make 3D perspective plots or contour plots. To begin with, we will make 3D perspective plots at several different perspectives in a panel of graphs. We also shrink the default margins. The results are in Figure 5-10.

```
par(mfrow = c(2, 2), mar = c(.1, .1, .1, .1))
vis.gam(mgam7,
  view = c("CESD11_W1", "SelfEsteem_W1"),
  theta = 210, phi = 40,
  color = "topo",
  plot.type = "persp")
vis.gam(mgam7,
  view = c("CESD11_W1", "SelfEsteem_W1"),
  theta = 150, phi = 40,
  color = "topo",
  plot.type = "persp")
vis.gam(mgam7,
  view = c("CESD11_W1", "SelfEsteem_W1"),
  theta = 60, phi = 40,
  color = "topo",
  plot.type = "persp")
vis.gam(mgam7,
  view = c("CESD11_W1", "SelfEsteem_W1"),
  theta = 10, phi = 40,
  color = "topo",
  plot.type = "persp")
```

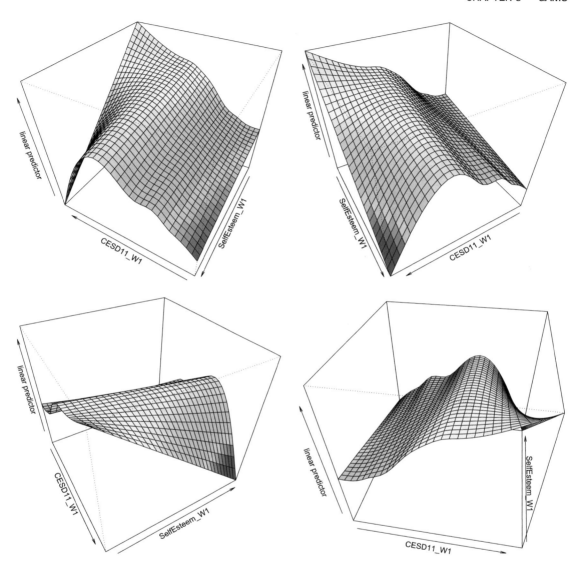

Figure 5-10. *3D perspective plots showing the result of a tensor product smooth between depression symptoms and self-esteem at wave 1 predicting depression at wave 2*

A plot easier to visualize in two-dimensions is a contour plot. Contour plots show the predictors on the x axis and y axis but use lines and colors to show the third dimension. Each line or contour represents the same predicted value, and the curves demonstrate how the same predicted value can be accomplished by varying combinations of the two predictors. An example contour plot is shown in Figure 5-11.

```
par(mfrow = c(1, 1), mar = c(5.1, 4.1, 4.1, 2.1))
vis.gam(mgam7,
  view = c("CESD11_W1", "SelfEsteem_W1"),
  color = "topo",
  plot.type = "contour")
```

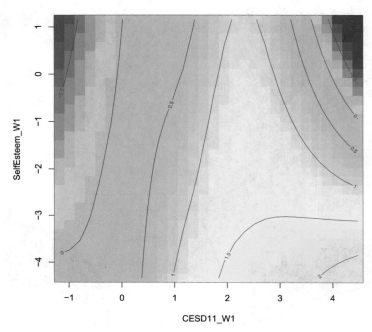

Figure 5-11. *Contour plot showing the result of a tensor product smooth between depression symptoms and self-esteem at wave 1 predicting depression at wave 2.*

If we want to attempt a rough decomposition of the variance, we can use tensor product interactions, using the `ti()` function. The results are shown in the following code, and they suggest that the interaction between depression symptoms and self-esteem does not provide much added value between the smooth terms for both depression symptoms and self-esteem.

```
mgam8 <- gam(CESD11_W2 ~ Sex +
                ti(CESD11_W1, k = 4) +
                ti(SelfEsteem_W1, k = 4) +
                ti(CESD11_W1, SelfEsteem_W1, k = 4^2),
```

```
        data = acl,
      family = gaussian())
```

summary(mgam8)

```
##
## Family: gaussian
## Link function: identity
##
## Formula:
## CESD11_W2 ~ Sex + ti(CESD11_W1, k = 4) + ti(SelfEsteem_W1, k = 4) +
##     ti(CESD11_W1, SelfEsteem_W1, k = 4^2)
##
## Parametric coefficients:
##               Estimate Std. Error t value Pr(>|t|)
## (Intercept)    -0.0156     0.0281   -0.55    0.579
## Sex(2) FEMALE   0.0681     0.0338    2.02    0.044 *
## ---
## Signif. codes:  0 '***' 0.001 '**' 0.01 '*' 0.05 '.' 0.1 '_' 1
##
## Approximate significance of smooth terms:
##                             edf Ref.df      F p-value
## ti(CESD11_W1)              1.72   2.08 260.86 < 2e-16 ***
## ti(SelfEsteem_W1)          1.00   1.00  22.77 1.9e-06 ***
## ti(CESD11_W1,SelfEsteem_W1) 21.77 31.08   1.03    0.43
## ---
## Signif. codes:  0 '***' 0.001 '**' 0.01 '*' 0.05 '.' 0.1 '_' 1
##
## R-sq.(adj) =  0.284   Deviance explained =   29%
## GCV = 0.74778   Scale est. = 0.74087   n = 2867
```

The final additional feature of the mgcv package we will examine is the ability to run some quick checks whether the maximum smooth allowed is restrictive. Although the degree of smoothness is learned, the parameter k controls the maximum allowed. Typically if the estimated degrees of freedom are very much lower than $k - 1$, it is unlikely that increasing k will have any benefit as the model already determined that a simple structure was adequate. However, this is not always true, and especially if the estimated degrees of freedom are close to $k - 1$, this may suggest that the artificial limit imposed resulted in an overly constrained model and that if we increased k we may get a different set of results. To examine this, we will go back to our earlier model examining depression symptoms and age predicting subsequent depression without any interactions. For convenience, we reproduce the model here. Notice that the estimated degrees of freedom for depression symptoms are close to $k - 1 = 4 - 1 = 3$, suggesting there may be some issues.

```
mgam5 <- gam(CESD11_W2 ~ Sex +
                s(CESD11_W1, k = 4) +
                s(AGE_W1, k = 4), data = acl,
        family = gaussian())

summary(mgam5)

##
## Family: gaussian
## Link function: identity
##
## Formula:
## CESD11_W2 ~ Sex + s(CESD11_W1, k = 4) + s(AGE_W1, k = 4)
##
## Parametric coefficients:
##                 Estimate Std. Error t value Pr(>|t|)
## (Intercept)     -0.0207     0.0272   -0.76    0.447
## Sex(2) FEMALE    0.0688     0.0342    2.01    0.044 *
## ---
## Signif. codes:  0 '***' 0.001 '**' 0.01 '*' 0.05 '.' 0.1 '_' 1
##
## Approximate significance of smooth terms:
##               edf Ref.df     F p-value
## s(CESD11_W1) 2.86   2.99 344.7  <2e-16 ***
## s(AGE_W1)    1.78   2.16   2.4   0.076 .
## ---
## Signif. codes:  0 '***' 0.001 '**' 0.01 '*' 0.05 '.' 0.1 '_' 1
##
## R-sq.(adj) =  0.272   Deviance explained = 27.3%
## GCV = 0.75508  Scale est. = 0.75333   n = 2867
```

To check whether we may want to increase k, we can use the gam.check() function. All it requires to begin with is a fitted GAM. Because gam.check() relies on some simulations, it can vary depending on the random seed. To ensure reproducibility, set the random seed, as we do next using set.seed(). Results are printed and also some plots are made, shown in Figure 5-12. The results suggest that k may not have been high enough for depression symptoms at wave 1. In general because of the smoothing, it is not required to guess k exactly correct, but it does need to be large enough that the functional form is not inappropriately restricted.

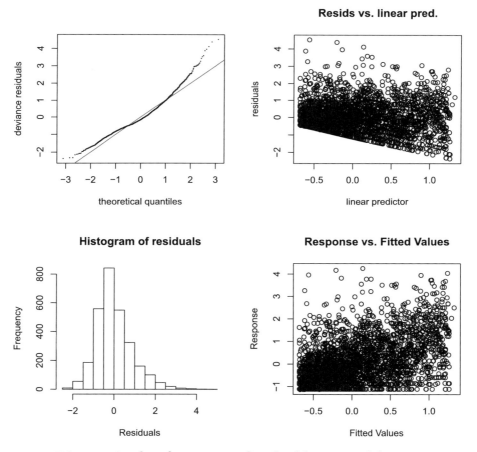

Figure 5-12. *Diagnostic plots from generalized additive model*

```
par(mfrow = c(2, 2))
set.seed(12345)
gam.check(mgam5)
```

```
##
## Method: GCV   Optimizer: magic
## Smoothing parameter selection converged after 9 iterations.
## The RMS GCV score gradient at convergence was 6.7e-07 .
## The Hessian was positive definite.
## Model rank = 8 / 8
##
## Basis dimension (k) checking results. Low p-value (k-index<1) may
## indicate that k is too low, especially if edf is close to k'.
##
##                   k'  edf k-index p-value
## s(CESD11_W1) 3.00 2.86    0.97   0.025 *
## s(AGE_W1)    3.00 1.78    0.99   0.230
## ---
## Signif. codes:  0 '***' 0.001 '**' 0.01 '*' 0.05 '.' 0.1 '_' 1
```

In light of the information from gam.check(), we might refit our model, increasing the k for depression symptoms. We choose a new value, $k = 20$. Now the results show estimated degrees of freedom much lower than $k - 1 = 20 - 1 = 19$. Although we did not need to, we increase the k for age as well to demonstrate the impact of changing the parameter.

```
mgam5b <- gam(CESD11_W2 ~ Sex +
                s(CESD11_W1, k = 20) +
                s(AGE_W1, k = 20), data = acl,
         family = gaussian())

summary(mgam5b)

##
## Family: gaussian
## Link function: identity
##
## Formula:
## CESD11_W2 ~ Sex + s(CESD11_W1, k = 20) + s(AGE_W1, k = 20)
##
```

```
## Parametric coefficients:
##                Estimate Std. Error t value Pr(>|t|)
## (Intercept)    -0.0202      0.0271   -0.75    0.456
## Sex(2) FEMALE   0.0680      0.0341    2.00    0.046 *
## ---
## Signif. codes:  0 '***' 0.001 '**' 0.01 '*' 0.05 '.' 0.1 '_' 1
##
## Approximate significance of smooth terms:
##                edf Ref.df      F p-value
## s(CESD11_W1) 11.68   13.99 76.43    <2e-16 ***
## s(AGE_W1)     1.69    2.13  1.87      0.14
## ---
## Signif. codes:  0 '***' 0.001 '**' 0.01 '*' 0.05 '.' 0.1 '_' 1
##
## R-sq.(adj) =  0.278   Deviance explained = 28.2%
## GCV = 0.7508  Scale est. = 0.74678   n = 2867
```

We plot the results in Figure 5-13, which shows the much greater flexibility in the trend for depression symptoms than previously included. Also notice that the trend for age is not meaningfully altered despite increasing the k. This occurs when a simpler fit is already adequate; thus, increasing k has little impact on the results as the additional flexibility was already penalized away.

```
par(mfrow = c(1, 2))
plot(mgam5b, se = TRUE, scale = 0)
```

Although brief, this section covers many of the basic uses and features of GAMs for continuous, normally distributed data. The next sections utilize the same ideas, but applied to other types of outcome data and assuming different distributions. However, the basic steps of model fitting, model comparison, and visualization tend to be similar.

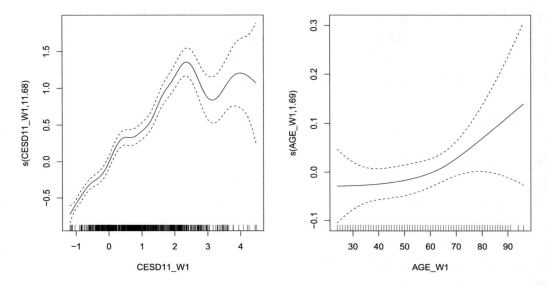

Figure 5-13. *Plot of model results for generalized additive model after increasing k for depression symptoms*

Binary Outcomes

Binary outcomes were already introduced in the chapters discussing GLMs. GAMs for binary outcomes rely on the same theory as the GLMs and work with the same families (Bernoulli/binomial). The novel aspect of GAMs for binary outcomes is the use of smoothing splines, which function the same as they did for continuous, normally distributed, outcomes, except that the smoothing splines are on the link scale, which in the case of binary outcomes typically is the logit (log odds).

To examine GAMs for binary outcomes, we will use smoking as an outcome, comparing current smokers (1) to former or never smokers (0). We begin with age as our predictor. The summary suggests there is some non-linearity in the association of age with smoking status.

```
library(VGAM)

##
## Attaching package: 'VGAM'

## The following object is masked from 'package:car':
##
##     logit
```

```
## The following objects are masked from 'package:rms':
##
##      calibrate, lrtest

## The following object is masked from 'package:mgcv':
##
##      s

## The following objects are masked from 'package:boot':
##
##      logit, simplex

acl$CurSmoke <- as.integer(acl$Smoke_W1 == "(1) Cur Smok")


mgam.lr1 <- vgam(CurSmoke ~ s(AGE_W1, df = 3),
            family = binomialff(link = "logit"),
            data = acl, model = TRUE)

summary(mgam.lr1)

##
## Call:
## vgam(formula = CurSmoke ~ s(AGE_W1, df = 3), family = binomialff(link =
## "logit"),
##      data = acl, model = TRUE)
##
##
## Number of linear predictors:    1
##
## Name of linear predictor: logit(prob)
##
## (Default) Dispersion Parameter for binomialff family:    1
##
## Residual deviance:   4173 on 3613 degrees of freedom
##
## Log-likelihood: -2087 on 3613 degrees of freedom
##
## Number of iterations:   5
```

```
##
## DF for Terms and Approximate Chi-squares for Nonparametric Effects
##
##                      Df Npar Df Npar Chisq P(Chi)
## (Intercept)          1
## s(AGE_W1, df = 3)    1        2           44  2e-10
```

To better understand the results, we can plot the smoothing spline.

```
par(mfrow = c(1, 1))
plot(mgam.lr1, se = TRUE,
     lcol = viridis(4)[1],
     scol = viridis(4)[2])
```

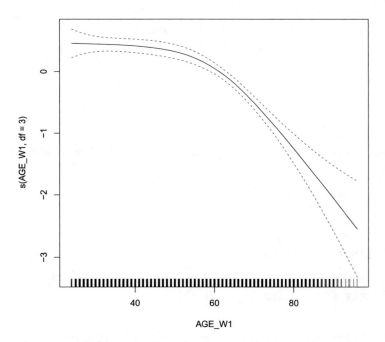

Figure 5-14. *Generalized additive model for age and current smoking status*

This shows that there is little change in younger ages but a steep decline around age 60 years. However, currently, the plot is on the link scale. We can get predictions on the probability scale for plotting in a more intuitive metric.

```
## generate new data for prediction
## use the whole range of age
newdat <- as.data.table(expand.grid(
  AGE_W1 = seq(
    from = min(acl$AGE_W1, na.rm=TRUE),
    to = max(acl$AGE_W1, na.rm=TRUE),
    length.out = 1000)))

newdat$yhat <- predict(mgam.lr1,
                       newdata = newdat,
                       type = "response")
```

Once we have a dataset of predictions, we can make our final plot using ggplot(). The results are shown in Figure 5-15. The figure shows a similar picture in this case to the results on the link scale. However, transformed, the results can be more easily interpreted.

```
ggplot(newdat, aes(AGE_W1, yhat)) +
  geom_line() +
  scale_y_continuous(labels = percent) +
  xlab("Age (years)") +
  ylab("Probability of Smoking") +
  coord_cartesian(xlim = range(acl$AGE_W1),
                  ylim = c(0, .4),
                  expand = FALSE)
```

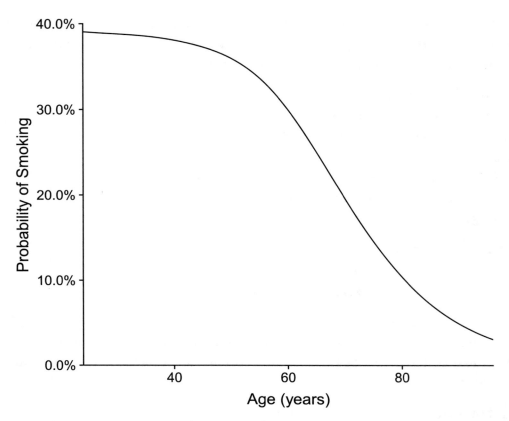

Figure 5-15. *Predicted probability of smoking across ages*

One limitation is that there is no built-in way to get confidence intervals for GAM predictions on new data from the VGAM package. One way to generate confidence intervals is to use bootstrapping. With such a simple model, this does not take too much time. We could also utilize parallel processing to speed it up, particularly for more complex models or if we were going to do a greater number of bootstrap resamples.

```
nboot <- 500

out <- matrix(NA_real_, ncol = nboot, nrow = nrow(newdat))

start.time <- proc.time()
set.seed(12345)
for (i in 1:500) {
  tmp <- vgam(CurSmoke ~ s(AGE_W1, df = 3),
          family = binomialff(link = "logit"),
```

```
                 data = acl[sample(nrow(acl), replace = TRUE)], model = TRUE)
  out[, i] <- predict(tmp,
                      newdata = newdat,
                      type = "response")
}
stop.time <- proc.time()

## time to bootstrap 500 times
stop.time - start.time
##    user  system elapsed
##   19.12    0.03   19.21
```

Now we can generate some summaries from the bootstrapped predictions. First, sometimes people compare the average of the bootstrap predictions to the actual model to see if there is systematic bias. The following code is just a quick check calculating the mean absolute difference. In this case it is very small.

```
mean(abs(newdat$yhat - rowMeans(out)))

## [1] 0.00031
```

Next we can calculate the confidence intervals, but taking the percentiles of the bootstrap samples.

```
newdat$LL <- apply(out, 1, quantile,
  probs = .025, na.rm = TRUE)

newdat$UL <- apply(out, 1, quantile,
  probs = .975, na.rm = TRUE)
```

Finally, we can remake our predicted probability plot, but now with confidence intervals added. The results are shown in Figure 5-16. The confidence intervals are slightly jagged due to the relatively small number of bootstrap samples drawn. Quantile-based confidence intervals tend to smooth out with larger number of samples, but even as is, they provide a relatively quick and very useful addition by indicating the degree of uncertainty in the predicted probability of smoking across the smooth spline for age.

```
ggplot(newdat, aes(AGE_W1, yhat)) +
  geom_ribbon(aes(ymin = LL, ymax = UL), fill = "grey80") +
  geom_line(size = 2) +
```

```
scale_y_continuous(labels = percent) +
xlab("Age (years)") +
ylab("Probability of Smoking") +
theme_tufte() +
coord_cartesian(xlim = range(acl$AGE_W1),
                ylim = c(0, .5),
                expand = FALSE)
```

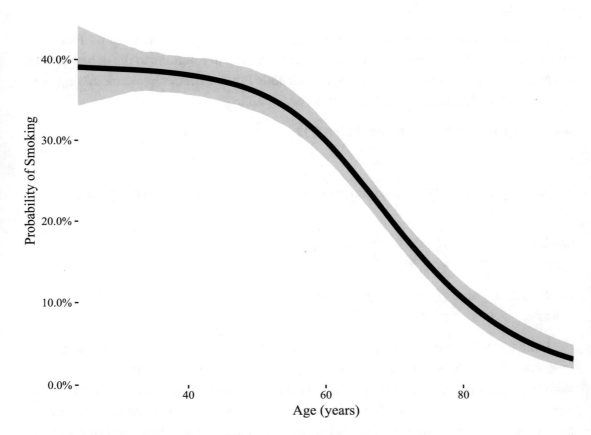

Figure 5-16. *Predicted probability of smoking across ages*

In the chapter on GLMs, we introduced a measure, the average marginal change in predicted probability as a more intuitive summary of a predictor when outcomes are categorical. Although this can work to deal with the non-linearity due to the link function, in the case of smoothing splines, it often does not make sense to produce such a measure as it not only incorporates differences due to starting probability but also the fact that the effect of the predictor is not linear. Thus a figure with confidence intervals as we produced in Figure 5-16 would often be the final result of a GAM for binary outcomes.

Unordered Outcomes

In the chapter on GLMs, we saw multinomial logistic regression models for unordered categorical outcomes with more than two levels. The procedures are similar for GAMs with unordered categorical outcomes. We first do some data management to generate a collapsed employment at wave 2 variable and then fit a GAM predicting this from a smooth spline of age. The summary reveals evidence of significant non-linearity for several contrasts. One interesting feature is that now instead of a single test for whether there is non-linearity in a smooth term, there are now $k - 1$ tests, where k is the number of categories in the outcome. In our case we have five categories; one serves as the reference and hence there are four tests of non-linearity.

```
acl[, EmployG_W2 := as.character(Employment_W2)]
acl[EmployG_W2 %in% c(
  "(2) 2500+HRS", "(3) 15002499",
  "(4) 500-1499", "(5) 1-499HRS"),
  EmployG_W2 := "(2) EMPLOYED"]
acl[, EmployG_W2 := factor(EmployG_W2)]

mgam.mr1 <- vgam(EmployG_W2 ~ s(AGE_W1, k = 5),
              family = multinomial(),
              data = acl, model = TRUE)

summary(mgam.mr1)

##
## Call:
## vgam(formula = EmployG_W2 ~ s(AGE_W1, k = 5), family = multinomial(),
##     data = acl, model = TRUE)
```

```
##
##
## Number of linear predictors:     4
##
## Names of linear predictors:
## log(mu[,1]/mu[,5]), log(mu[,2]/mu[,5]), log(mu[,3]/mu[,5]), log(mu[,4]/
mu[,5])
##
## Dispersion Parameter for multinomial family:    1
##
## Residual deviance:   5261 on 11450 degrees of freedom
##
## Log-likelihood: -2631 on 11450 degrees of freedom
##
## Number of iterations:   8
##
## DF for Terms and Approximate Chi-squares for Nonparametric Effects
##
##                          Df  Npar  Df  Npar  Chisq  P(Chi)
## (Intercept):1             1
## (Intercept):2             1
## (Intercept):3             1
## (Intercept):4             1
## s(AGE_W1, k = 5):1        1         3           16    0.0
## s(AGE_W1, k = 5):2        1         2           83    0.0
## s(AGE_W1, k = 5):3        1         2           71    0.0
## s(AGE_W1, k = 5):4        1         2            5    0.1
```

We could again plot the results, by default on the link scale, and again we now get four plots instead of one because we have a plot for each level of the categorical outcome. The results are in Figure 5-17. Some comparisons show a greater degree of non-linearity than do others and obviously the trends are quite different. GAMs for multinomial logistic regression allow the shape and flexibility of the smoothing spline to vary across all the levels of the outcome.

```
par(mfrow = c(2, 2))
plot(mgam.mr1, se = TRUE,
    lcol = viridis(4)[1],
    scol = viridis(4)[2])
```

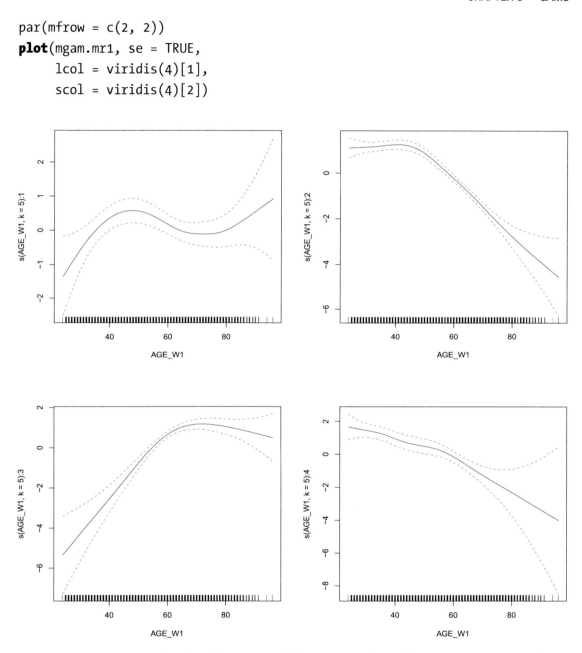

Figure 5-17. *Generalized additive model for age and employment status as a five-level unordered categorical outcome resulting in four distinct effects of age*

Again we might prefer to generate predicted probabilities rather than plots of the logits. We generate the predictions as usual. Here we combine the predicted probabilities with our dataset using the cbind() function because, rather than one vector of predicted probabilities, a matrix is returned as there is a probability of membership in each level of the outcome. Then we melt the data into a long dataset for plotting. The new dataset has three variables, one being age, another the level of the outcome, and the third the actual predicted probability.

```
## generate new data for prediction
## use the whole range of age
newdat <- as.data.table(expand.grid(
  AGE_W1 = seq(
    from = min(acl$AGE_W1, na.rm=TRUE),
    to = max(acl$AGE_W1, na.rm=TRUE),
    length.out = 1000)))

newdat <- cbind(newdat, predict(mgam.mr1,
                newdata = newdat,
                type = "response"))

newdatlong <- melt(newdat, id.vars = "AGE_W1")

summary(newdatlong)
```

```
##      AGE_W1              variable          value
## Min.    :24   (1)  DISABLED:1000   Min.    :0.00
## 1st Qu. :42   (2)  EMPLOYED:1000   1st Qu.:0.03
## Median  :60   (6)  RETIRED :1000   Median :0.08
## Mean    :60   (7)  UNEMPLOY:1000   Mean    :0.20
## 3rd Qu. :78   (8)  KEEP HS :1000   3rd Qu.:0.28
## Max.    :96                        Max.    :0.84
```

Finally, we can make a plot of the results using ggplot(). The results are shown in Figure 5-18. These findings highlight something well known but not necessarily well captured by a linear model: people tend to retire after age 60 years. Because there is a relatively narrow age window around when many (not all, but many) people retire, the models are decidedly not linear. Instead they are relatively flat, briefly change greatly, and return to relatively flat. Given what we know about retirement ages, we might consider adding even more flexibility or a purposeful piecewise model around that age as the smoothing splines will still try to smooth what in fact may be a relatively discrete process. Nonetheless, even without such additional efforts, we can see the value of a GAM in capturing this rapid transition relatively well, despite us providing no guidance to the model where such a transition might occur.

```
ggplot(newdatlong, aes(
  AGE_W1, value,
  colour = variable, linetype = variable)) +
  geom_line(size = 2) +
  scale_color_viridis(discrete = TRUE) +
  scale_x_continuous("Age (years)") +
  scale_y_continuous("Probability", label = percent) +
  coord_cartesian(ylim = c(0, 1), expand = FALSE) +
  theme_tufte() +
  theme(legend.position = c(.2, .5),
        legend.key.width = unit(2, "cm"))
```

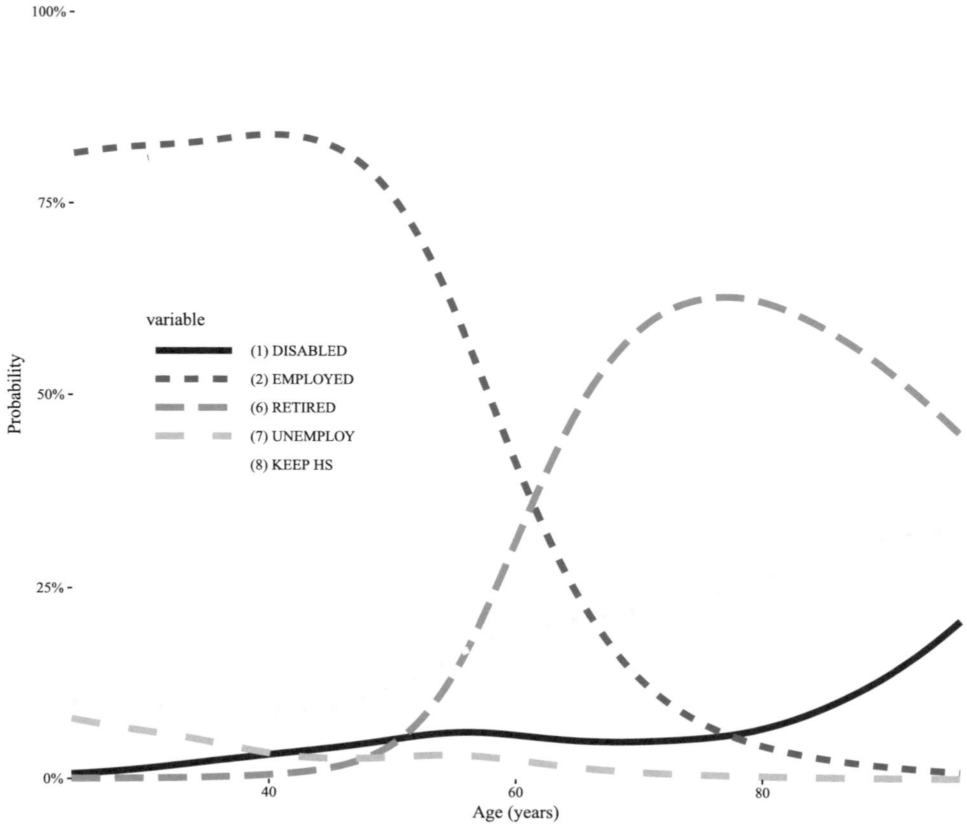

Figure 5-18. *Predicted probability of employment status across ages*

Count Outcomes

Count outcomes were already introduced in the chapters discussing GLMs. GAMs for count outcomes rely on the same theory as the GLMs and work with the same families (Poisson, negative binomial). The novel aspect of GAMs for count outcomes is the use of smoothing splines, which function the same as they did for continuous, normally distributed, outcomes, except that the smoothing splines are on the link scale, which in the case of count outcomes typically is the natural logarithm. In the GLM chapter, we saw the limitations of the Poisson distribution where overdispersion often may occur and not be captured adequately. Thus here we will go directly to GAMs using a negative binomial distribution that can allow overdispersion. Although we have only been looking at a single predictor, we are not limited to this. In the following example, we examine the number of chronic conditions and predict this from sex and a smoothing spline for age. The summary shows strong evidence of non-linearity of age.

```
## negative binomial regression model
mgam.nbr1 <- vgam(NChronic12_W2 ~ Sex + s(AGE_W1, k = 5),
            family = negbinomial(),
            data = acl, model = TRUE)

summary(mgam.nbr1)

##
## Call:
## vgam(formula = NChronic12_W2 ~ Sex + s(AGE_W1, k = 5), family =
negbinomial(),
##      data = acl, model = TRUE)
##
##
## Number of linear predictors:   2
##
## Names of linear predictors: loge(mu), loge(size)
##
## Dispersion Parameter for negbinomial family:   1
##
## Log-likelihood: -3636 on 5727 degrees of freedom
##
```

```
## Number of iterations:     8
##
## DF for Terms and Approximate Chi-squares for Nonparametric Effects
##
##                    Df Npar Df Npar Chisq P(Chi)
## (Intercept):1     1
## (Intercept):2     1
## Sex               1
## s(AGE_W1, k = 5)  1         3         112      0
```

To better understand the results, we can plot the smoothing spline, shown in Figure 5-19. There we can see that women tend to report more chronic conditions than do men and that the number of conditions reported increases with age rapidly when younger, but the increase slows at older ages (Figure 5-19).

```
par(mfrow = c(1, 2))
plot(mgam.nbr1, se = TRUE,
     lcol = viridis(4)[1],
     scol = viridis(4)[2])
```

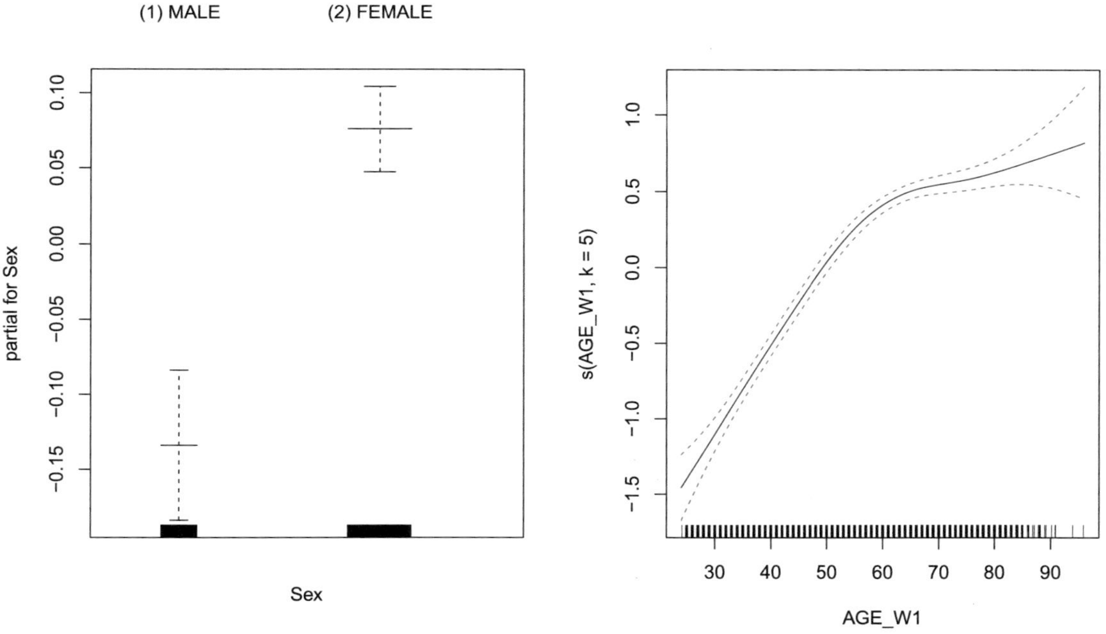

Figure 5-19. *Generalized additive model for sex and age and number of chronic conditions*

215

As with binary and multinomial logistic regression, currently, the plot is on the link scale. We can get predictions on the raw count scale for plotting in a more intuitive metric.

```
## generate new data for prediction
## use the whole range of age and sex
newdat <- as.data.table(expand.grid(
  Sex = levels(acl$Sex),
  AGE_W1 = seq(
    from = min(acl$AGE_W1, na.rm=TRUE),
    to = max(acl$AGE_W1, na.rm=TRUE),
    length.out = 1000)))

newdat$yhat <- predict(mgam.nbr1,
                       newdata = newdat,
                       type = "response")
```

Once we have a dataset of predictions, we can make our plot using ggplot(). The results are shown in Figure 5-20. Although the effect of sex is constant on the link scale, on the original response scale, it increases in absolute magnitude as the predicted scores increase; hence, the gap between women and men is predicted wider at older ages. However, note that at present, there is no interaction between age and sex. Thus the effect of age is assumed the same for both women and men in Figure 5-20.

```
ggplot(newdat, aes(AGE_W1, yhat, colour = Sex)) +
  geom_line(size = 2) +
  scale_color_viridis(discrete = TRUE) +
  xlab("Age (years)") +
  ylab("Number Chronic Conditions") +
  theme_tufte() +
  coord_cartesian(xlim = range(acl$AGE_W1),
                  ylim = c(0, 2.5),
                  expand = FALSE) +
  theme(legend.position = c(.2, .8),
        legend.key.width = unit(1, "cm"))
```

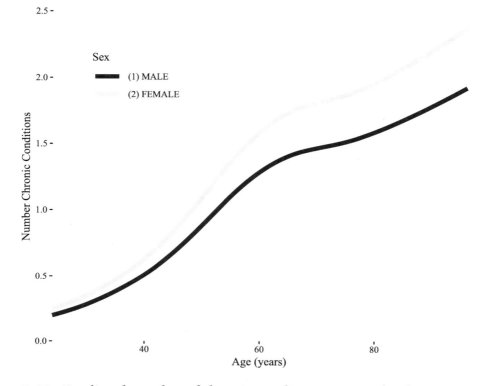

Figure 5-20. *Predicted number of chronic conditions across ages by sex*

We might wonder whether if women indeed report more chronic conditions at a younger age, whether they plateau quicker than do men. That would suggest an interaction. As interactions with smooth splines are not currently supported in VGAM, we detach the package and switch to mgcv. The model can be fit using the gam() function as shown next.

```
detach("package:VGAM")
library(mgcv)

mgam.nbr2 <- gam(NChronic12_W2 ~ Sex + s(AGE_W1, k = 10, by = Sex),
          family = nb(), data = acl)

summary(mgam.nbr2)

##
## Family: Negative Binomial(20719.179)
## Link function: log
```

```
##
## Formula:
## NChronic12_W2 ~ Sex + s(AGE_W1, k = 10, by = Sex)
##
## Parametric coefficients:
##              Estimate Std. Error z value Pr(>|z|)
## (Intercept)   -0.2674     0.0395   -6.77  1.3e-11 ***
## Sex(2) FEMALE  0.2661     0.0477    5.58  2.4e-08 ***
## ---
## Signif. codes:    0 '***' 0.001 '**' 0.01 '*' 0.05 '.' 0.1 '_' 1
##
## Approximate significance of smooth terms:
##                         edf Ref.df Chi.sq p-value
## s(AGE_W1):Sex(1) MALE   4.36   5.36    295  <2e-16 ***
## s(AGE_W1):Sex(2) FEMALE 4.11   5.10    447  <2e-16 ***
## ---
## Signif. codes:  0 '***' 0.001 '**' 0.01 '*' 0.05 '.' 0.1 '_' 1
##
## R-sq.(adj) =  0.248   Deviance explained = 25.6%
## -REML = 3649.7  Scale est. = 1          n = 2867
```

Now we again generate predictions, using the same code as we would had the model been fitted with vgam().

```
## generate new data for prediction
## use the whole range of age and sex
newdat <- as.data.table(expand.grid(
  Sex = levels(acl$Sex),
  AGE_W1 = seq(
    from = min(acl$AGE_W1, na.rm=TRUE),
    to = max(acl$AGE_W1, na.rm=TRUE),
    length.out = 1000)))

newdat$yhat <- predict(mgam.nbr2,
                   newdata = newdat,
                   type = "response")
```

Finally, we again plot the results, which are shown in Figure 5-21. However, the results reveal that far from women plateauing faster, if anything (albeit the results may not be reliably different), men plateau sooner, and even at older ages women are predicted to report even more chronic conditions.

```
ggplot(newdat, aes(AGE_W1, yhat, colour = Sex)) +
  geom_line(size = 2) +
  scale_color_viridis(discrete = TRUE) +
  xlab("Age (years)") +
  ylab("Number Chronic Conditions") +
  theme_tufte() +
  coord_cartesian(xlim = range(acl$AGE_W1),
                  ylim = c(0, 2.7),
                  expand = FALSE) +
  theme(legend.position = c(.2, .8),
        legend.key.width = unit(1, "cm"))
```

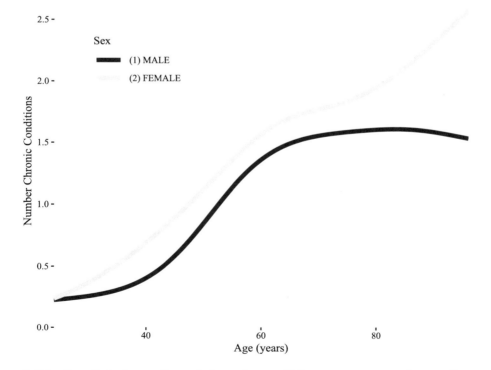

Figure 5-21. *Predicted number of chronic conditions across ages by sex from an interaction model*

As with the binary logistic GAM, if we wanted we could use bootstrapping to generate confidence intervals around the predictions. Note that this takes somewhat longer than the example of a binary logistic GAM, perhaps due to differences in software but also because this is a far more complex model with two predictors and the smooth spline for age varying as a function of sex. As it takes longer to run, this is an example where it would likely be advantageous to utilize parallel processing if this was being run for actual analysis as you would likely use at least a few thousand bootstrap samples.

```
nboot <- 500

out <- matrix(NA_real_, ncol = nboot, nrow = nrow(newdat))

start.time <- proc.time()
set.seed(12345)
for (i in 1:500) {
  tmp <- gam(NChronic12_W2 ~ Sex + s(AGE_W1, k = 10, by = Sex),
             family = nb(),
             data = acl[sample(nrow(acl), replace = TRUE)])
  out[, i] <- predict(tmp,
                      newdata = newdat,
                      type = "response")
}
stop.time <- proc.time()

## time to bootstrap 500 times
stop.time - start.time

##     user  system elapsed
##   167.18    0.73  168.08
```

Now we can generate some summaries from the bootstrapped predictions. First, sometimes people compare the average of the bootstrap predictions to the actual model to see if there is systematic bias. The following code is just a quick check calculating the mean absolute difference. In this case it is very small.

```
mean(abs(newdat$yhat - rowMeans(out)))

## [1] 0.0094
```

Next we can calculate the confidence intervals, but taking the percentiles of the bootstrap samples.

```
newdat$LL <- apply(out, 1, quantile,
  probs = .025, na.rm = TRUE)

newdat$UL <- apply(out, 1, quantile,
  probs = .975, na.rm = TRUE)
```

Finally, we can remake our predicted count plot, but now with confidence intervals added. The results are shown in Figure 5-22.

```
ggplot(newdat, aes(AGE_W1, yhat)) +
  geom_ribbon(aes(ymin = LL, ymax = UL, fill = Sex), alpha = .2) +
  geom_line(aes(colour = Sex), size = 2) +
  scale_color_viridis(discrete = TRUE) +
  scale_fill_viridis(discrete = TRUE) +
  xlab("Age (years)") +
  ylab("Number Chronic Conditions") +
  theme_tufte() +
  coord_cartesian(xlim = range(acl$AGE_W1),
                  ylim = c(0, 4),
                  expand = FALSE) +
  theme(legend.position = c(.2, .8),
        legend.key.width = unit(2, "cm"))
```

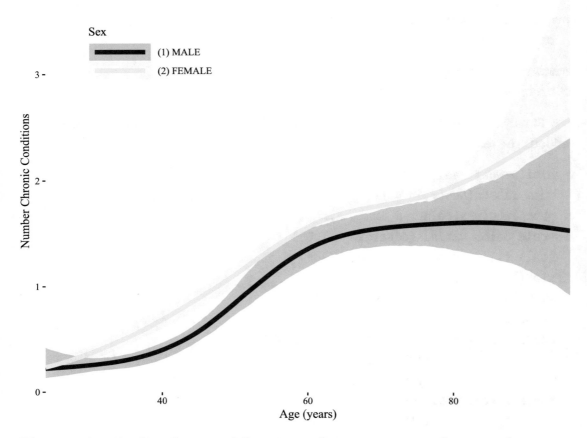

Figure 5-22. *Predicted count of chronic conditions across ages by sex with bootstrapped confidence intervals*

Overall there are some distinctions between women and men in the shape of the trend across age, but there also is considerable uncertainty in the predictions, particularly at the oldest ages, suggesting that perhaps that apparent increase past 80 in women may or may not be a very reliable trend. Likewise although men are predicted to be relatively stable after age 80 years, they have a wide confidence interval that could include a noticeable increase. In this case there simply are not enough data to be confident. Indeed, if we look at the number of people aged 80 or more at wave 1 who completed reports of chronic conditions at wave 2, we see there are just 27 men and 73 women, not much data, especially for men, on which to draw strong inference about the trends past 80 years, hence the wide confidence interval.

```
xtabs(~Sex + I(AGE_W1 > 80), data = acl[!is.na(NChronic12_W2)])
```

```
##                I(AGE_W1 > 80)
## Sex              FALSE  TRUE
##   (1) MALE        1010    27
##   (2) FEMALE      1757    73
```

5.3 Summary

This chapter provided a brief introduction to polynomials, splines, and smoothing splines for regression and then introduced a flexible class of models, generalized additive models (GAMs). It showed how GAMs can be used to extend parametric generalized linear models (GLMs) to capture unknown functional forms for predictors. In particular, GAMs are often useful when there are continuous predictors, and there is a concern that the association with the outcome may be inadequately captured by a linear or polynomial trend, or there is too little information to speculate as to the degree of polynomial trend. In these cases and when there is a sufficiently large sample size, GAMs excel and allow one to capture and model these unknown trends. This chapter also showed how the results could be checked and the basics of visualizing and presenting results from GAMs including how bootstrapping can be used to derive uncertainty estimates and how interactions can be plotted and visualized. A brief overview of some of the key functions used is presented in Table 5-1.

Table 5-1. *Listing of Key Functions Described in This Chapter and Summary of What They Do*

Function	What It Does
vgam()	Vector generalized additive models from the VGAM package to fit semi-parametric models that can include both standard parametric terms like generalized linear models but also smooth splines for one or more terms.
s()	Function to indicate which predictor variable should have a smoothing spline applied. Note that this function occurs by the same name in both the VGAM package and in the mgcv package, but in the VGAM package, the argument to control flexibility is df, whereas in the mgcv package, the same argument is k.
gam()	Generalized additive models from the mgcv package; see the preceding vgam() as they are similar.
plot()	When applied to a generalized additive model, typically produces a plot of each predictor variable with the parametric or smoothing spline results.
linearHypothesis()	Function to test linear hypotheses from the car package. Allows us to test hypotheses about the parametric terms from a vgam() model.
predict()	Generic function with methods in place for generalized additive models in R to generate predicted scores from the original or new data based on a model. Plots of predicted values often are used when presenting generalized additive models due to smooth splines being difficult to summarize verbally.
vis.gam()	Perspective (3D) and contour plots for models fit by the gam() function in the mgcv package. Especially useful for visualizing interactions.
gam.check()	Function to check whether the maximum flexibility allowed is (probably) sufficient or whether the flexibility parameter, k, should be increased.

CHAPTER 6

ML: Introduction

Machine learning (ML) is a rather amorphous, in the authors' opinions at least, toolkit of computer aided statistics. While our eventual target will be support vector machines, classification and regression trees, and artificial neural networks using some recent R packages, at their heart machine learning is simply pattern recognition of various flavors.

The twin foci of this introductory chapter, understanding sample structures and introducing parallel processing, seek to lay the groundwork for machine learning. Every workflow and project is different, so our intentions are to dive into what will happen to the data in Chapter 6, take a step back in the following chapter once our feet are wet and properly clean and set up and reduce our data, and finally do some genuine machine learning in the third act.

If you, gentle reader, are familiar with train/validation/test datasets, bootstrap, and parallel/multi-core processing, please feel free to skip to the next chapter. If, on the other hand, you are a more classically trained statistician, from the comfortable world of lookup tables one of the authors fondly remembers, then be prepared to think about data in a rather different way than before.

In any case, in this chapter, we pretend data come to us clean, small, and tidy—a word chosen because we use `tibbles` and other elements of the newish uberpackage named `tidyverse` that combines several packages [111] useful for data munging. It will be more useful perhaps later in our machine learning quest. For now, take a look at the set up and library calls, and we introduce the other packages used in this chapter.

```
library(checkpoint)
checkpoint("2018-09-28", R.version = "3.5.1",
  project = book_directory,
  checkpointLocation = checkpoint_directory,
  scanForPackages = FALSE,
  scan.rnw.with.knitr = TRUE, use.knitr = TRUE)
```

© Matt Wiley and Joshua F. Wiley 2019
M. Wiley and J. F. Wiley, *Advanced R Statistical Programming and Data Models*,
https://doi.org/10.1007/978-1-4842-2872-2_6

```
library(knitr)
library(tidyverse)
library(rsample)
library(data.table)
library(boot)
library(parallel)
library(foreach)
library(doParallel)

options(width = 70, digits = 3)
```

As always we use several packages and briefly discuss the new additions here. The rsample package [52] allows for simple resampling, and the boot package handles generic bootstrap for us [26]. The parallel package provides functions to run computations on multiple cores in parallel, thereby reducing time to results [76]. The foreach package is a variant of the well-known for loop that lends itself to parallelization [63]. Finally (for now), the doParallel package allows the foreach to actually use parallelization [24].

6.1 Training and Validation Data

We start with the iris sample and a word about data. The iris dataset has 150 perfect observations. In machine learning, one risk is the machine learns too well our sample data and is then less accurate during real-world testing. This is called overtraining or overfitting. In other words, the metrics we might use to estimate likely error between the model and real life may be too optimistic and too small. One solution is to split the data into two pieces. One, a training piece, is shown to the machine. The other, a test piece, is kept in reserve to later estimate how well the model might work on new or "wild" data. Nothing comes free of trade-offs or costs, and with this technique the challenge is all else equal, the machine ought to do better the more data provided. Common ratios include 80/20, 75/25, or 70/30, with the bulk of the data in the training sample, while the smaller set is reserved for testing. Indeed, to discourage even human bias, performing the split prior to doing exploratory data analysis is perhaps recommended.

It is worth taking a moment here to discuss this theoretical framework a bit. Technically, if we follow this through to the logical conclusion, it makes sense to have three pieces of data. One set, the training set, would be the bulk of the data and would be used for exploratory data analysis as well as model training. The second piece, a smaller set, would be used to validate the model and see how well it was doing. As we might be selecting from several models, this validation set, not used to train the model, could be used to decide between a linear fit and a quadratic fit or between artificial neural networks and random forests. Since this validation data would essentially still be used in model selection, it is still technically part of the fitting process and thus not truly able to give a good sense about real-world performance with new data. The final, third piece would have been kept aside until the very end and only used to estimate the error of the model when faced with real-world, new data.

This of course requires enough data to sacrifice. For our examples in machine learning, we are not ignoring this concern. Rather, we are showing the mechanics behind any single calculation or technique under the local hypothesis that the technique being studied has already been determined as the best way forward through the literature or some other process. Failure to reserve a final, third test set can be one reason machine learning techniques fail to perform well when faced with new data (not the only reason, naturally).

In the case of our sample, it perhaps makes sense to use an 80/20 split. We use the function set.seed(5) from base R to fix the random number generator that the rsample resampling function uses. This allows exact duplication of values by the interested reader. The function used is initial_split() and takes arguments of the data to be split as well as a proportion of the data to set into the training set. Given our particulars, out of iris's 150 rows, we reserve 29 to be our test set and keep 80% for model building and training.

```
set.seed(5)
case_data <- initial_split(data = iris, prop = 0.8)
case_data
```

```
## <121/29/150>
```

This resample is indeed a random pull of the data—which is important because if the data were left sorted in some fashion, it could unduly influence the model, especially because we split the data. We have not chosen to use any strata in this resample split. If that were important to data, then those strata would each be subject to a split based on the given proportion. From the case_data object, we pull out both our training and testing data through relevant function calls from the rsample package. Additionally, we use glimpse() from the tidyverse to take a look at our training dataset.

```
data_train <- training(case_data)
data_test <- testing(case_data)
glimpse(data_train)

## Observations: 121
## Variables: 5
## $ Sepal.Length <dbl> 5.1, 4.9, 4.7, 4.6, 5.0, 5.4, 4.6, 5.0, 4.4,...
## $ Sepal.Width  <dbl> 3.5, 3.0, 3.2, 3.1, 3.6, 3.9, 3.4, 3.4, 2.9,...
## $ Petal.Length <dbl> 1.4, 1.4, 1.3, 1.5, 1.4, 1.7, 1.4, 1.5, 1.4,...
## $ Petal.Width  <dbl> 0.2, 0.2, 0.2, 0.2, 0.2, 0.4, 0.3, 0.2, 0.2,...
## $ Species      <fct> setosa, setosa, setosa, setosa, setosa, seto...
```

Noting we have four columns of type double, we do see there is a factor column for species. These, upon further inspection, turn out to be three species of iris flower.

```
unique(data_train$Species)

## [1] setosa     versicolor virginica
## Levels: setosa versicolor virginica
```

In order to focus on the ways samples may be used with ML and the computer environment to support reasonably fast ML, we fit a simple linear model to our numeric data to showcase three different ways to think about sample data structure. The objective is to provide a go-to example to act as a sort of lens through which to understand these three methods.

Linear models have been fit before in this text, so we only note this model is a linear fit where three of the four numeric variables available to us in our dataset are used to predict the fourth variable Petal.Length.

```
length.lm = lm(Petal.Length ~ Sepal.Length +
               Sepal.Width + Petal.Width,
            data = data_train)
length.lm

##
## Call:
## lm(formula = Petal.Length ~ Sepal.Length + Sepal.Width + Petal.Width,
##     data = data_train)
##
```

```
## Coefficients:
##   (Intercept)  Sepal.Length   Sepal.Width   Petal.Width
##        -0.274         0.723        -0.630         1.466
```

summary(**length**.lm)

```
##
## Call:
## lm(formula = Petal.Length ~ Sepal.Length + Sepal.Width + Petal.Width,
##     data = data_train)
##
## Residuals:
##     Min     1Q  Median      3Q     Max
## -1.0349 -0.1699 -0.0061  0.1976  0.5751
##
## Coefficients:
##               Estimate Std. Error t value Pr(>|t|)
## (Intercept)    -0.2735     0.3091   -0.88     0.38
## Sepal.Length    0.7230     0.0609   11.87  < 2e-16 ***
## Sepal.Width    -0.6298     0.0715   -8.81  1.3e-14 ***
## Petal.Width     1.4661     0.0700   20.93  < 2e-16 ***
## ---
## Signif. codes:  0 '***' 0.001 '**' 0.01 '*' 0.05 '.' 0.1 '_' 1
##
## Residual standard error: 0.299 on 117 degrees of freedom
## Multiple R-squared:  0.971,  Adjusted R-squared:  0.97
## F-statistic: 1.32e+03 on 3 and 117 DF,  p-value: <2e-16
```

The residual standard error (RSE) is 0.299, and RSE can be thought of as a kind of average between the y-height of the regression line and the actual y-height of the points in the training set. Values closer to zero are preferable to values further away from zero (in terms of telling us something of efficacy of our model). Formally, the equation [47], which is a really great text for a deeper theory behind all this, is given as follows:

$$RSE = \sqrt{\frac{\sum_{i=1}^{n}(y_i - \hat{y}_i)^2}{n-4}}$$

(6.1)

Translated into R, we show the code here:

```r
sqrt(
  sum(
    (fitted(length.lm)-data_train$Petal.Length)^2
  )/(nrow(data_train)-4)
)
```

```
## [1] 0.299
```

In both cases, note that the division portion has the size of the dataset less the number of variables (in our case 4). While this is kind of R to provide us this value for free, mean squared error (MSE) is perhaps a more common way to measure goodness or quality of fit. These are essentially the same equation, as, at their heart, they both involve the y-heights generated by the model less the y-heights found in the data given the paired input values. Since R has already stored these residuals for us for the model, the code looks a bit different, but a quick experiment will show that the third line of the preceding code and the argument inside the following mean() are the same.

```r
mse_train<- mean(length.lm$residuals^2)
mse_train
```

```
## [1] 0.0862
```

Both these measures are in the units of the original data, and both measures share a key flaw. They are measured on training data, and the model specifically trained to perform well on these same data. This is an example of in-sample accuracy measurement, and it is unlikely to be a great estimator of how our model might perform on wild or real-world data. Rather than calculate MSE (or indeed any type of goodness of fit metric) from training data, the purpose of having both training and testing data is precisely to allow us to estimate model performance on data that was not used in training. We turn naturally to our data_test.

If we measure the same sort of y value differences on our test data, what we see is the numbers are a bigger in both cases. This gives us some evidence that we have a model guilty of overfitting. It is also a great motivator about how important it may be to consider models not just against in-sample data but against out-of-sample data too.

```
sqrt(
  sum(
    (predict(length.lm, data_test)-data_test$Petal.Length)^2
  )/(nrow(data_test)-2)
)
```

`## [1] 0.41`

```
mse_test <- mean((predict(length.lm, data_test) -
                  data_test$Petal.Length)^2)
mse_test
```

`## [1] 0.156`

This train vs. test process is key. The goal of any good model is to have small residuals. Indeed, residuals of the training data are often minimized as part of the model coefficient selection, so rather than look at MSE from data inside our training sample, we look at test data MSE, data from outside our sample. This can help us understand how our model is likely to perform. Keep in mind, as mentioned at the beginning, we are not presently in the business of determining if this model is good or not. We are truly using our test data to understand how our model might hold up to real-world use upon deployment. If we were going to modify our model based on the testing data, again, we would need to actually have reserved three sets of data.

Remember, MSE is in squared original data units. So, while the MSE is higher on test data vs. training data—and indeed is almost double—it may still be good enough to proceed. Sometimes actionable data of any sort is better than nothing, and the fact that the test MSE is higher may simply warn our user to be cautious when using model data to inform decisions. It also depends on the type of data being used. College enrollment trends, for example, might be more tolerant of error than health results. In the case of these flower data, these data are in centimeters squared, and our test data suggests we may be off by about half a centimeter, while the training data claimed to be off by only perhaps a quarter centimeter.

We have a model, and we have an idea on how it might perform in the real world. We did, however, lose the value of 29 data points. It would be nice if we could have our out-of-sample estimate of MSE and still have all 150 data points. Part of the loss is not just that our model might do better with 20% more data. It is that the specific break of data, while random, might still be a bit extreme, by which we mean our choice of where to split the data, on a small dataset, even though random, may well have influenced our end result.

Consider our test vs. train data on just two dimensions. In the top row of graphs, one sees that our input of just Sepal.Length vs. our output of Petal.Length has some different ranges in training vs. testing. The second row of Q-Q plots confirms these samples do look slightly different in Figure 6-1.

```
par(mfrow=c(2,2))
plot(data_train$Sepal.Length, data_train$Petal.Length)
plot(data_test$Sepal.Length, data_test$Petal.Length)
qqnorm(data_train$Petal.Length,
      xlab = "Theoretical Quantiles Train")
qqnorm(data_test$Petal.Length,
      xlab = "Theoretical Quantiles Test")
```

Even a quick six-figure summary shows the test data are just a bit different than our sample data:

```
summary(data_train$Sepal.Length)
```

```
##    Min. 1st Qu.  Median    Mean 3rd Qu.    Max.
##    4.3     5.1     5.7     5.8     6.3     7.9
```

```
summary(data_test$Sepal.Length)
```

```
##    Min. 1st Qu.  Median    Mean 3rd Qu.    Max.
##    4.40    5.20    6.10    6.02    6.50    7.70
```

So while we perhaps have a better idea than just looking at training data MSE what our real-world MSE might be, we are at the mercy of the set.seed(5) data pull. Two risks we run doing this are potentially high variance based on sample split and, all else equal, given a choice between 121 and 150 observations, we'd expect a model trained on 150 observations to be somewhat better.

Our goal in this section was to motivate the need to have separate train vs. test data. In this overly simple case, our test data perhaps gave us some insight into how our trained model might perform in the real world. In general though, we sacrifice a lot by having missing data. How might we do better, particularly for smaller samples?

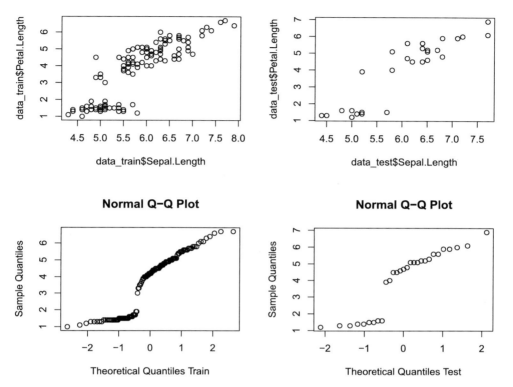

Figure 6-1. *2D and Q-Q plots of some data*

6.2 Resampling and Cross-Validation

A solution for this can be what is called cross-validation. Instead of doing only one train/
test split, we do this multiple times.

There are a few ways of going about this process. Rather than splitting our data into
two pieces, let us instead imagine 5 pieces where each holds 30 observations of our iris
dataset. We merge the first four bins together to form our training set while keeping
bin 5 reserved as test. Fit the model to the training set, and calculate MSE on the test
set. Repeat the process five times, where the next we reserve only bin 4 as our test set,
merging 1-3 and 5 as our training set and so on. We get five different MSE values from
each of the bins in turn. If we average those, we now have the what is called the cross-
validation MSE (sometimes called CV).

While there are perhaps packages that do this sort of calculation for us, it can help to
get our hands dirty once in a while to see what is going on. First we randomize our iris
sample into a variable we call `crossData` without replacement (this makes sure our bins
are not ordered in any way). We do this via the `sample_n()` function from the `tidyverse`

package. By using the %>% pipe operator, we do not have to write the dataset as the first argument of our function. Once our data are shuffled, we add a column to break it all into five bins. Five was a number chosen simply because it is conveniently small, divides nicely into 150, and we like it. Additionally, a `store` variable is invoked to hold the MSEs.

```
crossData <- iris %>%
  sample_n(nrow(iris), replace = FALSE)
crossData <- add_column(crossData,
  Bin = cut(1:150, breaks = 5, labels = c(1:5)))
store <- tibble(Fold=1:5, MSE=NA_integer_)
```

Now we use a for loop to cycle through our five bins. We do this five times because we have five bins, and each time we leave out a single bin to reserve as test data as described. In the parlance of cross-validation, the number of folds is our K number, and this is called K-Fold (or in our case 5-fold). Each pass of this for loop represents one "fold" of our data. We systematically proceed through our data to train our linear model, compute the MSE on the reserved test-bin, store that MSE in our `store` variable, and move to the next fold.

```
for(i in 1:5){
  data_train<-crossData %>% filter(Bin != i)
  data_test<-crossData %>% filter(Bin == i)
  lengthFold.lm = lm(Petal.Length ~ Sepal.Length +
                     Sepal.Width + Petal.Width,
                   data = data_train)
  store[i,]$MSE <- mean((predict(lengthFold.lm, data_test) -
                        data_test$Petal.Length)^2)
}
```

Now we get our mean standard error from all five folds. Notice this MSE is from our test portion of the folds, and thus this perhaps gives us a decent idea of how we might expect a model trained on all the data to perform in real life on new incoming data.

```
mse_k <- mean(store$MSE)
mse_k
```

```
## [1] 0.109
```

In this case, the MSE is approximately 0.109, which is less than it was estimated back when we only had a single set of train/test data in the prior section.

When doing K-Folds and cross-validation, the model one uses is not the individual fold models but rather a model trained on all data, and we see the MSEs are fairly close for the out-of-sample estimate vs. the in-sample estimate. So while our cross-validation MSE is a bit larger than our full dataset MSE, they aren't impossibly far away from each other, at least not too far for something like petal lengths perhaps.

```
lengthFold.lm <- lm(Petal.Length ~ Sepal.Length +
                    Sepal.Width + Petal.Width,
                    data = iris)
lengthFold.lm
```

```
##
## Call:
## lm(formula = Petal.Length ~ Sepal.Length + Sepal.Width + Petal.Width,
##     data = iris)
##
## Coefficients:
##  (Intercept)  Sepal.Length   Sepal.Width   Petal.Width
##       -0.263         0.729        -0.646         1.447
```

```
mse_ALL <- mean(lengthFold.lm$residuals^2)
mse_ALL
```

```
## [1] 0.099
```

Notice in the stored MSE values from each fold (where the model was fitted on other bins so this is truly out-of-sample testing for each iteration of the for loop), we have quite a wide spread from 0.075 to 0.18—which just shows the risk of a single test/train dataset! This goes back to our earlier warning, that smaller datasets run bit of a risk when portioning off too much to the test datasets.

```
store
```

```
## # A tibble: 5 x 2
##    Fold    MSE
##   <int>  <dbl>
## 1     1  0.180
```

```
## 2      2 0.109
## 3      3 0.0822
## 4      4 0.0746
## 5      5 0.101
```

Now, there is no reason we could not have done just 4 folds or even 100 (with some code edits) of course, hence the general name K-Fold, and K can be any number one chooses from 1 (which would be no data split), 2 (of which our first example is a special, weighted case), up to n (which is also a special case called Leave One Out Cross-Validation or LOOCV) . One last thing to notice, well, notice and then forget for a section or so, is something about our for loop. While we ran it in serial, namely, 1, 2, 3, 4, 5, in order, there was no reason we had to run it that way. In other words, iteration 2 does not depend on iteration 1. If we were fitting a complex model or fitting on more data, our for loop could well take some time to run each pass. All the same, each run is independent as long as we get our table of fold-MSEs. Again, notice this, and forget it right away for now.

K-Fold seems a decent technique, and it is. We were able to train a model on a full dataset, yet we still gained some sense of how our model might perform on data from outside the sample. However, while this may be all well and good for the linear model we are using, depending on the actual machine learning model(s) we truly deploy, there may not be an obvious connection between the K-Fold model and the model used on a full dataset. Or at least, not so much as that one might not want to cope with the fact that the model we are deploying on the full dataset might well be constructed on rather different coefficient weights than any single K-Fold model. How might we sort through this?

Notice also that while the K-Fold model works, it does require more computing power than our simple train/test model did. We went from fitting a model just once to fitting the model six times. In terms of work, that is a fairly big increase potentially. So, as we consider ways to allow our model to train on a whole dataset and get our error estimates from a completely trained model, recognize that each improvement depends on increased machine technology.

6.3 Bootstrapping

We now turn our attention to another technique, the bootstrap. In the K-Fold or cross-validation technique we just used, while at the end of the day we did in fact get our "true" MSE estimate from out-of-sample data, each model was built using less than a full sample of the dataset, which means each model was perhaps a bit weak. Thus, our error estimate, while better than pass one, may still be somewhat stronger than it needs to be.

Bootstrap tackles the problem a different way. Assuming our sample, the original `iris` sample was a well-drawn sample from the population of iris flowers, we could suppose that if we resampled from `iris`, with replacement, and did that many, many times, we ought to see an overall group that looks pretty close to the population.

First, we write a function that is the statistic of interest in our bootstrapping. We already know what MSE is from before. Notice we fit the model each time. Our function will get a dataset, compute the linear model, and then calculate the in-sample MSE and return that value. Presumably we will collect it elsewhere, and in fact we will!

```
mse <- function(data, i) {
  lengthBoot.lm <- lm(Petal.Length ~ Sepal.Length +
                      Sepal.Width + Petal.Width,
                      data=data[i,])
  return(mean(lengthBoot.lm$residuals^2))
}
```

Now, using the function boot() from library(boot), we run the bootstrap. What is happening is this function is resampling, with replacement, from our iris data sample. In fact, it does this 10,000 times. Each time, that resample is sent to our mse() function, and the returned value for that model's MSE is stored in bootResults.

```
bootResults <- boot(data=iris, statistic=mse, R=10000)
bootResults

##
## ORDINARY NONPARAMETRIC BOOTSTRAP
##
##
## Call:
## boot(data = iris, statistic = mse, R = 10000)
##
```

```
##
## Bootstrap Statistics :
##      original    bias    std. error
## t1*    0.099 -0.00263       0.0129
```

We may view the results both numerically and graphically. This shows the overall distribution of all these in-sample MSEs in Figure 6-2.

```
plot(bootResults)
```

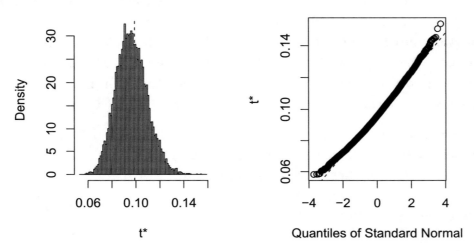

Figure 6-2. *Bootstrap results.*

From this distribution, we see the 95% confidence interval (note 0.95 is also the default) of the MSEs is somewhere between 0.08 and 0.13.

```
boot.ci(bootResults, conf = 0.95, type="bca")
```

```
## BOOTSTRAP CONFIDENCE INTERVAL CALCULATIONS
## Based on 10000 bootstrap replicates
##
## CALL :
## boot.ci(boot.out = bootResults, conf = 0.95, type = "bca")
##
## Intervals :
## Level        BCa
## 95%    ( 0.0794,  0.1343 )
## Calculations and Intervals on Original Scale
```

Just as for the K-Fold method, for the actual model this is attached to, we would use our linear fit of the entire dataset `lengthFull.lm` for variable coefficients. This simply gives us a sense of where we might expect real-world MSE to live.

We might expect this confidence interval to be a trifle too low, because it was run on in-sample mean standard errors. One could imagine an improvement to the bootstrap where one used any leftover data points after the resample was done to calculate an out-of-sample bootstrap MSE. Of course, that sort of estimate would likely be slightly too high because those are the data points removed from the training. While each booted linear model was trained on 150 values, it is not likely that all 150 of the original values were pulled—in fact, that is why one gets that normal distribution shown in the histogram. The distribution is precisely because not all values were pulled and so MSE varied. Using leftover data values of course gets us to the same place as the K-Fold, in that we would expect the MSE of such a bootstrap technique to be a bit larger perhaps than real-world data might experience compared to a train on the full model. Although, by doing the bootstrap so many times, we do perhaps get a sense of how our model works. Clever researchers develop new techniques fairly often, and currently the `.632 bootstrap` is worth exploring. However, we stay for now in our basic bootstrap. This introductory chapter is all about having three simple structures to think about when performing machine learning.

Another point of interest for both K-Fold and bootstrap is it took a bit of time to run. Likely the K-Folds for this dataset are so small it would be tough to spot. However, the 10,000 bootstraps probably did become noticeable. Now, not too long, maybe only 6 seconds, but we did just pull 10,000 samples of size 150–1,500,000 total points!

What if our original data had 1,000 points? Or 10,000? That could well slow down our ability to use bootstrap. At least, until we got a bit more clever with how we might use computers.

6.4 Parallel Processing and Random Numbers

R, at least as far as compute environment goes, has two main features. It lives in memory, as opposed to hard disk. Thus, any discussion of having millions or billions of data points vs. hundreds or thousands needs to give consideration to the size of random access memory. Still, thanks to modern graphics greed, most computers have a fair bit of that at their disposal. The other feature is that, by default, R works on a single processor. Most modern computers have at least two. As we already loaded the `parallel` library, go ahead and run the code snippet on your system and compare results:

```
detectCores()
```

```
## [1] 12
```

```
detectCores(logical = TRUE)
```

```
## [1] 12
```

```
detectCores(logical = FALSE)
```

```
## [1] 6
```

The computer we wrote this text on has four physical cores, each of which has two logical "cores." Now, this is not a book about computer hardware, so let us ignore the nuances of physical vs. logical cores. Indeed, depending on your operating and hardware system, you may well not get a difference between any of those three. In any case, provided the number returned for you is more than one, then bootstrapping and K-Fold can run faster if more processors are thrown into the workpool.

Recall also that we said, back in K-Fold, that our for loop was precisely the same sort of operation each time and that iteration c was not dependent on iteration a to be calculated. Same thing for the bootstrap. We are pulling 10,000 random pulls and feeding those into our function mse(), and if we were doing a more complex model than a simple linear fit, or if our data got much bigger, we could find ourselves waiting for a fair bit of time. The methods of this section work very well provided the same process is being done independently. If the process is not independent, that is beyond the scope of this text (although there are ways to adapt this process, of course).

Now, there are nuances between multi-core vs. cluster vs. parallel. If your data live in the world of truly big data, those nuances may be meaningful to you. But then, we hazard memory may also become an issue, and those are also beyond the scope of this text. However, if our 6-second estimate for the 10,000 resamples from our bootstrap is true, and we get two cores instead of one going, we ought to be able to cut execution time in half.

This is not quite true in practice, because there is some overhead cost to starting and stopping a multi-core environment. However, recall our analysis was just a simple linear model and then a mean. Those overhead costs are mostly fixed, so if the model fit were more exotic, and the bootstrap took a day or half a day, then cutting that process in roughly half starts to become meaningful. And, as our system has several cores, we can in fact do better than half! One word of caution: whatever the number of cores in your system, leave at least one free for the system to use—otherwise everything grinds to a halt until R is done. Most likely, you will hear the fan come on—you are going to do some compute-work after all!

In our following code, we will run on just 2 cores, so we divide our 10,000 boot runs in half to 5,000. For simplicity's sake, we build a function `runP()` to handle the parallel process run. Next, we use the `makeCluster()` function to create two parallel copies of R on which to run our function. Our copies of R need to have the same environment our machine generally has, so we use the `clusterEvalQ()` function to set our `cl` cluster to have the library structure we need. While we use `checkpoint()` to make sure the version is correct, each copy of R does not need any more than the `boot` library. Being stingy here saves overhead time, so be judicious in what you pass along. From there, we also export using `clusterExport()` some already built functions from our global environment. The last bit of cluster overhead we do is to set our random number stream so that our results match yours via `clusterSetRNGStream()`.

From here, most of the rest ought to be familiar from bootstrapping, so we take a moment to dissect the actual parallel call—remember the previous steps were overhead. Recall the `apply()` function from base R applies functions over an array of some sort. The function `parLapply()` is the same, except the first argument links to the cluster we just set up, the second argument just says we will run our function twice, and the third is of course our bootstrapping 5,000 runs function we built earlier. The `do.call()` function is a base function also, and it just is calling the combine function `c()` on our parallel function so that the two results are merged into one bootstrap object. This is done so that the familiar `boot.ci()` works just as before. Of course, we do take a moment to shut our cluster down via `stopCluster()`.

```
## notice 10000/2 = 5000
runP <- function(...) boot(data=iris, statistic=mse, R=5000)

## makes a cluster with 2 cores as 10000/5000 = 2
cl<-makeCluster(2)

## passes along parts of the global environment
## to each node / part of the cluster
## again, base is a file path variable to our book's path
## set book_directory <- "C:/YourPathHere/"
clusterExport(cl, c("runP", "mse", "book_directory", "checkpoint_directory" ))
```

```
## creates the library and some environment on
## each of the parts of the cluster
clusterEvalQ(cl, {

library(checkpoint)
  checkpoint("2018-09-28", R.version = "3.5.1",
  project = book_directory,
  checkpointLocation = checkpoint_directory,
  scanForPackages = FALSE,
  scan.rnw.with.knitr = TRUE, use.knitr = TRUE)

    library(boot)
  })

## [[1]]
##  [1] "boot"          "checkpoint"    "RevoUtils"     "stats"
##  [5] "graphics"      "grDevices"     "utils"         "datasets"
##  [9] "RevoUtilsMath" "methods"       "base"
##
## [[2]]
##  [1] "boot"          "checkpoint"    "RevoUtils"     "stats"
##  [5] "graphics"      "grDevices"     "utils"         "datasets"
##  [9] "RevoUtilsMath" "methods"       "base"

## similar to set.seed() except for clusters
clusterSetRNGStream(cl, 5)

## uses the parLapply() function which works on windows too
pBootResults <- do.call(c, parLapply(cl, seq_len(2), runP))

#stop the cluster
stopCluster(cl)

# view results
pBootResults

##
## ORDINARY NONPARAMETRIC BOOTSTRAP
##
```

```
##
## Call:
## boot(data = iris, statistic = mse, R = 5000)
##
##
## Bootstrap Statistics :
##     original    bias    std. error
## t1*    0.099 -0.00295       0.0128
```

```
## get 95% confidence interval of the MSEs
## (note 0.95 is the default)
boot.ci(pBootResults, conf = 0.95, type="bca")
```

```
## BOOTSTRAP CONFIDENCE INTERVAL CALCULATIONS
## Based on 10000 bootstrap replicates
##
## CALL :
## boot.ci(boot.out = pBootResults, conf = 0.95, type = "bca")
##
## Intervals :
## Level         BCa
## 95%   ( 0.0795,  0.1352 )
## Calculations and Intervals on Original Scale
```

And that is it. By our calculation, it was not quite half as much time as using one processor, but that is okay. Notice much of this can be modified depending on the environment one has. Since we have more cores, we could increase the number of R instances activated and decrease the number of bootstrapping runs in our runP() function accordingly. Alternately, we could have had more than one cycle of runP() run on our two cores, by editing seq_len() to four instead of two. In that case we get our bootstrap function to run four times, twice on each R instance, and we get 20,000 bootstrap runs. So you can see, it is very customizable and flexible based on available hardware and data nuances. Results in Figure 6-2 are similar to those in Figure 6-3:

```
plot(pBootResults)
```

It is worth taking a moment though to admit the two figures are slightly different—although with luck your machine matches ours in each case. The reason is due to the random numbers and the way they are generated. To make research reproducible, it is helpful to set.seed() or clusterSetRNGStream(). This not only helps our output match yours on various machines, it helps our output today match our output tomorrow. Random numbers are generated from a starting point or seed, and that is what these two techniques do. However, in the case of a cluster, while we want to have the same seed at a global level, for each processor we do want a slightly different starting point so that we truly get 10,000 unique resamples rather than a double copy of 5,000. That is achieved by using the cluster variant of seed generator, and it handles behind the scenes the process rather cleverly, so that just the right sort of reproducible randomness occurs. All the same, it is a different randomness than that used earlier, so the numbers will be a trifle different.

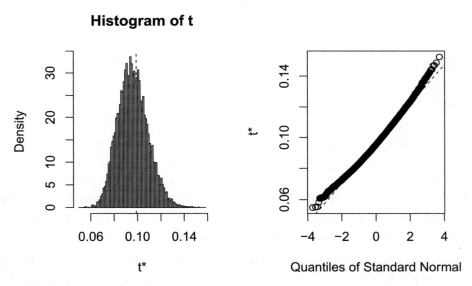

Figure 6-3. *Bootstrap results*

Now in this section, we saw how to build a cluster for any process and merge results from the cluster. We used our boot() function (which these days has some parallelization built into it admittedly). While the boot process could be handcoded via a for loop, we chose to use a package. This is often possible in R, with many packages and libraries out on CRAN. This constantly updating and improving ecosystem is one of the reasons for the checkpoint package. Code that works with one version of a library may well not work next month or year. For anyone who does work with constantly changing data, a feature of machine learning where real-world data comes in and needs analyzing, it is quite useful to both lock code into specific points in time so that it continues to work in the future and to occasionally revisit code and improve the writing to make use of newer or updated libraries.

foreach

Our last section is about the `foreach()` function and eponymous library. Now, while we introduce this in the context of the K-Fold `for()` function introduced earlier, this will of course work for any loop provided each iteration is independent. However, generally, there are often ways of vectorizing loop functions on data in R, and these will be faster than this process. Additionally, for this specific case, due to the overhead of creating the clusters, this does not actually save time as we have only 150 bits of data being processed on a simple linear model just five times. Our K-Fold was computationally much easier than our bootstrap!

Much of the code that follows is familiar. We start with the code that sets up our cluster. The only new function here is `registerDoParallel()` which connects our `foreach()` function to the right framework. As before, we chose to do this on two cores. Since we are not using `boot`, we do not call that library. We do, however, call the `tidyverse` library due to our use of pipe on our tibbles.

```
cl <- makeCluster(2)
registerDoParallel(cl)

clusterExport(cl, c("book_directory", "checkpoint_directory"))

clusterEvalQ(cl, {
  library(checkpoint)
  checkpoint("2018-09-28", R.version = "3.5.1",
  project = book_directory,
  checkpointLocation = checkpoint_directory,
  scanForPackages = FALSE,
  scan.rnw.with.knitr = TRUE, use.knitr = TRUE)

  library("tidyverse")
  })
```

```
## [[1]]
##  [1] "forcats"       "stringr"      "dplyr"       "purrr"
##  [5] "readr"         "tidyr"        "tibble"      "ggplot2"
##  [9] "tidyverse"     "checkpoint"   "RevoUtils"   "stats"
## [13] "graphics"      "grDevices"    "utils"       "datasets"
## [17] "RevoUtilsMath" "methods"      "base"
##
## [[2]]
##  [1] "forcats"       "stringr"      "dplyr"       "purrr"
##  [5] "readr"         "tidyr"        "tibble"      "ggplot2"
##  [9] "tidyverse"     "checkpoint"   "RevoUtils"   "stats"
## [13] "graphics"      "grDevices"    "utils"       "datasets"
## [17] "RevoUtilsMath" "methods"      "base"
```

The main code of interest is here. Again, this is almost precisely what happened back when we first did our K-Fold, and that is the point. The only change is the foreach() function and arguments. We again run our function through folds one to five. We do use a new argument, .combine and set that to our combine function, familiar from base R and of course our earlier do.call() function. Then we use the %dopar% to invoke the parallel processing via the registered back ends we set up in the preceding setup code. The packages handle the rest, sending these five runs out to two cores and combining the results.

```
k <- foreach(i=1:5, .combine = c) %dopar% {
  data_train <- crossData %>% filter(Bin != i)
  data_test <- crossData %>% filter(Bin == i)
  lengthFold.lm <- lm(Petal.Length ~ Sepal.Length +
                   Sepal.Width + Petal.Width,
                   data = data_train)
  mean((predict(lengthFold.lm, data_test) -
       data_test$Petal.Length)^2)
}
```

Notice that, with five folds to compute and only two processors, there is a bit of a mismatch. Depending on our loop, it is conceivable that one or more of these iterations might be more or less compute intensive. In this case, not likely, of course. All the same, it is possible to set up parallel processing in such a way that an interaction that would take a long time gets sent to one core first, while the rest go to another core. The default value is that they are run first in, first out, in order.

```
stopCluster(cl)
mse_Pk<-mean(k)
mse_Pk
```

```
## [1] 0.109
```

We see that our MSE from before in our for loop 0.109 matches our MSE from now 0.109. The two procedures are identical when run on the same data. Indeed, this can be wise, when getting code to run more quickly via foreach() for the first (or even last) time to make sure one is not making a silly error. The authors can only add their cold comfort, if our readers ever find themselves in such a scenario, that we were and are there more often than desired.

6.5 Summary

That completes the setup for machine learning. With three good methods to train data and then get some estimates on real-world performance, we now have the mental maps to understand how our data are going to be formatted and used. With a couple of good multi-core computing packages and code in hand, we have the technological tools to make our machine learning not take too long.

Indeed, the parallel compute techniques—particularly foreach()—are very easy to implement and can save quite a bit of time. Now, there are plenty of ways around a for loop. And vectorized computations are one of those. As a personal example, one author works with fairly large data daily (perhaps 200,000 rows and 70 columns or so). A basic sort of data munge—after pulling from the database—would take at least half a day. The code had been written a couple years back, and the scenario was not originally time-sensitive. The general use case was to run over the weekend. Operations change, however, and the code started needing to be run more frequently. While writing this text, your author decided to clean up the code. With vectorization of some procedures, and parallelization of others, on a small, four-core office computer, the time went from half a day to a bit less than 2 hours!

This is perhaps a great way to look at this chapter. It does set us up to do some pretty awesome things with machine learning. However, these techniques can be mixed and matched to make life and data easier and more friendly. Just like some old code that worked, yet was now just a bit too slow for convenience, it is possible to gain some pretty significant breakthroughs with some very minor changes!

As always, we close by calling out the main functions in Table 6-1.

Table 6-1. *Listing of Key Functions Described in This Chapter and Summary of What They Do*

Function	What It Does
boot()	From boot library and runs bootstrap
boot.ci()	Calculates confidence interval on bootstrap output
clusterEvalQ()	From parallel library sets up the environment to be copied to each cluster instance
clusterExport()	Exports values from global environment for use by each cluster environment
clusterSetRNGStream()	Creates a random number seed and properly percolates to each cluster
detectCores()	Identifies how many cores are available to use in R
do.call()	Base function that calls a function and passes arguments to it via list
filter()	tidyverse function to return certain rows
fitted()	Extracts the fitted y values from linear model
foreach()	A parallel version (mostly) of for
glimpse()	tidyverse function that shows column names horizontally followed by first few values
initial_split()	rsample function that splits data into train and test cases
lm()	Creates a linear model based on provided relation
makeCluster()	Makes a cluster
parLapply()	The apply function except suited for parallel computing
predict()	Shows predicted y values from linear model

(continued)

Table 6-1 (*continued*)

Function	What It Does
registerDoParallel()	doParallel package operator that registers the foreach() function with a parallel cluster
seq_len()	Creates a sequence of a certain length
set.seed()	Creates a random number seed for reproducible results
stopCluster()	Stops the cluster created by makeCluster
testing()	rsample function that extracts the test data
training()	rsample function that extracts the train data
unique()	Returns only unique values

CHAPTER 7

ML: Unsupervised

This chapter focuses on unsupervised machine learning, which typically deals with unlabelled data. The objective is to somehow sort these data into similar groups based on common feature(s). Often, although not always, unsupervised machine learning also is used as a type of dimension reduction. For example, if you get a dataset with hundreds or thousands of features, but only a few thousand cases, you may wish to first utilize unsupervised learning to distil the large number of features into a smaller number of dimensions that still capture most of the information from the larger set. Unsupervised machine learning also makes a good final step of the exploratory data analysis phase. Part of the sorting or clustering in unsupervised machine learning can be leveraged to understand how many "unique" groups or dimensions your data have. Imagine a dataset that is comprised of various indicators from several distinct geographic regions. One might expect an unsupervised grouping technique to indicate something about the geographic regions. Or, one might discover that physically distant locations have several highly common features.

Whether labelled or unlabelled, data often come to us messy. Thus, substantial effort often goes into the pre-processing phase. Often, data do not fit the format expected by machine learning algorithms. Generally, what is expected is a series of columns, where the first few columns are key columns in some sense (perhaps geographic regions and year some measurements were collected), while later columns indicate observations of specific features or variables. It can take manipulation to get the data into the correct format, where each row is a unique observation at a certain point in time. We will see some of this data manipulation in the example data we examine in this chapter.

As always we use several packages and briefly discuss the new additions here. The `readxl` package [112] allows for fast reading of our example data from an Excel data file, and the ape package handles dendrogram options for us [72]. The `MASS` package provides a function to handle non-linear dimension reduction [99]. The `matrixStats` package provides functionality for matrices and vectors [9]. Lastly, the `viridis` package works to provide a better color palette—particularly for accessibility purposes for the color blind [33].

© Matt Wiley and Joshua F. Wiley 2019
M. Wiley and J. F. Wiley, *Advanced R Statistical Programming and Data Models*,
https://doi.org/10.1007/978-1-4842-2872-2_7

```
library(checkpoint)
 checkpoint("2018-09-28", R.version = "3.5.1",
   project = book_directory,
   checkpointLocation = checkpoint_directory,
   scanForPackages = FALSE,
   scan.rnw.with.knitr = TRUE, use.knitr = TRUE)

library(ggplot2)
library(cowplot)
library(viridis)
library(scales)
library(readxl)
library(data.table)
library(ape)
library(MASS)
library(matrixStats)

options(width = 70, digits = 2)
```

One method discussed in this chapter is principal component analysis (PCA). Although R has some built-in functions for PCA, more extensive options are available in the package, pcaMethods, which is part of Bioconductor, an alternative to the CRAN package repository, with many R packages for bioinformatics. We can install pcaMethods using the code that follows. Note that

```
source("https://bioconductor.org/biocLite.R")
biocLite("pcaMethods")
```

Once pcaMethods is installed, we can load it as we load any other R package.

```
library(pcaMethods)
```

7.1 Data Background and Exploratory Analysis

The sample data used comes from the World Bank and is licensed under CC-BY 4.0 and has been modified to only include certain columns of data from certain regions [2]. The data are in Gender_StatsData_worldbank.org_ccby40.xlsx. In particular, we reduced the data to some key bits of information, in a range of recent years, such that there were no missing variables. The file used is available for download from the publisher's web site or the GitHub repository for this text.

It should be noted that `read_excel()` defaults to `stringsAsFactors = FALSE` which is generally the way one would want to read in data. Should factors be required, it is quite possible to control that through suitable function calls later on.

```
## Note: download Excel file  from publisher website first
dRaw <- read_excel("Gender_StatsData_worldbank.org_ccby40.xlsx")
dRaw <- as.data.table(dRaw) # convert data to data.table format.
```

Missing variables can be handled in different ways. The simplest approach to missing data is to remove all missing information. The price is the reduction of our data to a very small subset of the total information available. Less costly methods to more elegantly handle missing data are discussed in our chapter on missing data. For now, simply note there is no missing data in our example data for this chapter.

Understanding the structure of data is a critical first step to any analysis. For example, at a minimum, it is important to know what types of data are present and how the data are organized. Using the `str()` function on our data tells us the structure has character and numeric data which are organized by year, from 1997 to 2014. Indeed, most data are numeric.

The `summary()` function yields a brief summary by column and shows our data are quite variable. At this point we start to realize our data are not optimal for machine learning format. In other words, each column is not a single, unique measure. Further, time information is not captured in a variable for year, it is captured in the variable names themselves.

Delving deeper in the labels of our data, we see through the `unique()` function that our data come from several large geographic regions and seem to revolve around certain specific indicators.

```
str(dRaw)
```

```
## Classes 'data.table' and 'data.frame':      99 obs. of  21 variables:
##  $ CountryName   : chr  "Sub-Saharan Africa" "Sub-Saharan Africa"
"Sub-Saharan Africa" "Sub-Saharan Africa" ...
##  $ Indicator Name: chr  "Adolescent fertility rate (births per 1,000 women
ages 15-19)" "Age dependency ratio (% of working-age population)" "Children
out of school, primary, female" "Children out of school, primary, male" ...
```

```
##   $ IndicatorCode : chr  "SP.ADO.TFRT" "SP.POP.DPND" "SE.PRM.UNER.FE"
"SE.PRM.UNER.MA" ...
##   $ 1997          : num  1.32e+02 9.17e+01 2.44e+07 2.03e+07 1.55e+01 ...
##   $ 1998          : num  1.31e+02 9.13e+01 2.44e+07 2.05e+07 1.53e+01 ...
##   $ 1999          : num  1.30e+02 9.09e+01 2.43e+07 2.08e+07 1.52e+01 ...
##   $ 2000          : num  1.28e+02 9.04e+01 2.37e+07 2.00e+07 1.49e+01 ...
##   $ 2001          : num  1.27e+02 9.04e+01 2.31e+07 1.95e+07 1.47e+01 ...
##   $ 2002          : num  1.26e+02 9.02e+01 2.28e+07 1.91e+07 1.44e+01 ...
##   $ 2003          : num  124 90 21938840 18230741 14 ...
##   $ 2004          : num  1.23e+02 8.97e+01 2.14e+07 1.79e+07 1.36e+01 ...
##   $ 2005          : num  1.21e+02 8.94e+01 2.06e+07 1.71e+07 1.32e+01 ...
##   $ 2006          : num  1.20e+02 8.94e+01 1.99e+07 1.67e+07 1.28e+01 ...
##   $ 2007          : num  1.18e+02 8.94e+01 1.94e+07 1.52e+07 1.23e+01 ...
##   $ 2008          : num  1.16e+02 8.92e+01 1.90e+07 1.52e+07 1.19e+01 ...
##   $ 2009          : num  1.15e+02 8.89e+01 1.92e+07 1.56e+07 1.15e+01 ...
##   $ 2010          : num  1.13e+02 8.85e+01 1.98e+07 1.61e+07 1.10e+01 ...
##   $ 2011          : num  1.11e+02 8.83e+01 1.92e+07 1.54e+07 1.07e+01 ...
##   $ 2012          : num  1.09e+02 8.80e+01 1.91e+07 1.55e+07 1.03e+01 ...
##   $ 2013          : num  1.07e+02 8.75e+01 1.91e+07 1.53e+07 1.00e+01 ...
##   $ 2014          : num  1.05e+02 8.69e+01 1.92e+07 1.56e+07 9.73 ...
##   - attr(*, ".internal.selfref")=<externalptr>

summary(dRaw)

##   CountryName          Indicator Name         IndicatorCode
##   Length:99            Length:99              Length:99
##   Class :character     Class :character       Class :character
##   Mode  :character     Mode  :character       Mode  :character
##
##
##
##        1997                  1998                   1999
##   Min.   :      6     Min.   :      6     Min.   :      6
##   1st Qu.:     16     1st Qu.:     15     1st Qu.:     15
##   Median :     71     Median :     71     Median :     71
##   Mean   : 949741     Mean   : 872971     Mean   : 817616
##   3rd Qu.:   4366     3rd Qu.:   4338     3rd Qu.:   3993
```

```
##   Max.    :24371987    Max.    :24437801    Max.    :24292225
##          2000                  2001                  2002
##   Min.    :        6    Min.    :        6    Min.    :        5
##   1st Qu.:        15    1st Qu.:        16    1st Qu.:        16
##   Median :        70    Median :        70    Median :        70
##   Mean    :   781078    Mean    :   736806    Mean    :   674889
##   3rd Qu.:      4224    3rd Qu.:      4275    3rd Qu.:      4672
##   Max.    :23672959    Max.    :23125633    Max.    :22795557
##          2003                  2004                  2005
##   Min.    :        5    Min.    :        5    Min.    :        5
##   1st Qu.:        16    1st Qu.:        15    1st Qu.:        16
##   Median :        70    Median :        71    Median :        71
##   Mean    :   651075    Mean    :   637985    Mean    :   659420
##   3rd Qu.:      5568    3rd Qu.:      6772    3rd Qu.:      8042
##   Max.    :21938840    Max.    :21350198    Max.    :20582825
##          2006                  2007                  2008
##   Min.    :        5    Min.    :        5    Min.    :        5
##   1st Qu.:        15    1st Qu.:        16    1st Qu.:        16
##   Median :        71    Median :        71    Median :        72
##   Mean    :   653180    Mean    :   597847    Mean    :   573176
##   3rd Qu.:      9166    3rd Qu.:     11168    3rd Qu.:     13452
##   Max.    :19904220    Max.    :19402096    Max.    :19015196
##          2009                  2010                  2011
##   Min.    :        5    Min.    :        5    Min.    :        5
##   1st Qu.:        16    1st Qu.:        16    1st Qu.:        16
##   Median :        72    Median :        72    Median :        72
##   Mean    :   569320    Mean    :   569669    Mean    :   561551
##   3rd Qu.:     12484    3rd Qu.:     12654    3rd Qu.:     13404
##   Max.    :19209252    Max.    :19774011    Max.    :19191406
##          2012                  2013                  2014
##   Min.    :        5    Min.    :        5    Min.    :        5
##   1st Qu.:        16    1st Qu.:        16    1st Qu.:        16
##   Median :        72    Median :        73    Median :        73
##   Mean    :   567238    Mean    :   592806    Mean    :   610288
##   3rd Qu.:     13047    3rd Qu.:     13574    3rd Qu.:     13852
```

```
##  Max.    :19068296   Max.    :19092876   Max.    :19207489
unique(dRaw$CountryName)
```

```
## [1] "Sub-Saharan Africa"          "North America"
## [3] "Middle East & North Africa"  "Latin America & Caribbean"
## [5] "European Union"              "Europe & Central Asia"
## [7] "East Asia & Pacific"         "Central Europe and the Baltics"
## [9] "Arab World"
```

```
unique(dRaw$IndicatorCode)
```

```
##  [1] "SP.ADO.TFRT"      "SP.POP.DPND"      "SE.PRM.UNER.FE"
##  [4] "SE.PRM.UNER.MA"   "SP.DYN.CDRT.IN"   "SE.SCH.LIFE.FE"
##  [7] "SE.SCH.LIFE.MA"   "NY.GDP.PCAP.CD"   "NY.GNP.PCAP.CD"
## [10] "SP.DYN.LE00.FE.IN" "SP.DYN.LE00.MA.IN"
```

Before we can use machine learning algorithms on the data, we need to reorganize the data to a format where each column has only data tied to one indicator.

We will use the IndicatorCodes as the new column names and remove the human-friendly descriptions, although it's good to keep those handy to interpret the column labels once we have reorganized the data. As the indicator name column has a space, we delineate the start and the end of the column name with tick marks (found on the tilde key of the keyboard) and remove that column entirely via assignment to null.

```
dRaw[,'Indicator Name':= NULL]
```

To transform the data, we use the melt() function on the raw data to collapse the columns of years down to a single variable named year. That way information about timing is captured in a variable, not in the names of different variables. This leaves all numeric value in a single column titled "value". It almost seems the situation is worse, because now to determine what any single value means, one must check both the indicator code column and the year column.

However, using dcast(), we can convert the data, "cast" into the correct structure where the indicator codes are columns of value variables, and our CountryName and Year columns indicate observations of country regions over time.

```
## collapse columns into a super long dataset
## with Year as a new variable
d <- melt(dRaw, measure.vars = 3:20, variable.name = "Year")
head(d)
```

```
##             CountryName  IndicatorCode Year    value
## 1: Sub-Saharan Africa     SP.ADO.TFRT 1997 1.3e+02
## 2: Sub-Saharan Africa     SP.POP.DPND 1997 9.2e+01
## 3: Sub-Saharan Africa SE.PRM.UNER.FE 1997 2.4e+07
## 4: Sub-Saharan Africa SE.PRM.UNER.MA 1997 2.0e+07
## 5: Sub-Saharan Africa SP.DYN.CDRT.IN 1997 1.6e+01
## 6: Sub-Saharan Africa SE.SCH.LIFE.FE 1997 5.7e+00
```

```
str(d)
```

```
## Classes 'data.table' and 'data.frame':     1782 obs. of  4 variables:
##  $ CountryName  : chr  "Sub-Saharan Africa" "Sub-Saharan Africa"
"Sub-Saharan Africa" "Sub-Saharan Africa" ...
##  $ IndicatorCode: chr  "SP.ADO.TFRT" "SP.POP.DPND" "SE.PRM.UNER.FE"
"SE.PRM.UNER.MA" ...
##  $ Year         : Factor w/ 18 levels "1997","1998",..: 1 1 1 1 1 1 1 1
1 1 ...
##  $ value        : num  1.32e+02 9.17e+01 2.44e+07 2.03e+07 1.55e+01 ...
##  - attr(*, ".internal.selfref")=<externalptr>
```

```
## finally cast the data wide again
## this time with separate variables by indicator code
## keeping a country and time (Year) variable
d <- dcast(d, CountryName + Year ~ IndicatorCode)
```

```
head(d)
```

```
##     CountryName Year NY.GDP.PCAP.CD NY.GNP.PCAP.CD SE.PRM.UNER.FE
## 1:   Arab World 1997           2299           2310        6078141
## 2:   Arab World 1998           2170           2311        5961001
## 3:   Arab World 1999           2314           2288        5684714
## 4:   Arab World 2000           2589           2410        5425963
## 5:   Arab World 2001           2495           2496        5087547
## 6:   Arab World 2002           2463           2476        4813368
```

```
##     SE.PRM.UNER.MA SE.SCH.LIFE.FE SE.SCH.LIFE.MA SP.ADO.TFRT
## 1:        4181176            8.1            9.7          57
## 2:        4222039            8.3            9.8          56
## 3:        4131775            8.5           10.0          55
## 4:        3955257            8.7           10.0          54
## 5:        3726838            8.8           10.1          53
## 6:        3534138            9.1           10.2          52
##     SP.DYN.CDRT.IN SP.DYN.LEOO.FE.IN SP.DYN.LEOO.MA.IN SP.POP.DPND
## 1:             6.8                69                65          79
## 2:             6.7                69                65          78
## 3:             6.6                69                66          76
## 4:             6.5                70                66          75
## 5:             6.4                70                66          73
## 6:             6.3                70                66          72
str(d)
```

```
## Classes 'data.table' and 'data.frame':    162 obs. of  13 variables:
##   $ CountryName     : chr  "Arab World" "Arab World" "Arab World"
"Arab World" ...
##   $ Year            : Factor w/ 18 levels "1997","1998",..: 1 2 3 4 5 6
7 8 9 10 ...
##   $ NY.GDP.PCAP.CD  : num  2299 2170 2314 2589 2495 ...
##   $ NY.GNP.PCAP.CD  : num  2310 2311 2288 2410 2496 ...
##   $ SE.PRM.UNER.FE  : num  6078141 5961001 5684714 5425963 5087547 ...
##   $ SE.PRM.UNER.MA  : num  4181176 4222039 4131775 3955257 3726838 ...
##   $ SE.SCH.LIFE.FE  : num  8.08 8.27 8.5 8.65 8.84 ...
##   $ SE.SCH.LIFE.MA  : num  9.73 9.82 9.97 10.02 10.12 ...
##   $ SP.ADO.TFRT     : num  56.6 55.7 54.9 54.2 53.3 ...
##   $ SP.DYN.CDRT.IN  : num  6.8 6.68 6.57 6.48 6.4 ...
##   $ SP.DYN.LEOO.FE.IN: num  68.7 69 69.3 69.6 69.8 ...
##   $ SP.DYN.LEOO.MA.IN: num  65 65.3 65.7 65.9 66.2 ...
##   $ SP.POP.DPND     : num  79.1 77.7 76.2 74.7 73.2 ...
##   - attr(*, ".internal.selfref")=<externalptr>
##   - attr(*, "sorted")= chr  "CountryName" "Year"
```

Now the data are in a format suitable for machine learning, where each row is an observation of a global region by year, and all information is captured in a variable, not in the variable (column) names. Data in this format are sometimes called "tidy" data, a concept described in depth by Hadley Wickham in a 2014 article [113].

As our column names themselves and our regional names may be graphed, there is value in shortening the length of those names for higher visual clarity. Additionally, some algorithms may have reserved characters (such as the full point or other punctuation characters). In this case, the column names of our data are assigned to x, and then the gsub function is used to remove all punctuation. Next, the function abbreviate is used to uniquely reduce the column names to four characters each. The names are assigned to our dataset via the names function. Lastly the country names themselves are shortened. We hope our readers forgive us the trade-off between graph legibility and comprehension. Notice in this case we use the left.kept option in an effort to enhance comprehension.

```
## rename columns with shortened, unique names
x<-colnames(d)
x<-gsub("[[:punct:]]", "", x)
(y <- abbreviate(x, minlength = 4, method = "both.sides"))

##    CountryName               Year    NYGDPPCAPCD    NYGNPPCAPCD    SEPRMUNERFE
##          "CntN"            "Year"         "NYGD"         "NYGN"         "SEPR"
##    SEPRMUNERMA    SESCHLIFEFE    SESCHLIFEMA      SPADOTFRT    SPDYNCDRTIN
##          "ERMA"         "SESC"         "FEMA"         "SPAD"         "SPDY"
## SPDYNLEOOFEIN SPDYNLEOOMAIN      SPPOPDPND
##          "FEIN"         "MAIN"         "SPPO"

names(d) <- y

## shorten regional names to abbreviations.
d$CntN<-abbreviate(d$CntN, minlength = 5,
                  method = "left.kept")
```

We briefly describe what each column of the data represents in Table 7-1. We show the original column name first, followed by a | and then our shortened abbreviation.

Table 7-1. *Listing of Columns in Gender Data*

Variable (Feature)	Description
CountryName | CntN	The abbreviated name of the geographic region or country group
Year | Year	The year each data came from
SP.ADO.TFRT | SPAD	Adolescent fertility rate (births per 1,000 women ages 15–19)
SP.POP.DPND | SPPO	Age dependency ratio (percent of working-age population)
SE.PRM.UNER.FE | SEPR	Children out of school, primary, female
SE.PRM.UNER.MA | ERMA	Children out of school, primary, male
SP.DYN.CDRT.IN | SPDY	Death rate, crude (per 1,000 people)
SE.SCH.LIFE.FE | SESC	Expected years of schooling, female
SE.SCH.LIFE.MA | FEMA	Expected years of schooling, male
NY.GDP.PCAP.CD | NYGD	GDP per capita (current US dollars)
NY.GNP.PCAP.CD | NYGN	GNI per capita, Atlas method (current US dollars)
SP.DYN.LE00.FE.IN | FEIN	Life expectancy at birth, female (years)
SP.DYN.LE00.MA.IN | MAIN	Life expectancy at birth, male (years)

Now that the data are in the proper structure, the summary() function will provide us some information about our various units of measure. An item of interest is that the range of data in each column seems to vary widely in some cases. Another item of note is that the Year column is not numeric, but rather a factor. While it may make sense to leave it a factor, it may not. For now, we convert Year to character string with the as. character() function.

```
summary(d)
##      CntN              Year          NYGD              NYGN
##  Length:162        1997   : 9   Min.   :   496   Min.   :   487
##  Class :character  1998   : 9   1st Qu.:  3761   1st Qu.:  3839
##  Mode  :character  1999   : 9   Median :  7458   Median :  7060
##                    2000   : 9   Mean   : 13616   Mean   : 13453
##                    2001   : 9   3rd Qu.: 19708   3rd Qu.: 19747
##                    2002   : 9   Max.   : 54295   Max.   : 55010
##                    (Other):108
```

```
##         SEPR                ERMA              SESC             FEMA
##   Min.    :  100024   Min.    :  109075   Min.   : 5.7   Min.    : 7.0
##   1st Qu.:  482710   1st Qu.:  563119   1st Qu.:10.3   1st Qu.:11.2
##   Median : 1338898   Median : 1195360   Median :13.3   Median :13.1
##   Mean    : 3992637   Mean    : 3360191   Mean   :12.8   Mean    :12.8
##   3rd Qu.: 3936040   3rd Qu.: 3339679   3rd Qu.:15.7   3rd Qu.:14.9
##   Max.    :24437801   Max.    :20766960   Max.   :17.3   Max.    :16.5
##
##         SPAD             SPDY             FEIN            MAIN             SPPO
##   Min.    : 11   Min.    : 5.0   Min.   :52   Min.   :48   Min.    :41
##   1st Qu.: 21   1st Qu.: 6.0   1st Qu.:72   1st Qu.:68   1st Qu.:49
##   Median : 38   Median : 8.1   Median :77   Median :70   Median :51
##   Mean    : 45   Mean    : 8.5   Mean   :74   Mean   :69   Mean    :57
##   3rd Qu.: 53   3rd Qu.:10.6   3rd Qu.:80   3rd Qu.:73   3rd Qu.:63
##   Max.    :132   Max.    :15.5   Max.   :84   Max.   :78   Max.    :92
##
```

```
str(d)
```

```
## Classes 'data.table' and 'data.frame':    162 obs. of  13 variables:
##  $ CntN: chr   "ArbWr" "ArbWr" "ArbWr" "ArbWr" ...
##  $ Year: Factor w/ 18 levels "1997","1998",..: 1 2 3 4 5 6 7 8 9 10 ...
##  $ NYGD: num   2299 2170 2314 2589 2495 ...
##  $ NYGN: num   2310 2311 2288 2410 2496 ...
##  $ SEPR: num   6078141 5961001 5684714 5425963 5087547 ...
##  $ ERMA: num   4181176 4222039 4131775 3955257 3726838 ...
##  $ SESC: num   8.08 8.27 8.5 8.65 8.84 ...
##  $ FEMA: num   9.73 9.82 9.97 10.02 10.12 ...
##  $ SPAD: num   56.6 55.7 54.9 54.2 53.3 ...
##  $ SPDY: num   6.8 6.68 6.57 6.48 6.4 ...
##  $ FEIN: num   68.7 69 69.3 69.6 69.8 ...
##  $ MAIN: num   65 65.3 65.7 65.9 66.2 ...
##  $ SPPO: num   79.1 77.7 76.2 74.7 73.2 ...
##  - attr(*, ".internal.selfref")=<externalptr>
```

```
d[, Year := as.character(Year)]
```

Next, we start to explore the relationship between our various features. We use the ggplot() function from the ggplot2 package to make some graphs of per capita gross national income as an input (x axis) and adolescent fertility as a response variable (y axis). One is in dollars per inhabitant, and the other is in births per 1,000 women ages 15–19. The plot_grid() function helps us make a panel of graphs to show just those two variables and another plot where we color the data points by year of data.

```
## ggplot2 plot object indicating x and y variables
p1 <- ggplot(d, aes(NYGN, SPAD))

## make a grid of two plots
plot_grid(
  ## first plot data points only
  p1 + geom_point(),
  ## data poins colored by year
  p1 + geom_point(aes(colour = Year)) +
    scale_colour_viridis(discrete = TRUE),
  ncol = 1
)
```

Looking at Figure 7-1, it seems that there are distinct groups. The next section turns to machine learning methods to try to determine exactly how many distinct groups there are. One imagines that not all of our regions are clearly distinct (e.g., there may be overlap between the European Union and Europe and Central Asia perhaps). Thus simply separating every group by geographic region may not be necessary. Instead, we will examine some algorithms that attempt to empirically determine how many clusters or groupings are needed to capture the observed data.

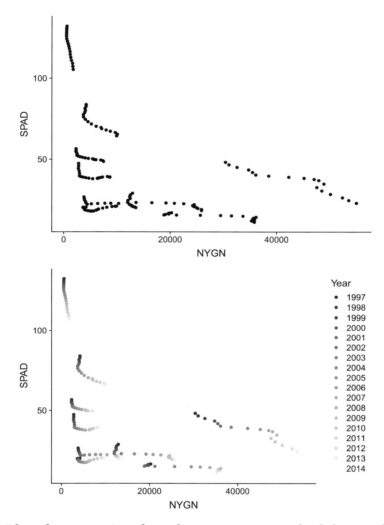

Figure 7-1. *Plot of gross national product per capita and adolescent fertility rate per 1,000 women*

7.2 kmeans

One of the more common grouping algorithms is kmeans. We can perform kmeans clustering in R using the aptly named kmeans() function from the stats package, which is one of the few packages that "ships" with base R.

kmeans() takes two main arguments: a dataset x that contains continuous numerical data and a center number k that tells the function how many clusters we wish to extract from the data. In a nutshell, what the algorithm does is create a good set of *k* center points such that the within-group sum of squares (a proxy for Euclidean distance) is minimized, given the constraint of the number of clusters we specified should be extracted.

The algorithm has some variability with how it starts, so if you want your analysis to be reproducible, it is important to set the random starting seed in R. We do this by using the set.seed() function with a value of 2468, which is what you should use if you want to reproduce our results.

Next, a variable wgss is initialized to contain the total within-group sum of squares. This is the metric the algorithm seeks to minimize for any given number of clusters.

The case k = 1 would be boring, so we start with two centers, and let our for() loop iterate the algorithm across 2 to 9 centers for our data of NYGN vs. SPAD. For each iteration, we store the total within-group sum of squares in our variable, and also plot the familiar graph of our data, this time coloring the dots by assigned cluster membership.

Finally, we add a scree plot in the very last tile to show how the number of clusters compares to the reduction of that within-group sum of squares. The final figure is in Figure 7-2.

```
set.seed(2468)
wgss <- vector("numeric", 8)
plots <- vector("list", 9)
p1 <- ggplot(d, aes(NYGN, SPAD))

for(i in 2:9) {
  km <- kmeans(d[, .(NYGN, SPAD)],
              centers = i)

  wgss[i - 1] <- km$tot.withinss

  plots[[i - 1]] <- p1 +
    geom_point(aes_(colour = factor(km$cluster))) +
    scale_color_viridis(discrete = TRUE) +
    theme(legend.position = "none") +
    ggtitle(paste("kmeans centers = ", i))
}

plots[[9]] <- ggplot() +
  geom_point(aes(x = 2:9, y = wgss)) +
  xlab("Number of Clusters") +
  ylab("Within SS") +
```

```
ggtitle("Scree Plot")
```

```
do.call(plot_grid, c(plots, ncol = 3))
```

In Figure 7-2, we see the various numbers of clusters possible. Keep in mind, in this case, we happen to know these data come from nine geographic regions. However, looking at the scree plot in the bottom right, we see that there is not a major improvement in the reduction of the total within-group sum of squares after 3 or 4. As an unsupervised learning technique, kmeans has shown us that these data are much less varied than might be expected based on geography alone.

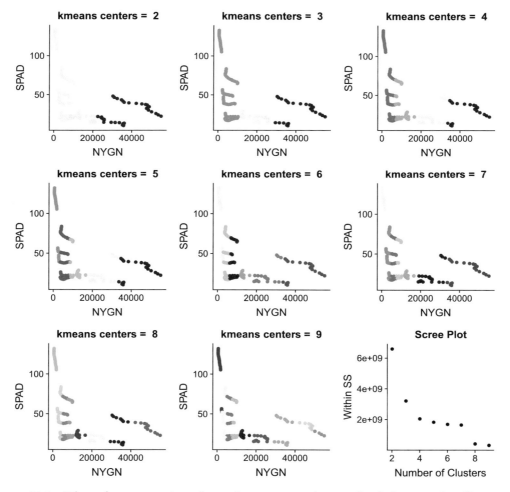

Figure 7-2. *Plot of gross national product per capita and adolescent fertility rate per 1,000 women for different numbers of k clusters*

Notice in Figure 7-2 that the main factor seems to be GNP. Because that is the farthest distance, it makes sense that it is the deciding factor. The goal of kmeans is to minimize distances between the center of each group and the points that comprise that group. Looking more closely at our data, we see the quite large difference in the scale between our two variables. That is, one ranges over a hundred or so and the other over thousands.

```
summary(d[,.(NYGN, SPAD)])
```

```
##       NYGN             SPAD
##  Min.   :   487   Min.   : 11
##  1st Qu.: 3839    1st Qu.: 21
##  Median : 7060    Median : 38
##  Mean   :13453    Mean   : 45
##  3rd Qu.:19747    3rd Qu.: 53
##  Max.   :55010    Max.   :132
```

Often, the scale of data may be relatively arbitrary and not indicate which variable should be given more weight in an analysis. One way to solve this type of problem is via centering and scaling. Centering data subtracts out the mean value of the data from each feature, so that every feature has mean equal to zero. Scaling divides our data out by the standard deviation of each feature so that there is a standard range as well. The scale() function in base R handles both these operations for us and works on matrix or data frame and data table style data.

```
x <- scale(d[,.(NYGN, SPAD)])
summary(x)
```

```
##       NYGN             SPAD
##  Min.   :-0.92    Min.   :-1.04
##  1st Qu.:-0.68    1st Qu.:-0.74
##  Median :-0.46    Median :-0.21
##  Mean   : 0.00    Mean   : 0.00
##  3rd Qu.: 0.45    3rd Qu.: 0.26
##  Max.   : 2.96    Max.   : 2.72
```

Now we can recreate the plots we made last time, but using our scaled data. This shows that now the y axis height is a distinguisher in the grouping (Figure 7-3).

```
set.seed(2468)
wgss <- vector("numeric", 8)
plots <- vector("list", 9)
p1 <- ggplot(d, aes(NYGN, SPAD))

for(i in 2:9) {
  km <- kmeans(x, centers = i)

  wgss[i - 1] <- km$tot.withinss

  plots[[i - 1]] <- p1 +
    geom_point(aes_(colour = factor(km$cluster))) +
    scale_color_viridis(discrete = TRUE) +
    theme(legend.position = "none") +
    ggtitle(paste("kmeans centers = ", i))
}

plots[[9]] <- ggplot() +
  geom_point(aes(x = 2:9, y = wgss)) +
  xlab("Number of Clusters") +
  ylab("Within SS") +
  ggtitle("Scree Plot")

do.call(plot_grid, c(plots, ncol = 3))
```

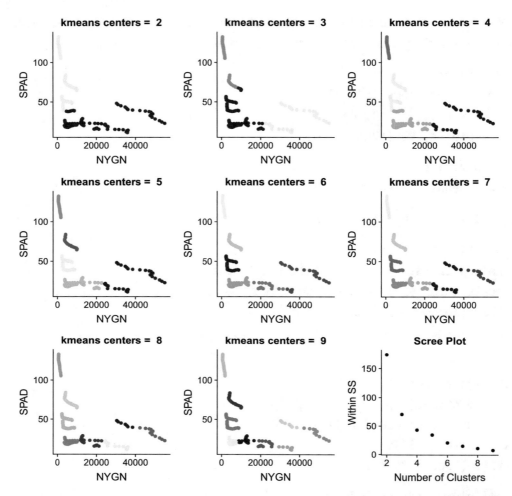

Figure 7-3. *Plot of gross national product per capita and adolescent fertility rate per 1,000 women for different numbers of k clusters*

By scaling our data, we make it more likely kmeans will spot groups based on y axis heights of fertility rates, although it is still clear in the left side of our graphs that there must be some factor besides GNP that accounts for various rates. Again, despite the scale, we see from the scree plot that there may well be fewer "logical" groups than the nine regions we selected. Indeed, if we think a bit about real-world geography, we might decide that six groups may well be the maximum that makes sense, and looking at the various kmeans plots and the scree plot might show us that there is relatively little improvement to our total within-group sum of squares values after 6, perhaps even four clusters may approximate the data well enough.

Still, supposing we decide to go with 6 as the number of centers, let us see what might be gained by changing the default iteration max from 10. While there are some slight differences in groupings, all we really notice is what our visual inspection told us anyway, that group on the far right is rather sparsely defined (and thus it switches groups as it lacks a strong signal). The iteration max controls how many times the algorithm runs through its sort procedure.

kmeans takes a set of k random starting points called centroids, and then every value (observation) is assigned cluster membership based on which centroid is the minimal Euclidean distance from that observation. From there, all the points in each cluster are used to calculate a new centroid, which is now at the center of that cluster. So now there are second-generation k centroids, and this means every point is again checked against every centroid and assigned cluster membership based on minimum Euclidean distance. If there are changes in group membership, then the points in each cluster are used to calculate the center of that group, and this becomes the new centroid, and we now have third-generation centroids. The process repeats until either there are no more changes in group membership or the iteration max is reached. In this case, it seems that, due to weak signal, we have some points that could go either way, and thus successive iterations cause some membership switches. The results are shown in Figure 7-4.

```
set.seed(2468)
plots <- vector("list", 9)
p1 <- ggplot(d, aes(NYGN, SPAD))

for(i in 6:14) {
  km <- kmeans(x, centers = 6, iter.max = i)

  plots[[i - 5]] <- p1 +
    geom_point(aes_(colour = factor(km$cluster))) +
    scale_color_viridis(discrete = TRUE) +
    theme(legend.position = "none") +
    ggtitle(paste("kmeans iters = ", i))
}

do.call(plot_grid, c(plots, ncol = 3))
```

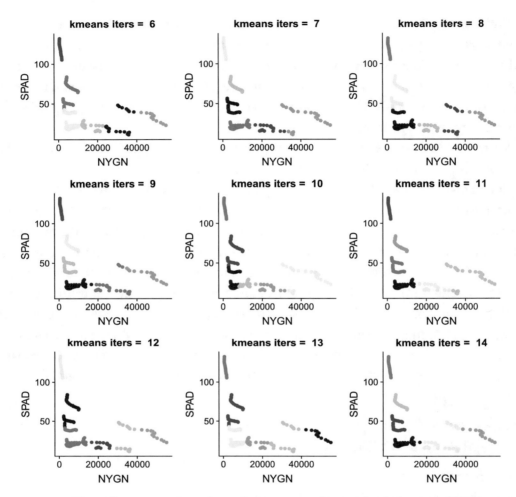

Figure 7-4. *Plot of gross national product per capita and adolescent fertility rate per 1,000 women for different numbers of iterations*

The last formal argument regularly modified in kmeans() is nstart. kmeans computes groups by measuring the distance of each point to the center of its group and attempts to make that distance measure smaller in each iteration. However, for the first generation, on start, it randomly sets k centers. Thus, nstart determines how many different randomizations compete to be the first-generation k centroids. The algorithm selects the optimal choice as the place to start the first generation and run the next iteration. Still, we do not detect any major changes in this case for this example, as shown in Figure 7-5.

```
set.seed(2468)
plots <- vector("list", 9)
p1 <- ggplot(d, aes(NYGN, SPAD))
```

```
for(i in 1:9) {
  km <- kmeans(x, centers = 6, iter.max = 10, nstart = i)

  plots[[i]] <- p1 +
    geom_point(aes_(colour = factor(km$cluster))) +
    scale_color_viridis(discrete = TRUE) +
    theme(legend.position = "none") +
    ggtitle(paste("kmeans nstarts = ", i))
}

do.call(plot_grid, c(plots, ncol = 3))
```

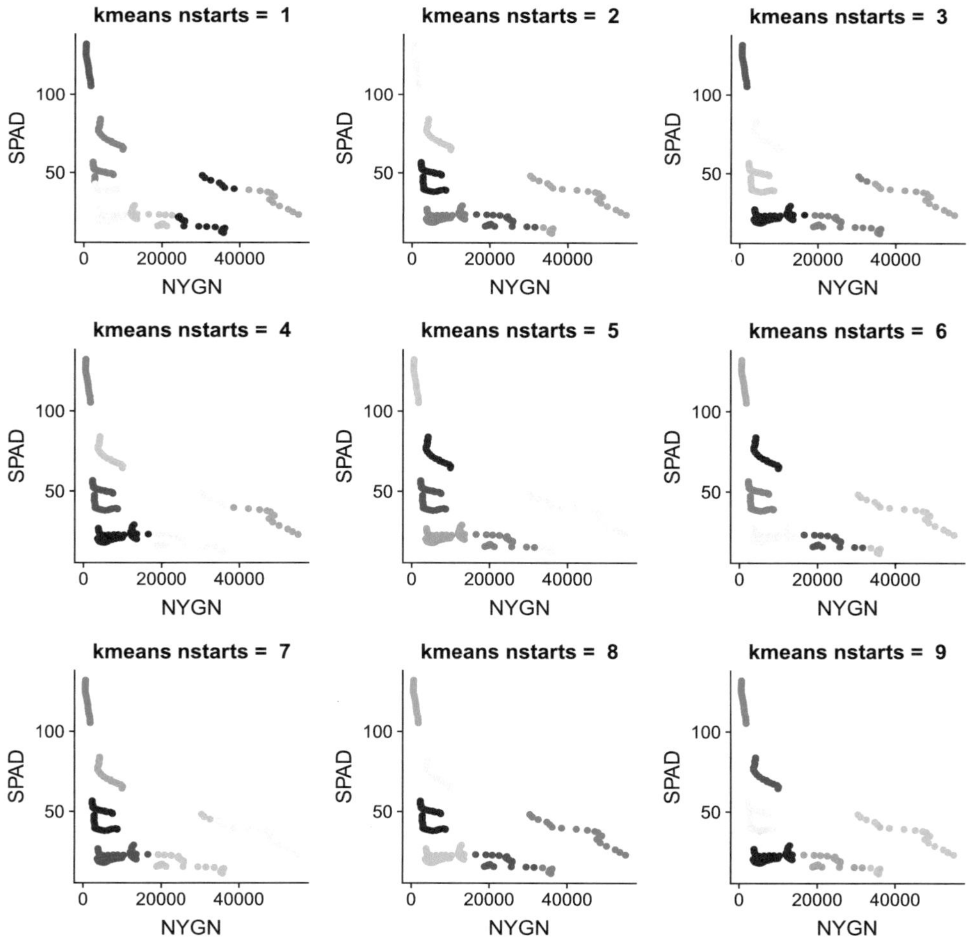

Figure 7-5. *Plot of gross national product per capita and adolescent fertility rate per 1,000 women for different nstart values*

While these examples were done in just two dimensions, we can perform kmeans clustering on higher dimensions of data. While it is not possible to visualize all 11 variables via a two-dimensional plot, we can still observe the scree plot and determine an optimal number of groups. We perform the same calculations as before, only this time only graphing the scree plot.

```
x <- scale(d[,-c(1,2)])
wgss<-0
set.seed(2468)
for( i in 1:11){
  km <- kmeans(x, centers = i)
  wgss[i]<-km$tot.withinss
}

ggplot() +
  geom_point(aes(x = 1:11, y = wgss)) +
  xlab("Number of Clusters") +
  ylab("Within SS") +
  ggtitle("Scree Plot - All Variables")
```

Based on the scree plot, Figure 7-6, we determine an optimal number of groups for our full dataset. In this case we choose perhaps 4 and now run the algorithm one last time and see what happens with the groupings. The result is stored in the kmAll variable, and we column bind those results to our data using cbind() along with our country names and the year.

```
kmAll <- kmeans(x, centers = 4, nstart = 25)
x <- cbind(d[, c(1,2)], x,
           Cluster = kmAll$cluster)
tail(x)
```

```
##       CntN Year  NYGD  NYGN SEPR ERMA SESC FEMA SPAD SPDY FEIN MAIN
## 1: Sb-SA 2009 -0.89 -0.87  2.4  2.3 -1.5 -1.5  2.2 1.11 -2.2 -2.2
## 2: Sb-SA 2010 -0.87 -0.87  2.5  2.4 -1.5 -1.5  2.1 0.95 -2.1 -2.1
## 3: Sb-SA 2011 -0.86 -0.86  2.4  2.3 -1.4 -1.4  2.1 0.81 -2.0 -2.0
## 4: Sb-SA 2012 -0.85 -0.84  2.4  2.3 -1.3 -1.4  2.0 0.68 -1.9 -1.9
## 5: Sb-SA 2013 -0.85 -0.84  2.4  2.3 -1.3 -1.3  1.9 0.56 -1.8 -1.8
## 6: Sb-SA 2014 -0.85 -0.83  2.4  2.3 -1.3 -1.4  1.9 0.45 -1.7 -1.7
```

```
##      SPPO Cluster
## 1:   2.2       1
## 2:   2.2       1
## 3:   2.2       1
## 4:   2.2       1
## 5:   2.1       1
## 6:   2.1       1
```

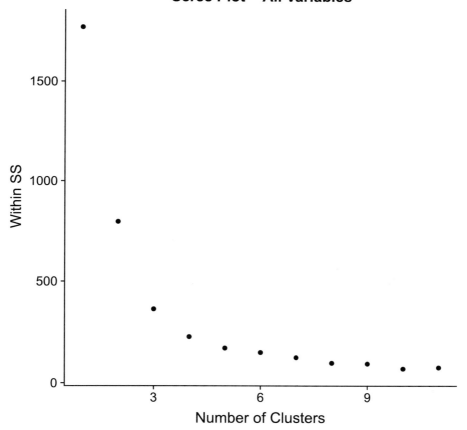

Figure 7-6. *Scree plot for all*

If we cross tabulate the frequency where cases belong to a country and a particular cluster, we can see that some regions are fairly consistent, while others seem to change cluster on occasion.

```
xtabs(~ CntN + Cluster, data = x)
```

```
##          Cluster
## CntN      1  2  3  4
##    ArbWr  0 18  0  0
##    CEatB  0  0  0 18
##    Er&CA  0  0  1 17
##    ErpnU  0  0 11  7
##    EsA&P  0 14  0  4
##    LtA&C  0 18  0  0
##    ME&NA  0 18  0  0
##    NrthA  0  0 18  0
##    Sb-SA 18  0  0  0
```

Does this mean our algorithm has failed? Firstly, as this is unsupervised learning, the answer is likely that even if specific geographic regions were not fully reproduced by the data, the empirical structure may still tell us something useful. Additionally, in this case, our data do have a time component. When we examine the results by time, we see the shift is consistent over time, with a region switching cluster membership.

```
unique(x[
  order(CntN, Year, Cluster),
  .(CntN, Year, Cluster)][
    CntN=="EsA&P"])
```

```
##         CntN Year Cluster
##  1: EsA&P 1997       2
##  2: EsA&P 1998       2
##  3: EsA&P 1999       2
##  4: EsA&P 2000       2
##  5: EsA&P 2001       2
##  6: EsA&P 2002       2
##  7: EsA&P 2003       2
##  8: EsA&P 2004       2
##  9: EsA&P 2005       2
## 10: EsA&P 2006       2
## 11: EsA&P 2007       2
## 12: EsA&P 2008       2
```

```
## 13: EsA&P 2009        2
## 14: EsA&P 2010        2
## 15: EsA&P 2011        4
## 16: EsA&P 2012        4
## 17: EsA&P 2013        4
## 18: EsA&P 2014        4
```

```
unique(x[
  order(CntN, Year, Cluster),
  .(CntN, Year, Cluster)][
    CntN == "ErpnU"])
```

```
##         CntN Year Cluster
##   1: ErpnU 1997        4
##   2: ErpnU 1998        4
##   3: ErpnU 1999        4
##   4: ErpnU 2000        4
##   5: ErpnU 2001        4
##   6: ErpnU 2002        4
##   7: ErpnU 2003        4
##   8: ErpnU 2004        3
##   9: ErpnU 2005        3
## 10: ErpnU 2006        3
## 11: ErpnU 2007        3
## 12: ErpnU 2008        3
## 13: ErpnU 2009        3
## 14: ErpnU 2010        3
## 15: ErpnU 2011        3
## 16: ErpnU 2012        3
## 17: ErpnU 2013        3
## 18: ErpnU 2014        3
```

As we have seen, kmeans() is a way to group similar cases together, and, while our data were labelled for pedagogical purposes, it is not required that our data have labels. Using a scree plot and for loop, reasonably optimal clusters can be determined to organize data into similar groups. It makes a good final step to exploratory data analysis in some ways, as the results may shed light on your organization's clients or on your study's participants. Keep in mind, though, that data which are not scaled may give unequal weight to certain dimensions and the algorithm's objective to minimize Euclidean distance may not yield the same obvious groupings a human eye might see.

7.3 Hierarchical Clusters

While kmeans started by picking k arbitrary centers, and assigning each point to the nearest center, and then iterating through that process a certain number of times, hierarchical clustering is different. Instead, each point is assigned to its own unique cluster—you have as many groups as you have points! Then, the distance between every point is determined, and the closest neighbor is found. Neighbors are then joined into a bigger group. This process repeats until the last two supergroups are joined into one single archgroup.

Because this relies on distance, the first step is to calculate the distance of each point from every other point. The distance function again defaults to the Euclidean distance. Thus, at its most basic, this again requires continuous numerical data. However, it is possible to write bespoke distance functions that allow for other types of "distances." As long as the output is a distance matrix where higher numbers indicate greater distance, the algorithm will process (although the efficacy of that processing is of course not guaranteed).

The first stage is to use the dist() function on our data. To build our intuition, we use just two dimensions of our data.

```
hdist <- dist(d[,.(NYGN, SPAD)])
str(hdist)
```

```
## 'dist' num [1:13041] 1.13 22.04 100.03 186.5 166.08 ...
##  - attr(*, "Size")= int 162
##  - attr(*, "Diag")= logi FALSE
##  - attr(*, "Upper")= logi FALSE
##  - attr(*, "method")= chr "euclidean"
##  - attr(*, "call")= language dist(x = d[, .(NYGN, SPAD)])
```

The resulting object encodes the relative distance between every point in our 2D dataset. This information is passed to the hclust() function, which creates our hierarchical cluster which can be plotted using a call to plot(). This is a dendrogram graph, and the heights of each segment show the distance between any two rows/observations of our data. The graph is in Figure 7-7.

```
hclust <- hclust(hdist)
plot(hclust)
```

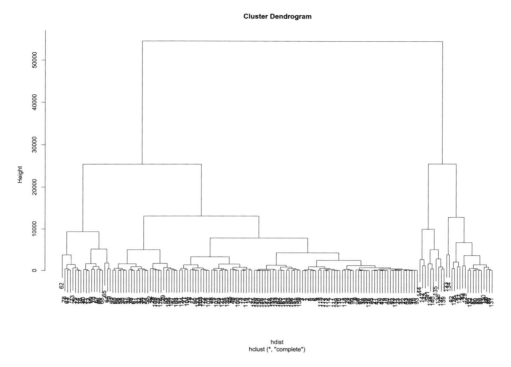

Figure 7-7. *Cluster dendrogram with row numbers*

Notice this uses row names, which in this case are numeric and are not overly helpful. We can set a key column so that we can understand this dendrogram a bit better. Although this is not the cleanest graph, one does note that it seems similar regions are often joined closely to each other, in Figure 7-8.

```
x <- d[, .(CntN, Year, NYGN, SPAD)]
x[, Key := paste(CntN, Year)]
x[, CntN := NULL]
x[, Year := NULL]

hdist <- dist(x[,.(NYGN, SPAD)])
hclust <- hclust(hdist)
plot(hclust, labels = x$Key)
```

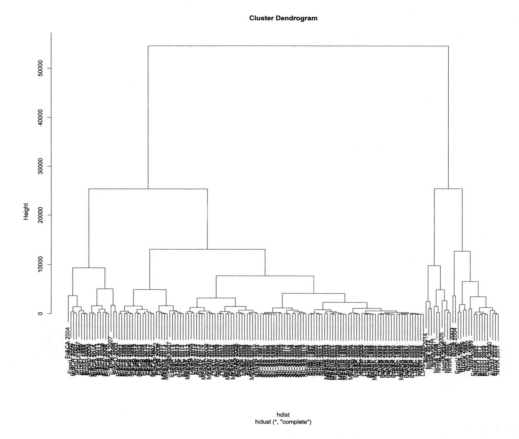

Figure 7-8. *Variations on dendrogram via ape package*

Now, while kmeans was perhaps simpler, it shows an end result where there are only as many groups as one chose. Thus, a scree plot was of value to determine how many clusters might make sense. Here, every row starts in its own group, and then each neighbor is systematically joined up into pairs.

The heights show us the distance between any two joined groups. Additionally, notice that, for a given height, we could understand how many "groups" our data fit.

```
plot(hclust, labels = x$Key)
abline(h = 30000, col = "blue")
```

At a height of h = 30,000, shown in Figure 7-9, we only have two groups. One contains North America (NrtA) and most of the European Union (ErpU) of the 2000s, while the other group contains the rest. A glance at the summary as well as a couple of data table calls on our parent data suggests that the main height (which is based on Euclidean distance, remember) is likely the GNP (NYGN). Note that this shows quite well why, once we are done with our exploratory work, scaling our data continues to be important. Without some effort on our part to ensure roughly equal data columns, something like GNP (NYGN) can drown out other distinguishing features.

```
summary(x)
```

```
##       NYGN              SPAD            Key
##   Min.   :   487   Min.    : 11   Length:162
##   1st Qu.:  3839   1st Qu.: 21   Class :character
##   Median :  7060   Median : 38   Mode  :character
##   Mean   : 13453   Mean    : 45
##   3rd Qu.: 19747   3rd Qu.: 53
##   Max.   : 55010   Max.    :132
```

```
d[, mean(NYGN), by = CntN][order(V1)]
```

```
##       CntN      V1
## 1: Sb-SA    953
## 2: ArbWr   4260
## 3: ME&NA   5045
## 4: EsA&P   5801
## 5: LtA&C   6004
## 6: CEatB   8531
## 7: Er&CA  19021
## 8: ErpnU  28278
## 9: NrthA  43188
```

```
d[, mean(SPAD), by = CntN][order(V1)]
```

```
##       CntN   V1
## 1: ErpnU   15
## 2: EsA&P   20
## 3: CEatB   23
## 4: Er&CA   23
## 5: NrthA   37
## 6: ME&NA   40
## 7: ArbWr   52
## 8: LtA&C   73
## 9: Sb-SA  120
```

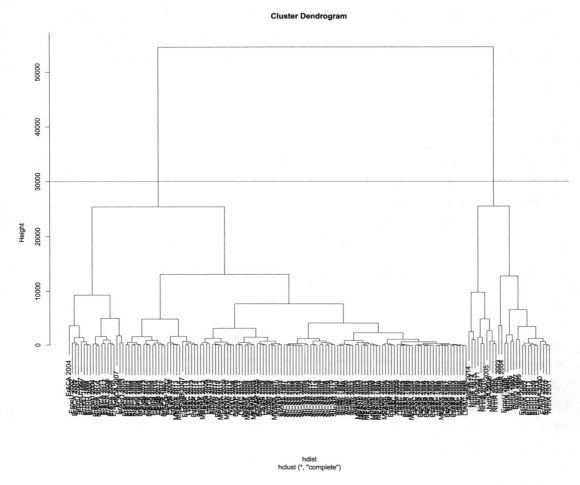

Figure 7-9. *Cluster dendrogram with country names and year and a height line*

Adjusting the height of our `abline()` allows us to change the way we view our data, and now it is in four clusters. Now, especially in book form, it is tough to get such a large chart to print cleanly, although our best attempt is in Figure 7-10. Try running the code on your own machine, and keep in mind we are still at an exploratory stage. This is helping to shape our understanding of similarities between countries and what the structure might be on these two dimensions.

```
plot(hclust, labels = x$Key)
abline(h = 20000, col = "blue")
```

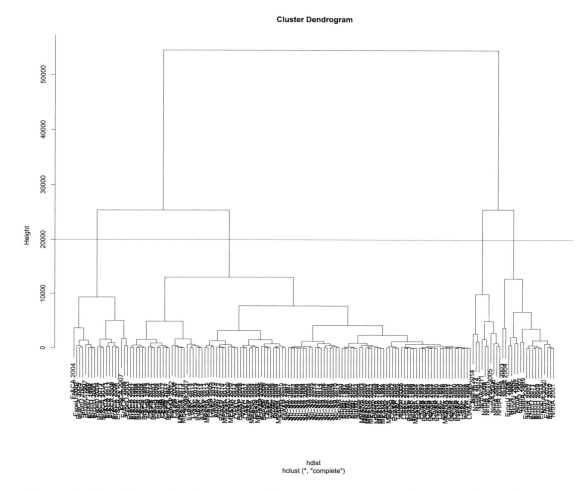

Figure 7-10. *Cluster dendrogram with country names and year and another height line*

Thinking about dimensions, recognize this model works well for visualizing the entire dataset—on all columns. It simply requires a small tweak to our code to not restrict the variables to just two. Of course, the height measurement will change drastically, as we now have a much larger space feeding our Euclidean distance. Additionally, the final two groupings join Sub-Saharan Africa (S-SA) to the rest of our regions, shown in Figure 7-11.

```
x <- copy(d)
x[, Key := paste(CntN, Year)]
x[, CntN := NULL]
x[, Year := NULL]
```

```
hdist <- dist(x[, -12])
hclust <- hclust(hdist)

plot(hclust, labels = x$Key)
```

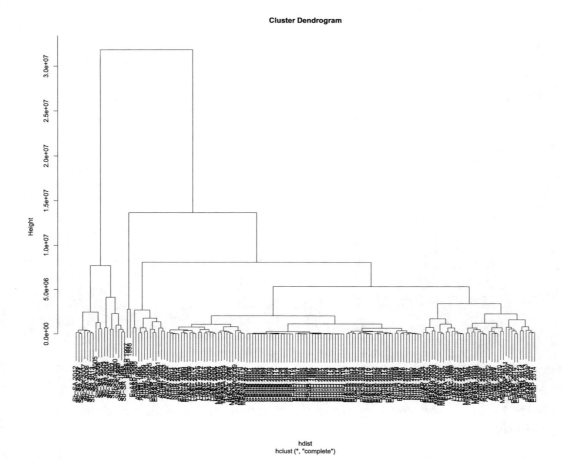

Figure 7-11. *Cluster dendrogram with country names and year and all dimensions of data*

Now, the hclust function takes a distance matrix as the first formal argument and a method type as the second. This second formal argument defaults to "complete" which uses the maximum distance between cluster points to determine group distance and then always groups the nearest groups based on that maximal distance. Other methods will yield different results, as they choose to use other methods (such as minimizing the iterative increase in within-cluster variance). Notice we do not refresh our hdist matrix here—we are simply changing which method is used. The result is in Figure 7-12.

```
hclust <- hclust(hdist, method = "ward.D2")
plot(hclust, labels = x$Key)
```

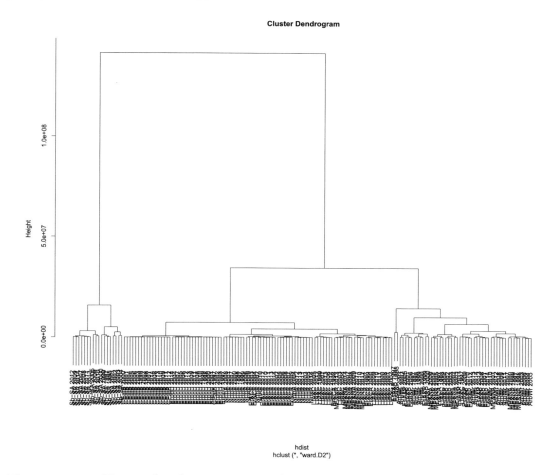

Figure 7-12. *Cluster dendrogram using the ward.D2 method*

It is worth noting that Euclidean distance could be swapped out for some other distance metric; indeed there are a couple of built-in options. Beyond the built-in functions, and without getting too esoteric, it may make sense to use some function that performs sentiment analysis on tweets, to determine how many groups of opinions there are about a particular trending hashtag. Of course, once that sentiment analysis was converted to some numeric number, it does perhaps make sense to use Euclidean distance on those. In any case, the point being, it is not required to that data start at a numeric level.

As mentioned earlier, and perhaps especially the more dimensions we add to our analysis, without scaling our data, the column(s) with the widest ranges will have outsized influence on our groupings. After scaling, using all the variables and replotting, the results are shown in Figure 7-13.

```
x <- scale(d[,-c(1,2)])
row.names(x) <- paste(d$CntN, d$Year)
hdist <- dist(x)
hclust <- hclust(hdist)

plot(hclust, labels = paste(d$CntN, d$Year))
abline(h = 6, col = "blue")
```

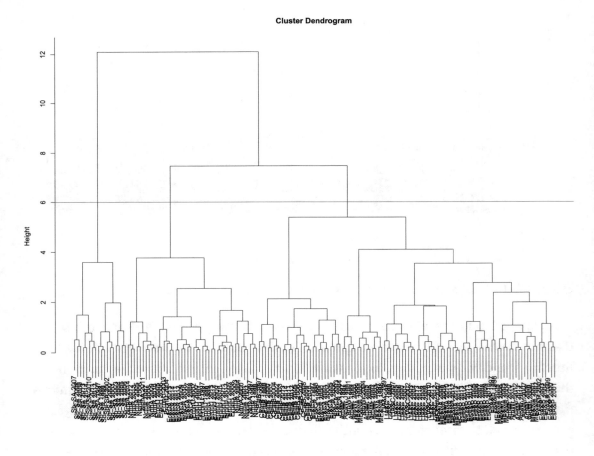

Figure 7-13. *Cluster dendrogram with scaling*

Now, while we may choose to cut our tree based on a certain height, that only shows us visually how our groups are arranged. We can also use the cutree() function to actually cluster our data at a certain height. In the case of $h = 6$, this cut our tree into three groups, which we can see that the function gave us.

```
cut_hclust <- cutree(hclust, h = 6)
unique(cut_hclust)
```

```
## [1] 1 2 3
```

Alternately, rather than cut at a specific height, the tree can be cut at certain number of clusters. Creating a copy of our data, we can record the cluster assignments in a new column titled cluster.

```
dcopy <- as.data.table(copy(d))
dcopy[, cluster:= NA_integer_]
```

```
dcopy$cluster <- cutree(hclust, k = 3)
```

```
tail(dcopy)
```

```
##        CntN Year NYGD NYGN    SEPR    ERMA SESC FEMA SPAD SPDY FEIN MAIN
## 1: Sb-SA 2009 1198 1186 1.9e+07 1.6e+07  8.2  9.4  115 11.5   58   55
## 2: Sb-SA 2010 1555 1287 2.0e+07 1.6e+07  8.3  9.4  113 11.0   58   55
## 3: Sb-SA 2011 1706 1412 1.9e+07 1.5e+07  8.4  9.6  111 10.7   59   56
## 4: Sb-SA 2012 1740 1631 1.9e+07 1.6e+07  8.6  9.7  109 10.3   60   57
## 5: Sb-SA 2013 1787 1686 1.9e+07 1.5e+07  8.9  9.9  107 10.0   61   57
## 6: Sb-SA 2014 1822 1751 1.9e+07 1.6e+07  8.7  9.7  105  9.7   61   58
##        SPPO cluster
## 1:       89       3
## 2:       89       3
## 3:       88       3
## 4:       88       3
## 5:       87       3
## 6:       87       3
```

To conclude our section on hierarchical clusters, we mention the ape package which has several visualization options to the dendrogram provided by default with plot(). Admittedly, it is easier to visualize models with fewer categories. Keep in mind, in the usual applications of unsupervised learning, a natural goal is to determine how many categories might reasonably exist. So our pedagogical example is going to be more than a little clunky, precisely because we are keeping a rather broad, fixed set of categories. We show the results in Figure 7-14.

```
plot(as.phylo(hclust), type = "cladogram")

plot(as.phylo(hclust), type = "fan")

plot(as.phylo(hclust), type = "radial")
```

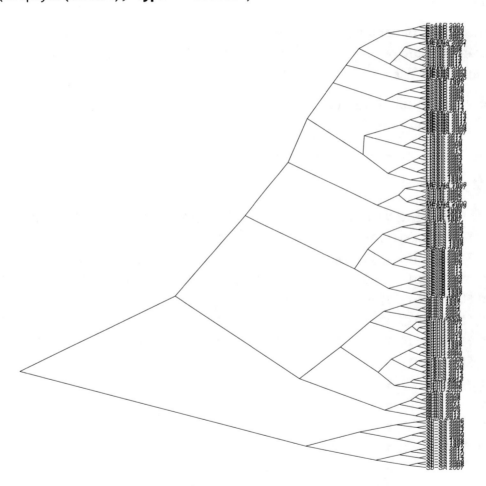

Figure 7-14. *Variations on dendrogram via ape package*

Part of the advantage of hierarchical clusters is that they can be a way to visualize similarity and dissimilarity between observations. As part of that process, cutting the tree into clear groups can be a helpful step. Now, based on our earlier kmeans analysis, we have a belief there might be four significant groups, so we attempt a visualization. Here, the second formal argument of cutree() is set to the number of clusters desired with k = 4. Additionally, the as.phylo() function is used to convert our hclust object into the type of object (a phylo object) for use by the ape package. We showcase a new type of graph—unrooted–as well as a label.offset which provides some distance between our labels and the graph (thereby creating some all-important space between characters of our human-friendly yet fit-on-graph unfriendly labels). Finally, we bring in our cutree() information in the tip.color argument and reduce the magnification of our text with cex. The end result is in Figure 7-15.

```
hclust4 <- cutree(hclust, k = 4)
plot(as.phylo(hclust), type = "unrooted", label.offset = 1,
     tip.color = hclust4, cex = 0.8)
```

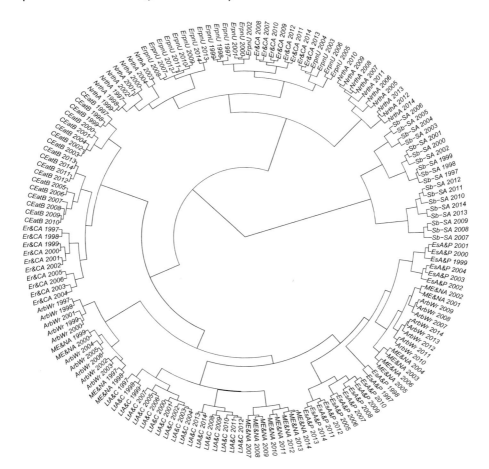

Figure 7-15. *Variations on dendrogram via ape package*

Notice that the plot itself is based on the original hierarchical clustering model—only the color of the tips is used to group our data into pairings (and yet those pairs make sense looking at this graph).

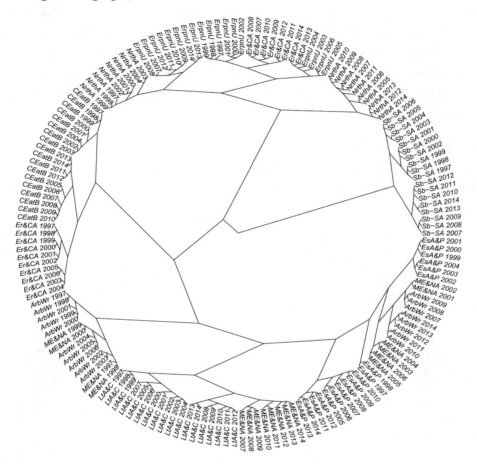

Figure 7-16. *Variations on dendrogram via ape package*

7.4 Principal Component Analysis

So far, in unsupervised learning, we have seen two techniques to understand how many groups or clusters there are in our data. One, kmeans, is told in advance how many groups it ought to find (based on perhaps scree plot analysis). The other, hclust, places each observation in single groups, and then eventually connects the dots, as it were, until there is only one group. It is left to the analyst to see which observations are closest, and which subgroups are closest. Principal component analysis (PCA) also attempts to determine groups after a fashion. PCA decomposes the data into the unique (i.e., uncorrelated, independent, orthogonal) components. For example, in our dataset, we may imagine that GDP and GNP are likely correlated at some level. In fact, they are highly correlated! These are essentially the same. Even their ranges and means are quite close, which can be seen in Figure 7-18.

```
cor(d$NYGD, d$NYGN)
```

```
## [1] 1
```

```
summary(d[,.(NYGD, NYGN)])
```

```
##          NYGD              NYGN
##  Min.    :  496    Min.    :  487
##  1st Qu.: 3761    1st Qu.: 3839
##  Median : 7458    Median : 7060
##  Mean    :13616    Mean    :13453
##  3rd Qu.:19708    3rd Qu.:19747
##  Max.    :54295    Max.    :55010
```

```
ggplot(d, aes(NYGD, NYGN)) +
  geom_point()
```

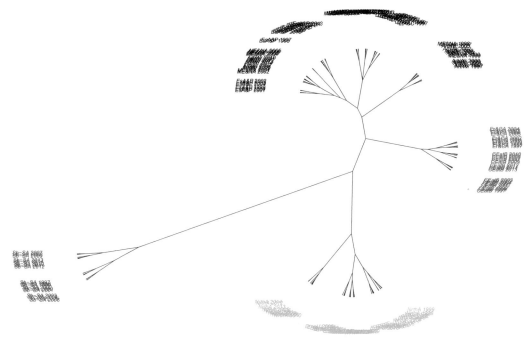

Figure 7-17. *Unrooted type on four clusters*

Visually, Figure 7-18 is essentially the line $y = x$. Principal component analysis can be thought of as a grouping operation. The goal of grouping is to see how many *truly* unique dimensions we have. In this case, while it may look like we have two different columns in the dataset, the fact is it seems reasonable to say we really only have one column (dimension) of unique information.

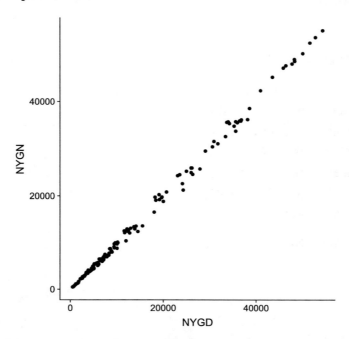

Figure 7-18. *A highly correlated plot—are there really two dimensions here?*

What principal component analysis (PCA) allows us to do is to determine we do not need both columns and to combine them into a single dimension in a principled way. This would simplify our feature space and, as we can see, would not result in much loss of information. On the other hand, if we were to look at some of our other variables, we see that perhaps it does not make sense to simply delete one column in favor of another. Even if they are highly correlated, it is still clear that sometimes, especially at a certain point, SE.SCH.Life.FE (SESC) has quite a bit of change with very little GDP (NYGD) change, shown in Figure 7-19. Clearly, there is more than one dimension of information available here.

```
ggplot(d, aes(NYGD, SESC)) +
  geom_point()

cor(d$NYGD, d$SESC)

## [1] 0.79
```

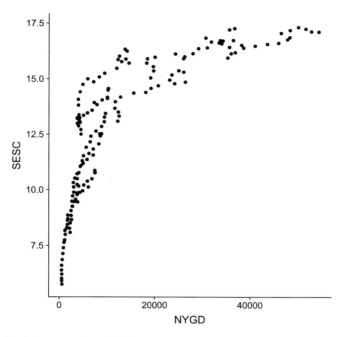

Figure 7-19. *A highly correlated plot—are there really two dimensions here?*

A related technique to PCA is factor analysis (FA). Both of these approaches can share an exploratory nature in that the true dimensionality of the data is unknown, although confirmatory FA exists where the goal is to test a hypothesized dimensionality. Further, both PCA and FA attempt to find a lower dimensional space that provides a reasonable approximation to the data. However, PCA and FA also have many differences. They come from different theoretical backgrounds, the interpretation of the principal components (PCA) and factors (FA) are not the same, and the fundamental assumptions about the data also differ. One notable difference is that FA typically focuses on the shared variance across variables, whereas PCA incorporates shared and unique variance. This difference is due to the different underpinnings and goals of PCA and FA. FA comes from the psychometric tradition and often is used with purpose-built tests, for example, to analyze different questions on an examination or multiple questions designed to measure IQ or to assess some psychological construct. In an exam, the goal of all the questions is to provide an index of a student's overall understanding of the course, which cannot be directly observed. The variance shared across questions is considered a better indicator of overall understanding than the variance unique to any given test question, which may represent a poorly worded question, poor teaching of one specific concept, or lack of understanding of a specific concept. In PCA, the goal is to typically reproduce a higher dimensional space with as few dimensions as possible. Thus if one question on an exam did not overlap much with any other, rather than considering that potentially a poor indicator of overall performance, PCA would consider it another unique dimension needed. In general, PCA is used more in machine learning as it builds less on any specific theory or belief that a set of items should have some shared overlap and because with PCA it often is easier to include enough components that the original data can almost perfectly be recovered, but from a smaller set of dimensions.

What PCA does is look at our data and break out the parts that move our data left and right vs. the parts that move it up and down—in other words, into the principal components. If you have a background in linear algebra, standard PCA is the eigenvalue decomposition of the covariance (or if standardized, correlation) matrix. In any case, we return to our go-to look at two columns of interest. Again reference the original plot of these two.

The pca() function has arguments both to scale data, setting the variance to one (unit variance) if set to "uv" and to center the data at zero if set to TRUE. The classical type of PCA is based on eigenvalues and the singular value decomposition (SVD), and hence for traditional PCA, we tell the pca() function to use method = "svd".

To begin, we collect just the raw data for GNP.PCAP vs. ADO.TFRT into our working dataset, x. Computing PCA is quite simple, and we go ahead and immediately scale and center our data and estimate the traditional PCA using singular value decomposition. The summary() function shows us how much of the total variance can be explained by each principal component individually, and another row shows the cumulative variance explained. In this case with only two variables, 100% of the variance, $R^2 = 1$, can be explained by two principal components.

```
x <- d[,.( NYGN, SPAD)]
res <- pca(x, method="svd", center=TRUE, scale = "uv")

summary(res)

## svd calculated PCA
## Importance of component(s):
##                    PC1    PC2
## R2              0.7213 0.2787
## Cumulative R2 0.7213 1.0000
```

To try to visualize our results, we can create a plot showing how the original data have been transformed to the biplot, graphed in Figure 7-20. Can you spot the orthogonal rotation? Of course, more than just a rotation is at play. The PC1 variable is the maximum variance explainable by a single straight line. The remainder of the variance is projected to PC2. The biplot() function shows the vectors of the original data values in the new space.

```
biplot(res, main = "Biplot of PCA")
```

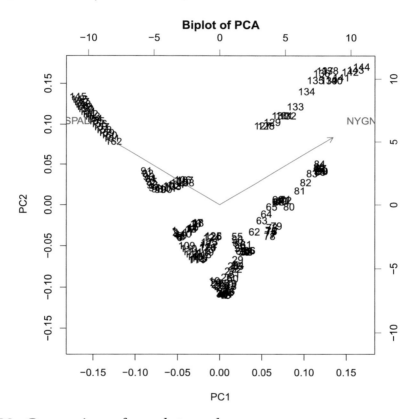

Figure 7-20. *Comparison of raw data and pca*

Next, we will examine what happens when we conduct a PCA using all the variables. We can make a scree plot to see how the accuracy improves as we add additional principal components, as a way of trying to decide how many components are needed to adequately represent the underlying dimensions in our data. We set an additional argument, nPcs, to the number of columns in our data, which is the maximum possible number of principal components we could need, although if the true dimensionality of our data is less, few components may adequately represent the data.

We make a plot of the proportion of raw information recovered from fewer dimensions, a sort of reverse scree plot, using the plot() function in Figure 7-21. What we see again looking at the scree plot is that there really are only four dimensions worth of unique data here, within even just one dimension able to capture more than half of all the variance across our 11 variables.

```
x <- d[, -c(1,2)]
res <- pca(x, method="svd", center=TRUE, scale = "uv",
          nPcs = ncol(x))

summary(res)

## svd calculated PCA
## Importance of component(s):
##                    PC1    PC2     PC3     PC4     PC5     PC6    PC7
## R2              0.7278 0.1614 0.06685 0.02554 0.01226 0.00447 0.0011
## Cumulative R2 0.7278 0.8892 0.95607 0.98161 0.99387 0.99834 0.9994
##                    PC8    PC9  PC10   PC11
## R2              0.00021 0.0002 1e-04 5e-05
## Cumulative R2 0.99965 0.9999 1e+00 1e+00

## reverse scree plot
ggplot() +
  geom_bar(aes(1:11, cumsum(res@R2)),
           stat = "identity") +
  scale_x_continuous("Principal Component", 1:11) +
  scale_y_continuous(expression(R^2), labels = percent) +
  ggtitle("Scree Plot") +
  coord_cartesian(xlim = c(.5, 11.5), ylim = c(.5, 1),
                  expand = FALSE)
```

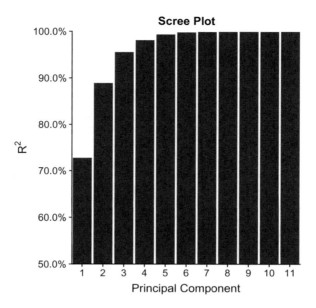

Figure 7-21. *Scree plot for traditional PCA on all features in the data*

We can also create biplots from this PCA, but can only plot two components at a time. The results for the first two components, which together explain most of the variance, are shown in Figure 7-22.

```
biplot(res, choices = c(1, 2))
```

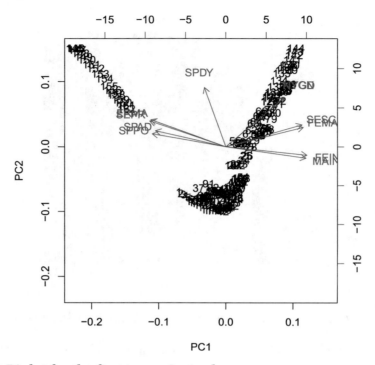

Figure 7-22. *Biplot for the first two principal components*

Incidentally, now that we have in the res variable our data, we can use that data (and as many columns of it as make sense such as the first four) instead of our original raw data. In particular, should we wish to perform supervised machine learning techniques, as is discussed in the next chapter, there may be advantages to using PCA data. For example, the results from our PCA indicate that we can with four components capture all but 2% of the variance in all 11 raw features.

The advantage of using principal components instead of raw features is that the principal components are orthogonal. The disadvantage is that, as each PCA component is comprised of potentially several of our raw variables, it becomes difficult to explain exactly what each feature means or represents anymore. However, if the predictive power of the model is improved or is the only goal, this may be a trade-off we are willing to make. In some cases, it also can speed up the model computation, as the later analysis only has to manage fewer features, reducing the memory and perhaps computation required for analysis.

We extract the principal component scores using the scores() function and look at the first few rows and then show using a correlation matrix that they indeed are linearly independent.

```
head(scores(res))
```

```
##           PC1  PC2  PC3   PC4  PC5    PC6   PC7    PC8     PC9     PC10
## [1,] -2.5 -1.2 0.74 -0.31 0.99 -0.075 -0.22 0.0046  0.0015 -0.00660
## [2,] -2.4 -1.2 0.72 -0.31 0.89 -0.057 -0.21 0.0075 -0.0035  0.00095
## [3,] -2.3 -1.3 0.70 -0.29 0.82 -0.047 -0.19 0.0121 -0.0046 -0.00521
## [4,] -2.2 -1.3 0.69 -0.29 0.75 -0.074 -0.18 0.0202 -0.0086 -0.01269
## [5,] -2.0 -1.3 0.66 -0.27 0.70 -0.095 -0.15 0.0290 -0.0149 -0.00243
## [6,] -1.9 -1.4 0.62 -0.25 0.64 -0.104 -0.13 0.0355 -0.0237 -0.00071
##           PC11
## [1,] -9.9e-03
## [2,]  4.7e-06
## [3,]  1.3e-02
## [4,]  1.1e-02
## [5,]  1.3e-02
## [6,]  1.3e-02
```

```
round(cor(scores(res)),2)
```

```
##      PC1 PC2 PC3 PC4 PC5 PC6 PC7 PC8 PC9 PC10 PC11
## PC1    1   0   0   0   0   0   0   0   0    0    0
## PC2    0   1   0   0   0   0   0   0   0    0    0
## PC3    0   0   1   0   0   0   0   0   0    0    0
## PC4    0   0   0   1   0   0   0   0   0    0    0
## PC5    0   0   0   0   1   0   0   0   0    0    0
## PC6    0   0   0   0   0   1   0   0   0    0    0
## PC7    0   0   0   0   0   0   1   0   0    0    0
## PC8    0   0   0   0   0   0   0   1   0    0    0
## PC9    0   0   0   0   0   0   0   0   1    0    0
## PC10   0   0   0   0   0   0   0   0   0    1    0
## PC11   0   0   0   0   0   0   0   0   0    0    1
```

Beyond the traditional PCA method of eigenvalues and singular value decomposition, the pca() function has several other methods available for a variety of other cases. One such approach is robustPca, which alters the algorithm to attempt to be more robust to extreme values or outliers. To see the effects of this, we will take our same data, pre-scale it using the prep() function, and then create two versions, one without outliers and one with outliers.

Next, we run four PCA models, traditional SVD PCA and the robust PCA on both the data without and with outliers. Finally, we plot the loadings, which are the values used to project the raw data onto the principal components, from the data without and with outliers using the regular and robust PCA methods. The results are in Figure 7-23. In that figure, you can clearly see how the loadings change dramatically in the traditional PCA, but in the robust PCA, the addition of outliers makes little difference on the loadings. In general, if you are concerned that your data may have extreme values or outliers, it is worth cleaning or removing them first, or at least comparing results from traditional and robust PCA methods, to ensure results are not too sensitive to the presence of a few extreme scores.

```
x <- d[, -c(1,2)]
x <- prep(x, center = TRUE, scale = "uv")

xout <- copy(x)
xout[1:5, "NYGD"] <- (-10)

res1 <- pca(x, method = "svd",
            center = FALSE, nPcs = 4)
res2 <- pca(xout, method = "svd",
            center = FALSE, nPcs = 4)

res1rob <- pca(x, method = "robustPca",
               center = FALSE, nPcs = 4)
res2rob <- pca(xout, method = "robustPca",
               center = FALSE, nPcs = 4)
plot_grid(
  ggplot() +
    geom_point(aes(
      x = as.numeric(loadings(res1)),
      y = as.numeric(loadings(res2)))) +
    xlab("Loadings, SVD, No Outliers") +
    ylab("Loadings, SVD, Outliers"),
  ggplot() +
    geom_point(aes(
      x = as.numeric(loadings(res1rob)),
      y = as.numeric(loadings(res2rob)))) +
    xlab("Loadings, Robust PCA, No Outliers") +
    ylab("Loadings, Robust PCA, Outliers"),
    ncol = 1)
```

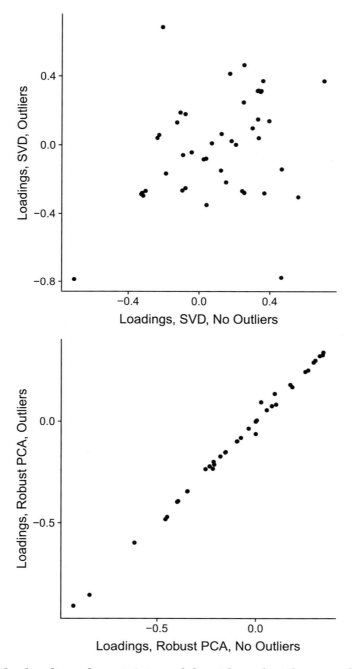

Figure 7-23. *Plot loadings from PCA models with and without outliers using traditional SVD PCA and robust PCA*

7.5 Non-linear Cluster Analysis

A frequent goal of PCA is to reduce the number of dimensions, thereby simplifying the model and improving predictive accuracy. One underlying assumption of PCA is the existence of sensible linear relationships in these fewer dimensions, in other words, that for n nominal dimensions, there exist some underlying orthogonal vectors to be projected upon without major loss of signal. If there is reason to suspect this is not the case, then non-linear methods may be in order.

While the first four lines of code are familiar from hierarchical clusters, the sammon() function from the MASS package takes those distances and attempts to map higher dimension (in our case 11 dimensions) down to k dimensions (in this example, we set k = 2 in order to graph). Our main objective in mentioning this is simply to observe that in the event of clearly non-linear relationships, PCA may be less effective than other methods that relax the linearity assumption.

```
x <- scale(d[, -c(1,2)])
row.names(x) <- paste(d$CntN, d$Year)
head(x)

##                NYGD  NYGN SEPR  ERMA SESC FEMA SPAD  SPDY  FEIN  MAIN
## ArbWr 1997 -0.81 -0.79 0.33 0.156 -1.5 -1.4 0.37 -0.66 -0.74 -0.62
## ArbWr 1998 -0.82 -0.79 0.31 0.164 -1.5 -1.3 0.34 -0.71 -0.70 -0.57
## ArbWr 1999 -0.81 -0.79 0.27 0.147 -1.4 -1.2 0.32 -0.75 -0.67 -0.53
## ArbWr 2000 -0.79 -0.79 0.22 0.113 -1.3 -1.2 0.30 -0.78 -0.63 -0.48
## ArbWr 2001 -0.80 -0.78 0.17 0.070 -1.3 -1.2 0.27 -0.81 -0.60 -0.44
## ArbWr 2002 -0.80 -0.78 0.13 0.033 -1.2 -1.1 0.24 -0.84 -0.57 -0.41
##                SPPO
## ArbWr 1997   1.6
## ArbWr 1998   1.4
## ArbWr 1999   1.3
## ArbWr 2000   1.2
## ArbWr 2001   1.1
## ArbWr 2002   1.0

sdist <- dist(x)

xSammon <- sammon(sdist, k = 2)
```

```
## Initial stress        : 0.04343
## stress after    7 iters: 0.03619

head(xSammon$points)

##             [,1] [,2]
## ArbWr 1997 -2.6 -1.2
## ArbWr 1998 -2.5 -1.2
## ArbWr 1999 -2.3 -1.3
## ArbWr 2000 -2.2 -1.3
## ArbWr 2001 -2.1 -1.4
## ArbWr 2002 -2.0 -1.4
```

The Sammon technique attempts to minimize a metric called `stress` that seeks to measure how effectively a higher dimensional object has been "crunched" into a lower dimensional space.

If the preceding code were adjusted to $k = 3$, we would expect stress to decrease. Figure 7-24 shows the result of using Sammon to reduce 11 dimensions to only 2.

```
plot(xSammon$points, type = "n")
text(xSammon$points, labels = row.names(x) )
```

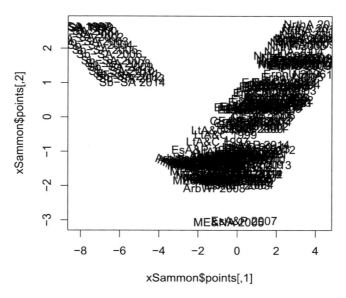

Figure 7-24. *Plot of Sammon points with text labels*

7.6 Summary

In this chapter we developed the idea of unsupervised machine learning. While the data subjected to such techniques is generally unlabelled, in this case we demonstrated with labelled data in order to gain some sense of what types of clusters or groupings we might expect. The frequently used techniques of kmeans and hclust were used with the standard Euclidean distance (thereby limiting us to numeric data). Nevertheless, these powerful techniques allow us to understand how many true groups there might be in a complex, multidimensional dataset. Finally, the concept of dimensional reduction was introduced, using principal component analysis as the most common method, while closing with a brief observation that the linearity assumption of PCA may be relaxed through more complex techniques. An overview of some of the functions used in this chapter and briefly what they do is in Table 7-2.

Table 7-2. *Listing of Key Functions Described in This Chapter and Summary of What They Do*

Function	What It Does
read_excel()	Takes a file path as a string and reads in an Excel file
as.data.table()	Converts a data.frame into a data.table object
str()	Shows the underlying structure of an object in R
summary()	Attempts a statistical summary of data or model objects
melt()	Melts wide data to long data
dcast()	Casts molten long data to a wide format
sort()	Sorts data in increasing order
unique()	Removes duplicates
plot()	Generic plotting of data and model results
set.seed()	Allows code to duplicate for pseudorandom algorithms
kmeans()	Runs the kmeans algorithm
scale()	Centers data by column at mean = 0 and adjusts to unit standard deviation
for()	Invokes the for loop operation over an iterator
cbind()	Binds data together by columns

(continued)

Table 7-2. (*continued*)

Function	What It Does
dist()	Creates a distance matrix that shows Euclidean distance between each element and every other element
hclust()	Creates a hierarchical cluster object
abline()	Draws a line on a plot object
row.names()	Allows access to the data.frame row name attribute
cutree()	Cuts a hierarchical tree object by either height or clusters
tail()	Shows the last six rows of data
copy()	Copies a data object rather than just assigning a new name by reference
as.phylo()	Creates a phylo object—for use with ape package graphing
cor()	Shows the correlation between items
pca()	Performs PCA reduction computations
scores()	Extracts principal component scores
loadings()	Extracts principal component loadings, used to project raw data into the principal component space
biplot()	Plots a PCA on PC1 and PC2 and shows original data vectors
sammon()	Non-linear dimension reduction algorithm; second formal is new dimensions
text()	Text labels on a plot object

CHAPTER 8

ML: Supervised

Machine learning (ML) and classification.

In the prior chapter, data was fed into an algorithm which then attempted to group common types of data. While we occasionally kept labels in order to gain a sense of how the algorithms performed, in real life, unsupervised machine learning is a form of exploratory analysis. Sometimes called pre-processing, it is often an initial phase that includes scaling and centering of data.

Often, the objective is not simply grouping, but the ability to use current data to predict the future. This frequent objective of machine learning is to train models with good predictive power. What is an acceptable level of predictive accuracy will vary from application to application. One of the authors works at a community college, where even small levels of predictive accuracy above chance for things such as a student's final letter grade may allow for more relevant, targeted interventions to occur (e.g., text messages about tutoring center hours or assignment to a mentor or coach). In that world, there is perhaps little risk with a false positive. Contrastingly, another author works with health outcome data, where false positives carry a higher level of likely risk to overall patient health.

For this chapter, new packages include `caret` [32] which contains useful functions for both preparing data and model development. The `kernlab` package is used for support vector machines [49]. In conjunction with `caret`, we will use `DALEX` [11] which will help explain models (and requires `spdep` [12]). In a similar fashion, `rattle` [119] allows additional visualizations. Finally, we also use the package `RSNNS` [10] for a multilayer perceptron model—a type of neural network.

A note about `caret` is it does not automatically install packages required for specific methods of analysis. Thus, as you step through the later analysis, expect to need to install some packages. To mitigate that, we include here the packages `ranger` [124], `e1071` [62], `gbm` [120], and `plyr` [110].

© Matt Wiley and Joshua F. Wiley 2019
M. Wiley and J. F. Wiley, *Advanced R Statistical Programming and Data Models*,
https://doi.org/10.1007/978-1-4842-2872-2_8

```
library(checkpoint)
checkpoint("2018-09-28", R.version = "3.5.1",
  project = book_directory,
  checkpointLocation = checkpoint_directory,
  scanForPackages = FALSE,
  scan.rnw.with.knitr = TRUE, use.knitr = TRUE)

library(ggplot2)
library(cowplot)
library(data.table)
library(readxl)
library(viridis)

library(RSNNS)
library(kernlab)
library(rpart)
library(rattle)
library(DALEX)
library(caret)
library(spdep)
library(ranger)
library(e1071)
library(gbm)
library(plyr)
set.seed(1234)

options(width = 70, digits = 2)
```

8.1 Data Preparation

For most applications of supervised learning, the data require some level of preparation.
We start this process using the same initial steps covered in the last chapter to read in
our sample dataset and convert it into the layout with which we are now familiar. Recall
the ' is the tick mark found on the tilde key above tab.

```
## Note: download Excel file  from publisher website first
dRaw <- read_excel("Gender_StatsData_worldbank.org_ccby40.xlsx")
dRaw <- as.data.table(dRaw) # convert data to data.table format.
```

```
dRaw[,'Indicator Name':= NULL]

## collapse columns into a super long dataset
## with Year as a new variable
data <- melt(dRaw, measure.vars = 3:20, variable.name = "Year", variable.
factor = FALSE)

## cast the data wide again
## this time with separate variables by indicator code
## keeping a country and time (Year) variable
data <- dcast(data, CountryName + Year ~ IndicatorCode)
rm(dRaw) #remove unneeded variable

#rename columns with shortened, unique names
x<-colnames(data)
x<-gsub("[[:punct:]]", "", x)
(y <- abbreviate(x, minlength = 4, method = "both.sides"))

##    CountryName            Year   NYGDPPCAPCD   NYGNPPCAPCD   SEPRMUNERFE
##          "CntN"          "Year"        "NYGD"        "NYGN"        "SEPR"
##    SEPRMUNERMA   SESCHLIFEFE   SESCHLIFEMA      SPADOTFRT   SPDYNCDRTIN
##          "ERMA"        "SESC"        "FEMA"        "SPAD"        "SPDY"
## SPDYNLEOOFEIN SPDYNLEOOMAIN     SPPOPDPND
##          "FEIN"        "MAIN"        "SPPO"

names(data) <- y

#shorten regional names to abbreviations.
data$CntN<-abbreviate(data$CntN, minlength = 5, method = "left.kept")
```

We again briefly describe what each column of the data represents in Table 8-1.

A note of warning: While many modern packages will perform all or some of the steps we seek to outline and visualize in this section, it helps to understand the value behind certain pre-processing techniques. The use of certain measures is dependent on both the model selected for use and the underlying data being fed into the model. For example, tree methods might work quite well with character or factor data, while a traditional linear regression requires numerical inputs. Thus, even in this chapter, it would not make sense to expect every model discussed to use our fixed dataset in quite the same way. Therefore, while we work our way through these measures, we shall preserve data as a holder of our working dataset from the last chapter. We will perform additional processing on copies thereof, and go back to data as seems reasonable.

Table 8-1. *Listing of Columns in Gender Data*

Variable (Feature)	Description
CountryName | CntN	The abbreviated name of the geographic region or country group
Year | Year	The year each data came from
SP.ADO.TFRT | SPAD	Adolescent fertility rate (births per 1,000 women ages 15–19)
SP.POP.DPND | SPPO	Age dependency ratio (percent of working-age population)
SE.PRM.UNER.FE | SEPR	Children out of school, primary, female
SE.PRM.UNER.MA | ERMA	Children out of school, primary, male
SP.DYN.CDRT.IN | SPDY	Death rate, crude (per 1,000 people)
SE.SCH.LIFE.FE | SESC	Expected years of schooling, female
SE.SCH.LIFE.MA | FEMA	Expected years of schooling, male
NY.GDP.PCAP.CD | NYGD	GDP per capita (current US dollars)
NY.GNP.PCAP.CD | NYGN	GNI per capita, Atlas method (current US dollars)
SP.DYN.LE00.FE.IN | FEIN	Life expectancy at birth, female (years)
SP.DYN.LE00.MA.IN | MAIN	Life expectancy at birth, male (years)

One Hot Encoding

We see there are nine unique categorical data types in the CntN (Regional Name) column. To convert nominal data to something applicable to regression or computing (in other words numeric), the basic technique is to create a set of new columns from the nominal categories which contain the value 0 if a row does not fit and a value of 1 if a row does fit. If we have nine values, we could create nine columns, one of each starting with ArbWr (ArabWorld) and ending with S-SA (Sub-Saharan Africa). In that case, there would be a single 1 in each row, the rest being left at 0. This is sometimes called one hot encoding. Alternatively, as we do have only nine options, we could encode only the first eight as described. If the first eight were all 0, this would still signal our model that a ninth option was the case. This method is sometimes called dummy coding. In practice, the two terms seem to be used almost interchangeably. Different models may have a preference for different layouts. As always, spend some time to understand the requirements of a particular model in terms of acceptable inputs.

```
d <- copy(data)
sort(unique(d$CntN))
```

```
## [1] "ArbWr" "CEatB" "Er&CA" "ErpnU" "EsA&P" "LtA&C" "ME&NA" "NrthA"
## [9] "Sb-SA"
```

We use the dummyVars function from caret to perform the one hot encoding for us. This function takes an input of a usual R formula, and here we ask it to operate on the entire dataset rather than only part of it. The function will ignore numeric data, but it will transform character and factor data. As we see from the str function, our year information is a character type. If left alone, the years would also be dummy coded. In this case, we go ahead and transform to numeric, hoping for a decent fit without. Admittedly, this is based at present on nothing more than a desire to keep total column counts low.

```
str(d)
```

```
## Classes 'data.table' and 'data.frame':      162 obs. of  13 variables:
##   $ CntN: chr  "ArbWr" "ArbWr" "ArbWr" "ArbWr" ...
##   $ Year: chr  "1997" "1998" "1999" "2000" ...
##   $ NYGD: num  2299 2170 2314 2589 2495 ...
##   $ NYGN: num  2310 2311 2288 2410 2496 ...
##   $ SEPR: num  6078141 5961001 5684714 5425963 5087547 ...
##   $ ERMA: num  4181176 4222039 4131775 3955257 3726838 ...
##   $ SESC: num  8.08 8.27 8.5 8.65 8.84 ...
##   $ FEMA: num  9.73 9.82 9.97 10.02 10.12 ...
##   $ SPAD: num  56.6 55.7 54.9 54.2 53.3 ...
##   $ SPDY: num  6.8 6.68 6.57 6.48 6.4 ...
##   $ FEIN: num  68.7 69 69.3 69.6 69.8 ...
```

```
##  $ MAIN: num   65 65.3 65.7 65.9 66.2 ...
##  $ SPPO: num   79.1 77.7 76.2 74.7 73.2 ...
##  - attr(*, ".internal.selfref")=<externalptr>

d[,Year:=as.numeric(Year)]
ddum <- dummyVars("~.", data = d)
d <- data.table(predict(ddum, newdata = d))
rm(ddum) #remove ddum as unneeded
str(d)

## Classes 'data.table' and 'data.frame':    162 obs. of  21 variables:
##  $ CntNArbWr: num   1 1 1 1 1 1 1 1 1 1 ...
##  $ CntNCEatB: num   0 0 0 0 0 0 0 0 0 0 ...
##  $ CntNEr&CA: num   0 0 0 0 0 0 0 0 0 0 ...
##  $ CntNErpnU: num   0 0 0 0 0 0 0 0 0 0 ...
##  $ CntNEsA&P: num   0 0 0 0 0 0 0 0 0 0 ...
##  $ CntNLtA&C: num   0 0 0 0 0 0 0 0 0 0 ...
##  $ CntNME&NA: num   0 0 0 0 0 0 0 0 0 0 ...
##  $ CntNNrthA: num   0 0 0 0 0 0 0 0 0 0 ...
##  $ CntNSb-SA: num   0 0 0 0 0 0 0 0 0 ...
##  $ Year     : num   1997 1998 1999 2000 2001 ...
##  $ NYGD     : num   2299 2170 2314 2589 2495 ...
##  $ NYGN     : num   2310 2311 2288 2410 2496 ...
##  $ SEPR     : num   6078141 5961001 5684714 5425963 5087547 ...
##  $ ERMA     : num   4181176 4222039 4131775 3955257 3726838 ...
##  $ SESC     : num   8.08 8.27 8.5 8.65 8.84 ...
##  $ FEMA     : num   9.73 9.82 9.97 10.02 10.12 ...
##  $ SPAD     : num   56.6 55.7 54.9 54.2 53.3 ...
##  $ SPDY     : num   6.8 6.68 6.57 6.48 6.4 ...
##  $ FEIN     : num   68.7 69 69.3 69.6 69.8 ...
##  $ MAIN     : num   65 65.3 65.7 65.9 66.2 ...
##  $ SPPO     : num   79.1 77.7 76.2 74.7 73.2 ...
##  - attr(*, ".internal.selfref")=<externalptr>
```

Scale and Center

The second stage of preparatory work is to scale and center our data for the reasons mentioned in the last chapter. In brief, those reasons are to prevent accidental overweighting of certain components that may have a longer range. We do not scale the first nine columns which are our dummy coded columns. We shall add those columns back in using cbind.

```
dScaled<-scale(d[,-c(1:9)])
dScaled<-as.data.table(dScaled)
d <- cbind(d[,c(1:9)], dScaled)
rm(dScaled) #remove d2 as unneeded
str(d)
```

```
## Classes 'data.table' and 'data.frame':   162 obs. of  21 variables:
##  $ CntNArbWr: num  1 1 1 1 1 1 1 1 1 1 ...
##  $ CntNCEatB: num  0 0 0 0 0 0 0 0 0 0 ...
##  $ CntNEr&CA: num  0 0 0 0 0 0 0 0 0 0 ...
##  $ CntNErpnU: num  0 0 0 0 0 0 0 0 0 0 ...
##  $ CntNEsA&P: num  0 0 0 0 0 0 0 0 0 0 ...
##  $ CntNLtA&C: num  0 0 0 0 0 0 0 0 0 0 ...
##  $ CntNME&NA: num  0 0 0 0 0 0 0 0 0 0 ...
##  $ CntNNrthA: num  0 0 0 0 0 0 0 0 0 0 ...
##  $ CntNSb-SA: num  0 0 0 0 0 0 0 0 0 0 ...
##  $ Year     : num  -1.633 -1.441 -1.249 -1.057 -0.865 ...
##  $ NYGD     : num  -0.813 -0.822 -0.812 -0.792 -0.799 ...
##  $ NYGN     : num  -0.793 -0.793 -0.795 -0.786 -0.78 ...
##  $ SEPR     : num  0.327 0.309 0.266 0.225 0.172 ...
##  $ ERMA     : num  0.1565 0.1643 0.1471 0.1134 0.0699 ...
##  $ SESC     : num  -1.52 -1.46 -1.39 -1.34 -1.28 ...
##  $ FEMA     : num  -1.35 -1.31 -1.25 -1.22 -1.18 ...
##  $ SPAD     : num  0.37 0.344 0.319 0.295 0.269 ...
##  $ SPDY     : num  -0.66 -0.708 -0.749 -0.784 -0.815 ...
##  $ FEIN     : num  -0.744 -0.704 -0.668 -0.635 -0.603 ...
##  $ MAIN     : num  -0.622 -0.573 -0.527 -0.484 -0.444 ...
##  $ SPPO     : num  1.55 1.45 1.34 1.24 1.13 ...
##  - attr(*, ".internal.selfref")=<externalptr>
```

Transformations

This scaling process does not remove outliers. It behooves us to check the data and see if there is significant non-normal behaviour as some machine learning techniques perform better on normal data. In those cases it can be helpful to perform other transformations (e.g., logarithmic scaling). As we see in Figure 8-1, there are some outliers in the data.

```
boxplot(d[,-c(1:9)], las = 2)
```

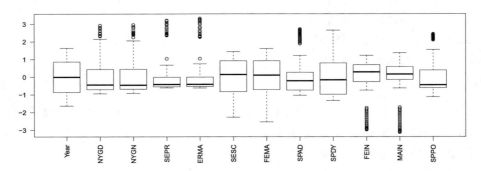

Figure 8-1. *Looking for significant data misshapes*

Looking more closely at the male children out of primary school in Figure 8-2 shows evidence of non-normal behaviour. The histogram is not normally distributed in a traditional bell-shape and the Q-Q plot poorly resembles a straight y = x line.

```
par(mfrow = c(1,2))
hist(d$ERMA, 100)
qqnorm(d$ERMA)

par(mfrow = c(1,1))
```

Indeed, if we conduct a hypothesis test using Shapiro-Wilk normality test, we see the p-value is significant, and thus we reject the null hypothesis of normal data.

```
shapiro.test(d$ERMA)

##
##      Shapiro-Wilk normality test
##
## data:  d$ERMA
## W = 0.6, p-value <2e-16
```

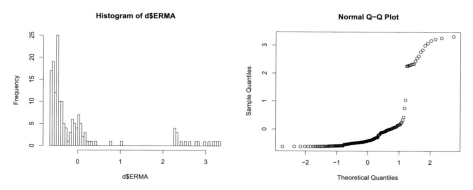

Figure 8-2. *Looking for significant data misshapes*

Based on the histogram and Q-Q from Figure 8-2, it may make sense to transform the data, perhaps using a logarithmic transformation. First, we observe that the range of the scaled and centered data naturally includes zero, which would be a challenge for a logarithmic transformation. The original data show this is not an issue for the underlying information. However, Shapiro-Wilk normality test shows that even the log transformation is still not in favor of matching a normal distribution.

```
range(d$ERMA)

## [1] -0.62   3.32

range(data$ERMA)

## [1] 1.1e+05 2.1e+07

shapiro.test( log(data$ERMA) )

##
##        Shapiro-Wilk normality test
##
## data:  log(data$ERMA)
## W = 1, p-value = 3e-04
```

Now, visually, we do see the log transformation does space out our data somewhat as seen in Figure 8-3.

```
par(mfrow = c(1,2))
hist(data$ERMA, 100)
hist( log (data$ERMA) , 100)

par(mfrow = c(1,1))
```

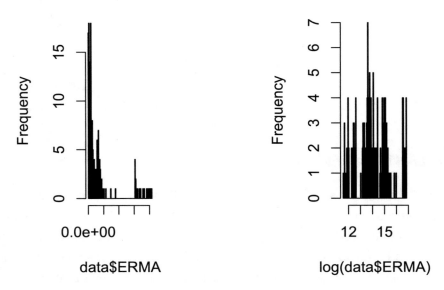

Figure 8-3. *Looking for significant data misshapes*

Is this spacing enough to warrant a transformation? One way to determine is to compare correlations between the original, raw data vs. the transformed data to the desired response variable. Our response/output variable will be SPAD, adolescent fertility. We create a copy of the dataset that includes both the ERMA in both logged and raw formats and observe the correlation. As the resulting correlation in the first column shows the correlation is stronger in the non-transformed column, our final choice will be to not perform a transformation.

```
d2 <- copy(data[,.(SPAD, ERMA)])
d2[, Log.ERMA := log(ERMA)]
cor(d2)

##           SPAD ERMA Log.ERMA
## SPAD      1.00 0.83     0.72
## ERMA      0.83 1.00     0.81
## Log.ERMA  0.72 0.81     1.00

rm(d2) #no longer needed
```

Technically, each variable under consideration ought to be inspected for normality (again, this is only because some of our models will perform better with normal predictors). Again, generally, when applying a specific model, it is important to understand expected inputs for optimal model performance. First, using `lapply`, we observe all the p-values are significant. Next, we use `sapply` to log transform our data, and use `colnames` to append a suffix. Finally, we compare and contrast the original data correlation to SPAD vs. the log transformed correlation. In some cases, the correlation is slightly better. In others, it is not.

```
lapply(data[,-c(1:2)], shapiro.test)

## $NYGD
##
##      Shapiro-Wilk normality test
##
## data:  X[[i]]
## W = 0.8, p-value = 3e-13
##
##
## $NYGN
##
##      Shapiro-Wilk normality test
##
## data:  X[[i]]
## W = 0.8, p-value = 1e-13
##
##
## $SEPR
##
##      Shapiro-Wilk normality test
##
## data:  X[[i]]
## W = 0.6, p-value <2e-16
##
##
## $ERMA
```

```
##
##     Shapiro-Wilk normality test
##
## data:  X[[i]]
## W = 0.6, p-value <2e-16
##
##
## $SESC
##
##     Shapiro-Wilk normality test
##
## data:  X[[i]]
## W = 0.9, p-value = 7e-06
##
##
## $FEMA
##
##     Shapiro-Wilk normality test
##
## data:  X[[i]]
## W = 1, p-value = 5e-05
##
##
## $SPAD
##
##     Shapiro-Wilk normality test
##
## data:  X[[i]]
## W = 0.8, p-value = 5e-13
##
##
## $SPDY
##
##     Shapiro-Wilk normality test
##
```

```
## data:  X[[i]]
## W = 0.9, p-value = 6e-08
##
##
## $FEIN
##
##      Shapiro-Wilk normality test
##
## data:  X[[i]]
## W = 0.8, p-value = 1e-13
##
##
## $MAIN
##
##      Shapiro-Wilk normality test
##
## data:  X[[i]]
## W = 0.8, p-value = 9e-14
##
##
## $SPPO
##
##      Shapiro-Wilk normality test
##
## data:  X[[i]]
## W = 0.8, p-value = 6e-13

dlog <- copy(data)
dlog <- sapply(dlog[,-c(1:2)], log)
dlog<-as.data.table(dlog)
colnames(dlog) <- paste(colnames(dlog), "LOG", sep = ".")

dlog<-cbind(data, dlog)
View(cor(dlog[,-c(1:2)]))
rm(dlog) #remove as we will not use.
```

For now, we proceed without log transformation in all cases. Depending on the analysis, depending on the model precision required, and even depending on the specific model chosen, we reserve the right to note which variables have a stronger correlation after transformation. Often, various levels of pre-processing become part of the model tuning process.

Train vs. Validation Data

We reach the point in our pre-processing where it comes time to discuss train vs. test data. Recall from the introduction chapter that the choice of train dataset is rather important. For any single model, it makes sense to choose a training set and a validation set (often an 80/20 split). This allows model creation on the training set and then some sense of "real" world accuracy on the validation set (ergo data the model has not seen). However, in the chapter, we intend to discuss several possible models. Should we choose one of those models, we once again (because we use the validation data to select the "best" model) are risking some level of overfit. Thus, we would need to have a final bit of test data held out in reserve from training and validation to use as our final "real" world estimate prior to our model being used in real life. This gets complicated, because a 60/20/20 split now leaves us only just more than half the data for training.

Of course, cross-validation is one way to avoid too little data left for training. We could perform a train/test split, 80/20, use cross-validation on our training data, and save ourselves the need for a formal validation set. This gets computationally expensive, especially on multiple models (or even multiple iterations of one model). Recall that, at the end, once a specific model is selected, the proper last step prior to use in the real world is indeed to retrain the model on all data—no more splits. Depending on the computational intensity of the model itself, this can come at quite a cost.

Due to our dataset having only 162 observations, and for simplicity, we shall proceed with only a standard 80/20 split. Our goal will be to use cross-validation in our models, so that the reserved test set may be used at the end.

Using set.seed(1234) to allow reproducibility, we introduce a new function to perform the split from the caret package createDataPartition. The first formal allows the function to understand how our data are stratified, and the second tells the ratio (in this case 80/20) of the split. Once the index is populated, we use that to separate our train and validation or test data. After splitting the data, we are ready for the final stage of pre-processing.

```
set.seed(1234)
index <- createDataPartition(data$CntN, p = 0.8, list = FALSE)
trainData <- data[index, ]
validationData <- data[-index, ]
```

We should note scaling and centering should technically be done separately for train vs. test vs. validation data. Indeed, the scale and center formula must be developed on train; otherwise there may be information leak between those sets. However, as we have been resetting our data, and as we intend to scale again in the upcoming section anyway, we hope our readers forgive us our ordering.

Principal Component Analysis

We saw principal component analysis in the last chapter. Advantages of dimensional reduction include reduced processing time, reduced overfit, and model simplification. Installing the pcaMethods package a second time is not required if you have already installed it in the unsupervised learning chapter. Should you need to install, remove the pound signs that comment-out the code, and run. In either case, the third line of code will for sure need to be run.

```
#source("https://bioconductor.org/biocLite.R")
#biocLite("pcaMethods")
library(pcaMethods)

## Loading required package: Biobase

## Loading required package: BiocGenerics

##
## Attaching package: 'BiocGenerics'

## The following objects are masked from 'package:Matrix':
##
##     colMeans, colSums, rowMeans, rowSums, which

## The following objects are masked from 'package:parallel':
##
##     clusterApply, clusterApplyLB, clusterCall, clusterEvalQ,
##     clusterExport, clusterMap, parApply, parCapply, parLapply,
##     parLapplyLB, parRapply, parSapply, parSapplyLB
```

```
## The following objects are masked from 'package:stats':
##
##      IQR, mad, sd, var, xtabs
## The following objects are masked from 'package:base':
##
##      anyDuplicated, append, as.data.frame, basename, cbind,
##      colMeans, colnames, colSums, dirname, do.call, duplicated,
##      eval, evalq, Filter, Find, get, grep, grepl, intersect,
##      is.unsorted, lapply, lengths, Map, mapply, match, mget,
##      order, paste, pmax, pmax.int, pmin, pmin.int, Position,
##      rank, rbind, Reduce, rowMeans, rownames, rowSums, sapply,
##      setdiff, sort, table, tapply, union, unique, unsplit,
##      which, which.max, which.min

## Welcome to Bioconductor
##
##      Vignettes contain introductory material; view with
##      'browseVignettes()'. To cite Bioconductor, see
##      'citation("Biobase")', and for packages
##      'citation("pkgname")'.

##
## Attaching package: 'pcaMethods'

## The following object is masked from 'package:stats':
##
##      loadings
```

Having already split our data, we prepare the data for PCA. Here again our choices are informed in part by our data and in part by our eventual model. For example, various types of forests and tree-based methods may not have issues with character or factor data. On the other hand, various types of regression will need the dummy coding mentioned earlier. PCA is not an especially good fit for data that is dummy coded. However, if a great deal of the data happens to be categorical, it may make sense to dummy code all categorical predictors first and then attempt various styles of dimension reduction, of which PCA (aka `svd`) is only one method.

On the other hand, our data have only one bit of fully categorical data–CntN. If our dataset were to have a handful of dummy coded columns, and then we reduced the other 10 variables down to 3 or 4, we'd still have gotten a decent reduction in dimension.

In general, PCA is the final pre-processing step applied to data. Certainly the numerical data ought to be scaled and centered prior to PCA. As for whether or not the categorical data ought also to go through PCA? For brevity we shall show how to exclude the categorical data manually. Should your data require a different or more nuanced technique, we hope we show enough of the process so you will adapt this code to your purpose.

Firstly, keeping in mind the known structure of our data, we verify that we have selected the correct rows of our training set with the str function.

The stats library comes by default with base R and prcomp is a function in that library. It performs the dimension calculation on our training data. Keep in mind that any model fit on our PCA training data would need to be given new data or validation data that had the same PCA process performed. To this end, after viewing the results of PCA via summary(pc), we use the predict function. The first formal is our PCA analysis, in this case stored in pc. The second is the new data, in this case our validationData. Of course, for a model being used in actual production, the new data would be literal new data we wished to then put into the model.

However, we know from the prior chapter there is more that may be done than "basic" PCA. Thus, we also show the steps required for using the pca function from the pcaMethods library. The specifics of that function are explained in the unsupervised machine learning chapter. For now we note that the results of that process work perfectly with predict, and we show a printout of just the first dimension from both methods to show that they are indeed the same.

The strength of course of pca is the ease of method conversion to any method listed by listPcaMethods() by the expediency of the method = " " formal.

```
#confirm structure
str(trainData[,c(3:8,10:13)])
```

```
## Classes 'data.table' and 'data.frame':    135 obs. of  10 variables:
##  $ NYGD: num  2299 2170 2314 2589 2495 ...
##  $ NYGN: num  2310 2311 2288 2410 2496 ...
##  $ SEPR: num  6078141 5961001 5684714 5425963 5087547 ...
##  $ ERMA: num  4181176 4222039 4131775 3955257 3726838 ...
```

```
## $ SESC: num   8.08 8.27 8.5 8.65 8.84 ...
## $ FEMA: num   9.73 9.82 9.97 10.02 10.12 ...
## $ SPDY: num   6.8 6.68 6.57 6.48 6.4 ...
## $ FEIN: num   68.7 69 69.3 69.6 69.8 ...
## $ MAIN: num   65 65.3 65.7 65.9 66.2 ...
## $ SPPO: num   79.1 77.7 76.2 74.7 73.2 ...
##  - attr(*, ".internal.selfref")=<externalptr>
```

```
#base R / traditional method
pc <- prcomp(trainData[,c(3:8,10:13)], center = TRUE, scale. = TRUE)
summary(pc)
```

```
## Importance of components:
##                          PC1    PC2    PC3    PC4     PC5      PC6
## Standard deviation      2.703  1.321 0.8175 0.4029 0.27986 0.16986
## Proportion of Variance 0.731  0.175 0.0668 0.0162 0.00783 0.00289
## Cumulative Proportion  0.731  0.905 0.9719 0.9881 0.99597 0.99886
##                          PC7     PC8    PC9    PC10
## Standard deviation      0.08417 0.04726 0.03535 0.02894
## Proportion of Variance 0.00071 0.00022 0.00012 0.00008
## Cumulative Proportion  0.99957 0.99979 0.99992 1.00000
```

```
pcValidationData1 <- predict(pc, newdata = validationData[,c(3:8,10:13)])
```

```
#scalable method using PcaMethods
pc<-pca(trainData[,c(1:8,10:13)], method = "svd",nPcs = 4, scale = "uv",
center = TRUE)
pc
```

```
## svd calculated PCA
## Importance of component(s):
##                   PC1    PC2     PC3     PC4
## R2              0.7306 0.1745 0.06683 0.01623
## Cumulative R2 0.7306 0.9051 0.97191 0.98814
## 10    Variables
## 135   Samples
## 0     NAs ( 0 %)
## 4     Calculated component(s)
```

```
## Data was mean centered before running PCA
## Data was scaled before running PCA
## Scores structure:
## [1] 135    4
## Loadings structure:
## [1] 10   4
```

```
summary(pc)
```

```
## svd calculated PCA
## Importance of component(s):
##                    PC1     PC2      PC3      PC4
## R2              0.7306 0.1745 0.06683 0.01623
## Cumulative R2 0.7306 0.9051 0.97191 0.98814
```

```
pcValidationData2 <- predict(pc, newdata = validationData[,c(3:8,10:13)])
```

```
#demonstration of how to access transformed validation data
pcValidationData1[,1]
```

```
##  [1] -1.11 -0.87 -0.77  0.96  1.08  1.67 -1.52 -1.17 -0.82  0.37  1.06
## [12]  1.67  1.94  2.67  3.36  0.76  0.86  1.00 -1.14 -0.49  0.40  2.66
## [23]  3.72  3.80 -7.52 -5.80 -4.95
```

```
pcValidationData2$scores[,1]
```

```
##  [1] -1.11 -0.87 -0.77  0.96  1.08  1.67 -1.52 -1.17 -0.82  0.37  1.06
## [12]  1.67  1.94  2.67  3.36  0.76  0.86  1.00 -1.14 -0.49  0.40  2.66
## [23]  3.72  3.80 -7.52 -5.80 -4.95
```

Having transformed our data, it would now be simple using cbind to fold in the one hot encoded CntN as well as our response variable SPAD.

As we conclude this section on pre-processing data, it is important to keep a few key principles in mind.

If a single model is to be used, then a train and validation split (often 80/20) is in order, unless cross-validation or bootstrapping is to be used. Keep in mind the latter two can be computationally expensive. If one is selecting from more than one model, then a train set will be needed to train the various models, a validation set will be needed to estimate model performance on unseen data, and finally a test set will be needed to estimate the final chosen model's performance on real-world data (otherwise we essentially have a second stage overfit at the validation step). Cross-validation can remove the need for a validation set, yet is even more computationally expensive when using multiple models. On the other hand, it is costly to use perhaps 70/20/10 for train, validation, and test. Whatever method one selects, it ought to be done first, to prevent information leak between sets.

Should your data have missing values, working through the chapter devoted to missing data first could provide some gains. Imputation normally is one of the first steps after the data is in the correct layout.

One hot or dummy coding for categorical data is a choice. Some models require only numerical data, others can deal with categorical data in a factor format. Just as each type of model may not be optimal for all types of data, certain types of data are not optimal for certain models. Thus, consider both data and model! For some models, keeping data categorical, yet in factor format, may give the model the information needed to both use the data effectively while not imposing an arbitrary linear effect on the data. Contrastingly, other models may not cope with factors, and thus there might be no choice except to dummy code.

In general, numerical data should be scaled and centered. Indeed, it may also be wise to remove columns with zero or near-zero variance. Columns that have near-zero variance may be retained and subjected to PCA later on in the pre-processing.

Along with basic centering and scaling, it can help to transform data to get a decent spread of data. While we only discussed logarithmic transformations, there are many possible transformations possible that may be of value for specific types and features of data and specific models. Again, if the first pass of a particular model is not accurate enough, more advanced mathematics may be in order.

8.2 Supervised Learning Models

Now that we understand some ways to pre-process data, our next task is to meet some models. Remember, supervised learning simply means we have a specific response variable in mind. For various types of response or predictor variables, different models may be more suited to better analyzing specific types of data. For example, if we have more categorical variables, then we may prefer classification-based supervised learning. On the other hand, if our data is mostly numeric, we may find more value in regression methods. Some models live in both worlds, or can be adapted to doing so. And, as we already saw in pre-processing, data itself can be mutated, for example, from categorical to numeric via dummy coding. Of course, it is a poor method that only works one way—and histogram bins are perhaps a great example of ways continuous, numeric data might be recoded to be categorical.

Something to keep in mind, as you proceed through these models, is what level of accuracy is required for your particular application. It may be reasonably achievable to find a model delivering better-than-chance results. Is simply better than chance enough? Is there time (both human and machine) to test more models? Most models have tuning parameters, additional inputs that allow more cross-validation or penalize additional model complexity. How much effort spent tuning is realistic?

In the following sections, we take a balanced approach. We prepare our data so it optimally interacts with a particular model, we assume a reasonable level of hardware and choose to use a sensible amount of compute time, and we discuss various tuning parameter options without necessarily exhausting those options. For any particular application of data, it may well make sense to spend more time on data collection, pre-processing, and final model tuning than it may on the model selection step. With that thought considered, and then stored safely in the back of our minds, we proceed.

Support Vector Machines

Support vector machines (SVM) are conceptually somewhat similar to k-nearest neighbors in that the idea is to find groupings of our data. This is done by finding boundary curves determined via identification of key data points—support vectors. The support vectors are those with some maximal separation (under some metric), and between these a boundary curve is drawn. As the boundary demarcation can be linear, polynomial, or more exotic, there are several ways to perform the basic SVM algorithm. The determination of the type of demarcation curve is called the kernel of the SVM. Naturally, more complex boundary kernels tend to be more computationally difficult. The caret package, along with some other packages, natively supports several variations for multiple styles of kernels including linear, exponential, polynomial, and radial. We shall restrict our discussion here to linear and polynomial.

Let us start with a simple dataset—not our full formal set—to understand visually how SVMs work. We use adolescent fertility rate and GNP per capita as the predictors, and our categorical response is going to be country name. We recycle our well-seen graph Figure 8-4 and note that for any single country grouping, there usually is a clear-cut linear separation possible. Admittedly, we do not expect miracles from the lower-left area/corner.

```
svmDataTrain <- trainData[,.(SPAD, NYGN)]
svmDataValidate <- validationData[,.(SPAD, NYGN)]

p1 <- ggplot(data = svmDataTrain,
             aes(x = NYGN, y = SPAD))
  ## data poins colored by country
  p1 + geom_point(aes(colour = trainData$CntN)) +
    scale_colour_viridis(discrete = TRUE)
```

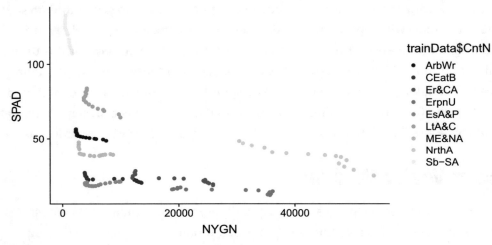

Figure 8-4. *Per capita GNP vs. adolescent fertility rate*

The caret package is powered by a function aptly named train. Take a moment to look at the following code layout.

The first formal is for the predictors, and the second formal is for the response. As of this writing, there are over 200 methods possible for the third formal. In this case, we are using a linear support vector machine, thus svmLinear. Because we intend to graph our results on the raw data, we choose to keep preProcess empty for now. In the future, it will take on such values as center, scale, and pca as well as various options for imputation. The formal argument metric = "Accuracy" signals we wish our summary output to let us know how accurate the trained regions of CountryName are compared with the actual countries. Again, think of svm as separating the data into various regions, using a linear boundary. We'll wish to know how often that boundary was drawn well. Lastly, the trainControl function is used to control how the model is trained. In this case, we will use cross-validation `cv` with `five` folds. There is actually a great deal of data contained in the `svm` variable. The printout shown is only a small summary. It does remind us there was no pre-processing, and as requested, tells us the estimated accuracy of our cross-validation. Keep in mind this accuracy is based on the cross-validation processes for each of the folds.

```
set.seed(12345)

  svm <- train(x = svmDataTrain,
            y = trainData$CntN,
            method = "svmLinear",
            preProcess = NULL,
            metric = "Accuracy",
            trControl = trainControl(method = "cv",
                                number = 5,
                                seeds = c(123, 234, 345, 456, 567, 678)
                                )
            )
svm

## Support Vector Machines with Linear Kernel
##
## 135 samples
##    2 predictor
##    9 classes: 'ArbWr', 'CEatB', 'Er&CA', 'ErpnU', 'EsA&P', 'LtA&C',
##               'ME&NA', 'NrthA', 'Sb-SA'
##
## No pre-processing
## Resampling: Cross-Validated (5 fold)
## Summary of sample sizes: 108, 108, 108, 108, 108
## Resampling results:
##
##    Accuracy  Kappa
##    0.82      0.8
##
## Tuning parameter 'C' was held constant at a value of 1
```

Next, we dig a little deeper into our model's accuracy. As we did in fact use cross-validation, that is considered a good estimate. First, we use the predict function, with two inputs. One is the new svm which contains our model. The other is the training data we used to train that model. Unlike with the cross-validation folds that only used 108 elements at a time, we use all 135 values. Next, we find when these predicted country names are the same as the correct country names found in our trainData variable. This creates Boolean values of TRUE and FALSE which equate to 1 and 0, respectively. Thus, the mean of those will give us the ratio of how often true occurs. Notice that, despite the estimated accuracy on unseen data of 0.84, on our training data we in fact see 0.85. This is part of what we mean by overfitting.

```
#predict the country name on our training data using our new model
predictOnTrain <- predict(svm, newdata = svmDataTrain)

mean( predictOnTrain == trainData$CntN)

## [1] 0.82
```

Now, it is not correct yet to use the validation data we held out. Technically, we must save that to the very end, not use it to inform our model selection at all, and then use it only to perform the final validation step after we've seen all the models. However, in light of showing what this process looks like in this first example sketch, it makes some pedagogical sense to show the last step. We perform the same calculation as with our training data (and indeed this prediction process is what we would do on genuinely new, real-world data). While the cross-validation was more conservative than the actual training data might suggest, it was not as conservative as it might have been.

Again, the 0.78 on fully unseen data by our model is an example of what is meant by overfitting of a model. The training data has biased our model in some way, and it does not behave so well with fully unseen validation data.

```
predictOnTest <- predict(svm, newdata = svmDataValidate)
mean(predictOnTest == validationData$CntN)

## [1] 0.81
```

Using the same process we used in the prior chapter, we plot the unseen validation data and compare the results of the correct answer with the prediction in Figure 8-5. Exactly where a linear separation might be expected to be difficult for our model to understand is indeed where the inaccuracies show up.

```
p1 <- ggplot(data = validationData,
             aes(x = NYGN, y = SPAD))

plot_grid(
  ## data poins colored by country
  p1 + geom_point(aes(colour = validationData$CntN, size = validation
  Data$CntN)) +
    scale_colour_viridis(discrete = TRUE),

  ## data poins colored by predicted country
  p1 + geom_point(aes(colour = predictOnTest, size = predictOnTest)) +
    scale_colour_viridis(discrete = TRUE),
ncol = 1
)

## Warning: Using size for a discrete variable is not advised.
## Warning: Using size for a discrete variable is not advised.
```

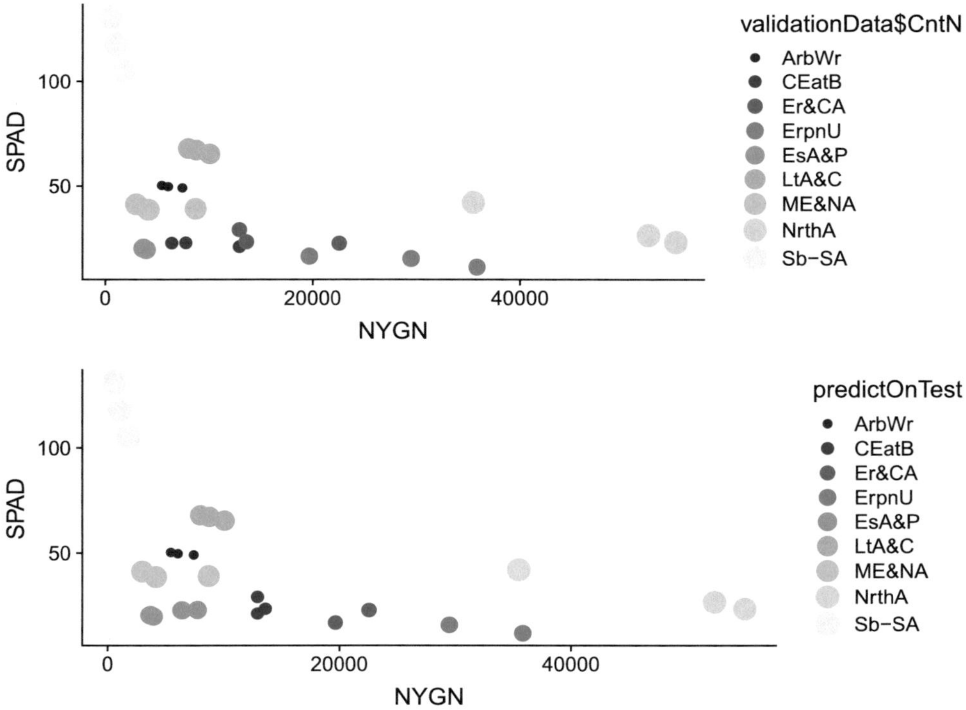

Figure 8-5. *Train vs. test predictions*

It is expected the preceding code generates two warnings—namely, that size is not recommended for discrete variables such as region names. This is a reasonable warning, and in real life we would never use such a technique (one great reason is humans tend to interpret size as important in and of itself, which in this case has no such meaning). We use it here only to help distinguish categories.

Now that we see and understand both the overall layout for `caret` model building, and how one might use validation data at the very end to understand which model to select, we put our validation data safely away, and will not use it again until the end, to see how our final chosen model might be expected to perform. We take a moment to also do some cleanup of our environment.

```
rm(p1)
rm(svm)
rm(svmDataTrain)
rm(svmDataValidate)
rm(pcValidationData1)
rm(pcValidationData2)
rm(predictOnTest)
rm(predictOnTrain)
rm(pc)
rm(d)
```

We are now ready to perform our SVM model on our entire `trainData` set. What we have left is one copy of our original `data` untouched, the `index` we used to split that data into training and validation sets, and the actual training and validation sets themselves.

We again build our model train data less our categorical country name column. For now, we recode the character year data as numeric and set our seeds for reproducibility. Other than including more variable columns, the only other difference is setting `preProcess = c("scale", "center", "pca")`. This defaults to a PCA `thresh = 0.95` which can be adjusted if desired by following the instructions in `?preProcess`. In this case, we stay with the default values. We see there is an improvement in estimated model accuracy.

```
# set up training & validation data
svmDataTrain <- trainData[,-1]
svmDataTrain[,Year:=as.numeric(Year)]
```

```
svmDataValidation <- validationData[,-1]
svmDataValidation[,Year:=as.numeric(Year)]
#run linear SVM on the full data set
set.seed(12345)
svmLinear <- train(x = svmDataTrain,
            y = trainData$CntN,
            method = "svmLinear",
            preProcess = c("scale", "center", "pca"),
            metric = "Accuracy",
            trControl = trainControl(method = "cv",
                               number = 5,
                               seeds = c(123, 234, 345, 456, 567, 678)
                               )
            )
svmLinear

## Support Vector Machines with Linear Kernel
##
## 135 samples
##   12 predictor
##    9 classes: 'ArbWr', 'CEatB', 'Er&CA', 'ErpnU', 'EsA&P', 'LtA&C',
##               'ME&NA', 'NrthA', 'Sb-SA'
##
## Pre-processing: scaled (12), centered (12), principal
##   component signal extraction (12)
## Resampling: Cross-Validated (5 fold)
## Summary of sample sizes: 108, 108, 108, 108, 108
## Resampling results:
##
##   Accuracy  Kappa
##   0.99      0.98
##
## Tuning parameter 'C' was held constant at a value of 1
```

It is important at this point to describe how linear SVM works. The algorithm is designed for binary classification. In this case, with multiple outcomes, it cycles through the data in a one-vs.-the-rest fashion. So this linear model is truly drawing a straight line to isolate one group at a time from the rest. The goal is to draw a line between the data such that the line is as far away from points on either side of the divide as possible. This helps a model predict where future data ought to go robustly. Via the tuning parameter C, we control how wide a margin is required (which if the data are too close together may make the line choose a different path). Higher values of C result in a higher training accuracy and a tighter margin—at the risk of overfitting.

When we change from linear to polynomial, it allows the line to become a polynomial curve. Additionally, this algorithm, by default, cycles through several values of C. Notice (after you run the following code) that, in general, for each set of fixed degree and scale, higher values of C produce a higher level of accuracy. The cost is that the margin about our curve shrinks as C increases, which translates to increased chance our model may be overfitting the data.

Next we make just one change in our model, from a linear model to a polynomial model. In caret this is readily accomplished by changing to method = "svmPoly". While the change in code is minor, the change in computational effort is less so. Notice svmPoly is a fully different model than the linear version. While the algorithm and mathematics are beyond the scope of this text, the process differs even at the linear degree level. While this does not currently appear to be as good as our linear model, if true, accuracy of 0.88 still may well be quite good.

```
#run polynomial SVM on the full data set
set.seed(12345)
svmPoly <- train(x = svmDataTrain,
          y = trainData$CntN,
          method = "svmPoly",
          preProcess = c("scale", "center", "pca"),
          metric = "Accuracy",
          trControl = trainControl(method = "cv",
                                   number = 5
                                   )
          )

svmPoly
```

```
## Support Vector Machines with Polynomial Kernel
##
## 135 samples
##  12 predictor
##   9 classes: 'ArbWr', 'CEatB', 'Er&CA', 'ErpnU', 'EsA&P', 'LtA&C',
##              'ME&NA', 'NrthA', 'Sb-SA'
##
## Pre-processing: scaled (12), centered (12), principal
##  component signal extraction (12)
## Resampling: Cross-Validated (5 fold)
## Summary of sample sizes: 108, 108, 108, 108, 108
## Resampling results across tuning parameters:
##
##   degree  scale  C     Accuracy  Kappa
##   1       0.001  0.25  0.76      0.73
##   1       0.001  0.50  0.76      0.73
##   1       0.001  1.00  0.76      0.73
##   1       0.010  0.25  0.76      0.73
##   1       0.010  0.50  0.76      0.73
##   1       0.010  1.00  0.76      0.73
##   1       0.100  0.25  0.76      0.73
##   1       0.100  0.50  0.78      0.75
##   1       0.100  1.00  0.81      0.79
##   2       0.001  0.25  0.76      0.73
##   2       0.001  0.50  0.76      0.73
##   2       0.001  1.00  0.76      0.73
##   2       0.010  0.25  0.76      0.72
##   2       0.010  0.50  0.76      0.72
##   2       0.010  1.00  0.76      0.72
##   2       0.100  0.25  0.79      0.76
##   2       0.100  0.50  0.81      0.79
##   2       0.100  1.00  0.87      0.85
##   3       0.001  0.25  0.76      0.73
##   3       0.001  0.50  0.76      0.73
##   3       0.001  1.00  0.76      0.73
```

```
##    3          0.010  0.25  0.76         0.72
##    3          0.010  0.50  0.76         0.72
##    3          0.010  1.00  0.76         0.72
##    3          0.100  0.25  0.80         0.78
##    3          0.100  0.50  0.86         0.84
##    3          0.100  1.00  0.90         0.88
##
## Accuracy was used to select the optimal model using the
##   largest value.
## The final values used for the model were degree = 3, scale = 0.1
##   and C = 1.
```

Notice that both of these models have fit our full training set perfectly. So, despite the lower estimated accuracy of the polynomial predictor, both models seem to well-fit our data.

```
predictOnTrainL <- predict(svmLinear, newdata = svmDataTrain)
mean( predictOnTrainL == trainData$CntN)
```

```
## [1] 1
```

```
predictOnTrainP <- predict(svmPoly, newdata = svmDataTrain)
mean( predictOnTrainP == trainData$CntN)
```

```
## [1] 0.98
```

Based on the accuracy level, if faced with these two options in real life, we would select the linear option. At that stage, having already made our choice, we would then run the validation data one last time on only our chosen model. We do so in the following code and have quite good results. Thus, we become fairly sure that based on our selected data, given a new observation in 2015 on the rest of our data points, we would be fairly confident in our ability to correctly sort to the correct region. Of course, we would be wise to keep track of our model's accuracy in the future.

```
predictOnTestL <- predict(svmLinear, newdata = svmDataValidation)
mean(predictOnTestL == validationData$CntN)
```

```
## [1] 1
```

Overall, it seems our dataset is fairly complete. In other words, these various geographic regions are well described in recent years by the set of predictors collected. This is not especially shocking, given the quite clear groupings seen in Figure 8-4. Indeed, realize that PCA would not have especially major changes for just those two variables. There's good vertical and horizontal separation.

That wraps up the section on support vector machines. In the upcoming sections, we will continue to use `caret`'s highly consistent structure to leverage other models. As our familiarity with the overall model structure grows, we attempt to add additional skillsets. Keep in mind the skills introduced later on may well be useful, in real life, to circle back to SVMs.

Classification and Regression Trees

Classification and regression trees (CART) are designed for numerical, continuous predictors and categorical response variables. This fits our dataset well. We refresh our train and validation data for CART, use `set.seed` for reproducibility, and use our now familiar `train` function. We add one new feature, `tuneLength = 10`, which controls the number of iterations on complexity. Increasing this value increases the overall computational time of the model; thus, for a larger dataset, gains in model accuracy would need to be weighted against time to train the model.

```
cartDataTrain <- copy(trainData[,-1])
cartDataTrain[,Year:=as.numeric(Year)]
cartDataValidation <- copy(validationData[,-1])
cartDataValidation[,Year:=as.numeric(Year)]

set.seed(12345)
cartModel <- train(x = cartDataTrain,
            y = trainData$CntN,
            method = "rpart",
            preProcess = c("scale", "center", "pca"),
            metric = "Accuracy",
            tuneLength = 10,
            trControl = trainControl(method = "cv",
                                number = 5
                                )
            )
```

cartModel

```
## CART
##
## 135 samples
##  12 predictor
##   9 classes: 'ArbWr', 'CEatB', 'Er&CA', 'ErpnU', 'EsA&P', 'LtA&C',
##              'ME&NA', 'NrthA', 'Sb-SA'
##
## Pre-processing: scaled (12), centered (12), principal
##  component signal extraction (12)
## Resampling: Cross-Validated (5 fold)
## Summary of sample sizes: 108, 108, 108, 108, 108
## Resampling results across tuning parameters:
##
##   cp     Accuracy  Kappa
##   0.000  0.84      0.83
##   0.014  0.84      0.83
##   0.028  0.84      0.83
##   0.042  0.85      0.83
##   0.056  0.85      0.83
##   0.069  0.85      0.83
##   0.083  0.85      0.83
##   0.097  0.76      0.73
##   0.111  0.73      0.70
##   0.125  0.11      0.00
##
## Accuracy was used to select the optimal model using the
##  largest value.
## The final value used for the model was cp = 0.083.
```

The model ends up selecting based on accuracy (as we have required) the largest complexity value before accuracy decreases (this is an example of the cross-validation helping us avoid overfitting the model).

The finalModel can be plotted using the generic plot function as seen in Figure 8-6. The tree does not have too many levels. Note that while our estimated accuracy is perhaps fairly high (and thus we may hope our model has a solid predictive value), the model is not so useful for understanding why a particular prediction should be made. So, while normally tree models might be considered fairly easy to interpret, using PCA has a net effect of reducing our ability to understand why our model might be true.

```
plot(cartModel$finalModel)
text(cartModel$finalModel, cex = 0.5)
```

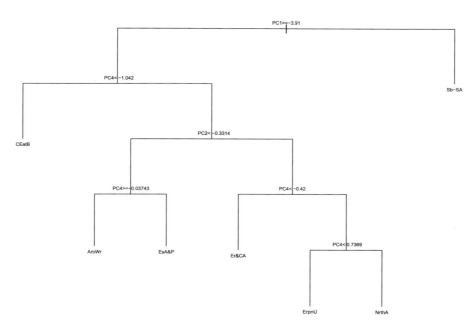

Figure 8-6. *Classification tree graph*

While the generic plot function can be helpful, the fancyRpartPlot often gives a cleaner look, although in this case it is admittedly tough to find the right level of zoom for the text, as seen in Figure 8-7. In any case, what is important to note is that while the accuracy of a model can be enhanced by PCA, the interpretability of the model becomes more complex. It would be helpful to develop a method to make interpretation better.

```
fancyRpartPlot(cartModel$finalModel, cex = 0.4, main = "")
```

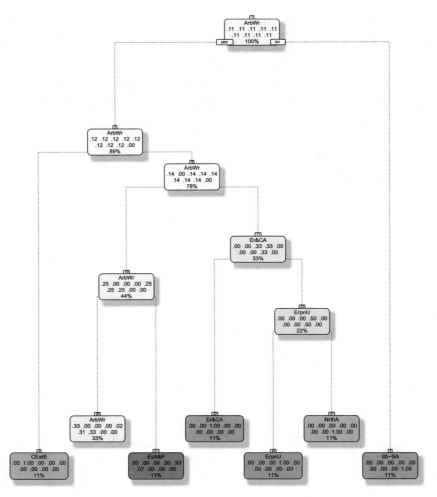

Rattle 2018−Nov−02 19:22:53 ceo

Figure 8-7. *Fancy classification tree graph*

Next, we repeat our measures of accuracy that we used for SVM. Our final model
is highly accurate on training data and indeed keeps that accuracy on fully unseen
data (well, almost keeps). Again, if we wish to have a true estimate of how a model
would perform on real-world data, as we continue to use cross-validation, the correct
methodology would be to wait until the very end of this chapter, select the most accurate
model via the cross-validation, and then finally run the `predictOnTestT` only on the
model selected. However, as a pedagogical method, there is value in seeing the results
on fully unseen data, so we ask our readers continue to forgive us.

```
predictOnTrainT <- predict(cartModel, newdata = cartDataTrain)
mean( predictOnTrainT == trainData$CntN)
```

```
## [1] 0.77
```

```
predictOnTestT <- predict(cartModel, newdata = cartDataValidation)
mean(predictOnTestT == validationData$CntN)
```

```
## [1] 0.67
```

The caret package has a function we have not introduced yet, called confusionMatrix. While the output is quite long in this case, we can see the results of the prediction vs. the validation reference. A bit of scrolling and a good memory show there is one case where the prediction of Arab World was incorrect—the actual data value was East Asia and Pacific. The confusion matrix can be helpful as it shows details beyond simple accuracy. It may allow the detection of a pattern that shows where the errors occur. If the model is meant to work on real-world data, that may make a case to collect additional information that might be expected to support classification between those data. In addition, it also provides a confidence interval.

```
confusionMatrix(predictOnTestT, as.factor(validationData$CntN))
```

```
## Confusion Matrix and Statistics
##
##            Reference
## Prediction ArbWr CEatB Er&CA ErpnU EsA&P LtA&C ME&NA NrthA Sb-SA
##      ArbWr     3     0     0     0     2     3     3     0     0
##      CEatB     0     3     0     0     0     0     0     0     0
##      Er&CA     0     0     2     0     0     0     0     0     0
##      ErpnU     0     0     1     3     0     0     0     0     0
##      EsA&P     0     0     0     0     1     0     0     0     0
##      LtA&C     0     0     0     0     0     0     0     0     0
##      ME&NA     0     0     0     0     0     0     0     0     0
##      NrthA     0     0     0     0     0     0     0     3     0
##      Sb-SA     0     0     0     0     0     0     0     0     3
##
## Overall Statistics
##
##                Accuracy : 0.667
##                  95% CI : (0.46, 0.835)
```

```
##        No Information Rate : 0.111
##        P-Value [Acc > NIR] : 1.15e-11
##
##                      Kappa : 0.625
##  Mcnemar's Test P-Value : NA
##
## Statistics by Class:
##
##                      Class: ArbWr Class: CEatB Class: Er&CA
## Sensitivity                 1.000        1.000       0.6667
## Specificity                 0.667        1.000       1.0000
## Pos Pred Value              0.273        1.000       1.0000
## Neg Pred Value              1.000        1.000       0.9600
## Prevalence                  0.111        0.111       0.1111
## Detection Rate              0.111        0.111       0.0741
## Detection Prevalence        0.407        0.111       0.0741
## Balanced Accuracy           0.833        1.000       0.8333
##                      Class: ErpnU Class: EsA&P Class: LtA&C
## Sensitivity                 1.000        0.333        0.000
## Specificity                 0.958        1.000        1.000
## Pos Pred Value              0.750        1.000          NaN
## Neg Pred Value              1.000        0.923        0.889
## Prevalence                  0.111        0.111        0.111
## Detection Rate              0.111        0.037        0.000
## Detection Prevalence        0.148        0.037        0.000
## Balanced Accuracy           0.979        0.667        0.500
##                      Class: ME&NA Class: NrthA Class: Sb-SA
## Sensitivity                 0.000        1.000        1.000
## Specificity                 1.000        1.000        1.000
## Pos Pred Value                NaN        1.000        1.000
## Neg Pred Value              0.889        1.000        1.000
## Prevalence                  0.111        0.111        0.111
## Detection Rate              0.000        0.111        0.111
## Detection Prevalence        0.000        0.111        0.111
## Balanced Accuracy           0.500        1.000        1.000
```

As you can see, in this dataset, a classification and regression tree worked quite well. While the PCA aspect perhaps complicates a perfect understanding of which input values drive which decisions, trees can often be succinctly graphed. In addition, if one wished, one could experiment with not using PCA. Using the original data (although perhaps normed and scaled) might yield an accurate enough model that was easier to interpret. On the other hand, a single tree might not be accurate enough in some cases. There are always trade-offs with models.

Random Forests

Random forests are a step forward with the idea of a tree. Thinking back to the last section, realize our tree was really good at identifying Central Europe (CEatB). While it made some errors with the Arab World, it was good at Central Europe (CEatB). What if, instead of training one tree, we trained a lot of trees? Of course, simply looping the CART algorithm would yield a cloned forest, and there is no need for that here. Instead, the conceptual first step in random forests is to randomly pick only certain predictor columns and randomly pick only certain observations from that column subset. This creates random subsets of the training dataset, and each of these subsets is used to train a tree. Once the trees are grown into a forest, the model is ready for prediction. Predictions on regression or numerical data would be an average of the predicted response values from each tree. Predictions on our classification or categorical data would be machine learning style democracy, where the majority vote wins (with an attached probability if requested).

We start off by setting up our now usual dataset copies. One of the highly useful features of `caret` continues to be the standardization of the `train` function, which allows us to quite easily work between these various models. For random forests, we use `method = "ranger"`. The model variable `num.trees` is set which controls the size of our forest, and we start off with only 20 trees in the forest. By default, the model does run a small grid search across some tuning parameters, which we discuss later in more depth. In this case, the model selected based on the cross-validation is estimated to have a high accuracy, with only 20 trees. Notice that the `finalModel` has several specific variables contained inside, and we do confirm there are only 20 trees in the model. In this case, that initial stage will create twenty random subsets, each used to train one of our twenty trees.

```
rfDataTrain <- copy(trainData[,-1])
rfDataTrain[,Year:=as.numeric(Year)]
rfDataValidation <- copy(validationData[,-1])
rfDataValidation[,Year:=as.numeric(Year)]

set.seed(12345)

rfModel <- train(x = rfDataTrain,
            y = trainData$CntN,
            method = "ranger",
            preProcess = c("scale", "center", "pca"),
            metric = "Accuracy",
            num.trees = 20,
            trControl = trainControl(method = "cv",
                                        number = 5
                                        )
            )

rfModel

## Random Forest
##
## 135 samples
##  12 predictor
##   9 classes: 'ArbWr', 'CEatB', 'Er&CA', 'ErpnU', 'EsA&P', 'LtA&C',
##              'ME&NA', 'NrthA', 'Sb-SA'
##
## Pre-processing: scaled (12), centered (12), principal
##   component signal extraction (12)
## Resampling: Cross-Validated (5 fold)
## Summary of sample sizes: 108, 108, 108, 108, 108
## Resampling results across tuning parameters:
##
##   mtry  splitrule   Accuracy  Kappa
##   2     gini        0.93      0.92
##   2     extratrees  0.96      0.96
##   3     gini        0.93      0.92
```

```
##   3      extratrees   0.96      0.96
##   4      gini         0.93      0.92
##   4      extratrees   0.99      0.98
##
## Tuning parameter 'min.node.size' was held constant at a value of 1
## Accuracy was used to select the optimal model using the
##  largest value.
## The final values used for the model were mtry = 4, splitrule
##  = extratrees and min.node.size = 1.
```

```
rfModel$finalModel$num.trees
```

```
## [1] 20
```

Naturally, we run our now usual checks to see how the final model performs on our test set and our validation set. The estimated accuracy from the reserved portions of the training data is perhaps a trifle high based on the validationData comparison.

```
predictOnTrainR <- predict(rfModel, newdata = rfDataTrain)
mean( predictOnTrainR == trainData$CntN)
```

```
## [1] 1
```

```
predictOnTestR <- predict(rfModel, newdata = rfDataValidation)
mean(predictOnTestR == validationData$CntN)
```

```
## [1] 1
```

The computation complexity of the random forest is in part a feature of the number of trees trained—and this indeed also starts to have a computational cost on future predictions as well (while not generally a huge burdan, predictions are not filtered through just a single tree). We next adjust our code to have 50 trees.

```
set.seed(12345)
rfModel <- train(x = rfDataTrain,
            y = trainData$CntN,
            method = "ranger",
            preProcess = c("scale", "center", "pca"),
            metric = "Accuracy",
            num.trees = 50,
```

```
                    trControl = trainControl(method = "cv",
                                            number = 5
                                            )
                    )
rfModel
```

```
## Random Forest
##
## 135 samples
##  12 predictor
##   9 classes: 'ArbWr', 'CEatB', 'Er&CA', 'ErpnU', 'EsA&P', 'LtA&C',
##              'ME&NA', 'NrthA', 'Sb-SA'
##
## Pre-processing: scaled (12), centered (12), principal
##  component signal extraction (12)
## Resampling: Cross-Validated (5 fold)
## Summary of sample sizes: 108, 108, 108, 108, 108
## Resampling results across tuning parameters:
##
##   mtry  splitrule   Accuracy  Kappa
##   2     gini        0.94      0.93
##   2     extratrees  0.99      0.99
##   3     gini        0.93      0.92
##   3     extratrees  0.98      0.97
##   4     gini        0.93      0.92
##   4     extratrees  0.98      0.97
##
## Tuning parameter 'min.node.size' was held constant at a value of 1
## Accuracy was used to select the optimal model using the
##  largest value.
## The final values used for the model were mtry = 2, splitrule
##  = extratrees and min.node.size = 1.
```

```
rfModel$finalModel$num.trees
```

```
## [1] 50
```

This increase has a net effect of providing a perfect match on our hold-out validation data. Whether it would hold up to additional, future real-world tests is another matter. It is important to realize each dataset will have various features that make it behave differently with different models. In this case, for now, the models are giving us highly accurate results. As mentioned, one of the authors works with student performance data, and in those cases accurate grade predictions can be more difficult.

```
predictOnTrainR <- predict(rfModel, newdata = rfDataTrain)
mean( predictOnTrainR == trainData$CntN)
```

```
## [1] 1
```

```
predictOnTestR <- predict(rfModel, newdata = rfDataValidation)
mean(predictOnTestR == validationData$CntN)
```

```
## [1] 1
```

One way to increase a given model's accuracy is through tuning. In the case of random forests, tuning parameters include num.trees, mtry, splitrule, and min.node.size. While num.trees has already been noted to control the number of trees, the other tuning parameters deserve some discussion. Based on our earlier exploration of PCA, it is reasonable to suppose that there continue to be four principal components of our data. In that case, when at any particular node, the random forest algorithm will randomly select some of those predictors and then decide which of those selected predictors will give the most sort gain (recall trees branch out to one of two child nodes). The mtry variable sets how many of our four predictors are chosen at random each time. Thus, one through four is our range, and becomes a way to tune our model. The splitrule allows selection of various nuances to the algorithm. In our case, as we are using the model to classify data, the options that make sense are gini and extratrees. As extratrees has been the specific method optimally selected so far, we will stick with that to streamline the tuning process. Finally, the min.node.size determines how many rows of data are allowed before the tree will stop growing. In the case of classification data, the default is one. However, our data has 135 observations of 9 regions which gives us 15 instances for each region. So while a minimal size larger than 15 would be less than ideal, it might make sense to try more than one. In order to tune the model with these variables, we use the expand.grid function from base R to create a data frame from all possible combinations of the ranges we chose. This is passed to the formal tuneGrid and the model run.

```
set.seed(12345)

rfModel <- train(x = rfDataTrain,
            y = trainData$CntN,
            method = "ranger",
            preProcess = c("scale", "center", "pca"),
            metric = "Accuracy",
            num.trees = 20,
            trControl = trainControl(method = "cv",
                                     number = 5
                                     ),
            tuneGrid = expand.grid(mtry = c(1, 2, 3, 4),
                                   splitrule = "extratrees",
                                   min.node.size = c(1, 5, 10, 15))
            )

rfModel

## Random Forest
##
## 135 samples
##  12 predictor
##   9 classes: 'ArbWr', 'CEatB', 'Er&CA', 'ErpnU', 'EsA&P', 'LtA&C',
##              'ME&NA', 'NrthA', 'Sb-SA'
##
## Pre-processing: scaled (12), centered (12), principal
##   component signal extraction (12)
## Resampling: Cross-Validated (5 fold)
## Summary of sample sizes: 108, 108, 108, 108, 108
## Resampling results across tuning parameters:
##
##    mtry  min.node.size  Accuracy  Kappa
##    1     1              0.95      0.94
##    1     5              0.94      0.93
##    1     10             0.88      0.87
##    1     15             0.77      0.74
##    2     1              0.98      0.97
```

##				
##	2	5	0.96	0.95
##	2	10	0.95	0.94
##	2	15	0.90	0.88
##	3	1	0.96	0.96
##	3	5	0.95	0.94
##	3	10	0.93	0.92
##	3	15	0.88	0.87
##	4	1	0.96	0.95
##	4	5	0.96	0.95
##	4	10	0.94	0.93
##	4	15	0.90	0.89

```
##
## Tuning parameter 'splitrule' was held constant at a value
##  of extratrees
## Accuracy was used to select the optimal model using the
##  largest value.
## The final values used for the model were mtry = 2, splitrule =
##  extratrees and min.node.size = 1.

rfModel$finalModel$num.trees

## [1] 20

rfModel$finalModel$mtry

## [1] 2

rfModel$finalModel$splitrule

## [1] "extratrees"

rfModel$finalModel$min.node.size

## [1] 1
```

Of note is that running the model on these more extensive options increased the time the model took to run by 1.9x on our system. We used the function system.time() as a wrapper around the original run with 20 trees and this was run with tuneGride to determine the time cost. As a general workflow, model tuning is a final sort of step, done once a specific model has been selected, for precisely this reason.

We run through the validation of our model one last time, and see that with the limit of 20 trees, the tuning did not increase the accuracy on our validation data.

```
predictOnTrainR <- predict(rfModel, newdata = rfDataTrain)
mean( predictOnTrainR == trainData$CntN)
```

```
## [1] 1
```

```
predictOnTestR <- predict(rfModel, newdata = rfDataValidation)
mean(predictOnTestR == validationData$CntN)
```

```
## [1] 1
```

Stochastic Gradient Boosting

Stochastic gradient boosting is a method of iteratively creating a "forest" of trees. The difference between this method and random forest lies in the iteration. After the first model is trained, the errors between predicted outputs and known training data outputs are calculated. The errors are then used as an additional response variable to the predictors in an effort to decrease the level of error. This process, done repeatedly, reduces the error rate. The technique can be highly effective and requires neither deletion/imputation of missing data nor scaling/centering/pca. Gains do tend to come with tradeoffs, and the model will almost certainly overfit when run without cross-validation. The iterative nature and the goal of error reduction may also be comparatively costly in both compute time and memory requirements.

This time, we are going to use a regression rather than a classification model. Toward that end, we dummy code our training and validation data for country and again choose to set our variable Year as a numeric value.

```
sgbDataTrain <- copy(trainData)
sgbDataTrain[,Year:=as.numeric(Year)]
sgbDataValidation <- copy(validationData)
sgbDataValidation[,Year:=as.numeric(Year)]

ddum <- dummyVars("~.", data = sgbDataTrain)
sgbDataTrain <- data.table(predict(ddum, newdata = sgbDataTrain))
sgbDataValidation <- data.table(predict(ddum, newdata = sgbDataValidation))
rm(ddum)
```

This time, our train function does have some differences. Because we are performing regression, `caret` does allow the first formal to be the regression function. In this case, we choose to let SPAD be the dependent variable and all other variables will be predictors. With the first formal no longer containing a mention of our dataset, we explicitly assert the data used in the second formal argument of the `train` function. The familiar scaling and centering are done, although we forgo principal component analysis for now. Additionally, as these are numeric data in a regression, we set `metric` = "RMSE" for detecting the optimal model via the root mean square error. We continue using cross-fold validation. This model prints a fair bit of text to the screen showing the many iterations. By setting `verbose` = `FALSE` we negate that, although we encourage the interested reader to change the setting to `TRUE`.

The `tuneGrid` options are set to default settings (and thus not strictly required, although for brevity we include them here). The `interaction.depth` will range from 1 to 3, so these trees range from a single node to grandchild nodes in "height". The value for `shrinkage` controls iteration moves where smaller values allow more fine-tuning yet perhaps take the model longer to find a "good enough" location. Familiar by now is `n.trees` controlling the max size of the forest and `n.minobsinnode` set to an equally familiar ten. As a general rule, only 150 trees might not be enough, and values over 10,000 are commonly found in the wild.

```
set.seed(12345)
sgbModel <- train(SPAD ~.,
                  data = sgbDataTrain,
            method = "gbm",
            preProcess = c("scale", "center"),
            metric = "RMSE",
            trControl = trainControl(method = "cv",
                                     number = 5
                                     ),
            tuneGrid = expand.grid(interaction.depth = 1:3,
                                   shrinkage = 0.1,
                                   n.trees = c(50, 100, 150),
                                   n.minobsinnode = 10),
            verbose = FALSE
            )
sgbModel
```

```
## Stochastic Gradient Boosting
##
## 135 samples
##  20 predictor
##
## Pre-processing: scaled (20), centered (20)
## Resampling: Cross-Validated (5 fold)
## Summary of sample sizes: 108, 108, 108, 107, 109
## Resampling results across tuning parameters:
##
##    interaction.depth  n.trees  RMSE  Rsquared   MAE
##    1                  50       14.6  0.81       11.5
##    1                  100      12.3  0.86        9.8
##    1                  150      11.1  0.88        8.8
##    2                  50       10.4  0.91        8.3
##    2                  100       7.8  0.94        5.7
##    2                  150       7.0  0.95        5.0
##    3                  50        8.9  0.93        6.7
##    3                  100       6.8  0.95        4.9
##    3                  150       6.2  0.96        4.4
##
## Tuning parameter 'shrinkage' was held constant at a value of
##   0.1
## Tuning parameter 'n.minobsinnode' was held constant at a
##   value of 10
## RMSE was used to select the optimal model using the smallest value.
## The final values used for the model were n.trees =
##   150, interaction.depth = 3, shrinkage = 0.1 and n.minobsinnode = 10.
```

The selected model has 150 trees and a minimum of 10 observations per node (contrast this with the earlier classification choice where the default observation minimum was 1). The RMSE of this model is 5.9 births per 1,000 women ages 15–19.

Avoiding PCA for this first attempt supports understanding what drives the model at a presumed cost of some level of accuracy and efficiency. Using the summary function, we see the most influential predictors. The largest influencer is SPPO, which shows the percentage of the working-age population. This is followed by SEPR which counts female children out of primary school, as well as SPDY which is the death rate per 1,000 people. In this case, the summary also generates a graph Figure 8-8, which is admittedly not optimal. Nevertheless, the data printout is useful, and the visualization can be helpful.

```
summary(sgbModel)
```

```
##                        var rel.inf
## SPPO              SPPO  31.337
## SEPR              SEPR  17.903
## SPDY              SPDY  11.753
## MAIN              MAIN   8.223
## NYGN              NYGN   6.945
## ERMA              ERMA   6.127
## NYGD              NYGD   5.158
## SESC              SESC   4.519
## `CntNLtA&C`  `CntNLtA&C`   3.912
## FEMA              FEMA   2.843
## Year              Year   0.504
## CntNNrthA    CntNNrthA   0.337
## CntNErpnU    CntNErpnU   0.250
## FEIN              FEIN   0.104
## `CntNME&NA`  `CntNME&NA`   0.049
## CntNArbWr    CntNArbWr   0.018
## `CntNSb-SA`  `CntNSb-SA`   0.016
## CntNCEatB    CntNCEatB   0.000
## `CntNEr&CA`  `CntNEr&CA`   0.000
## `CntNEsA&P`  `CntNEsA&P`   0.000
```

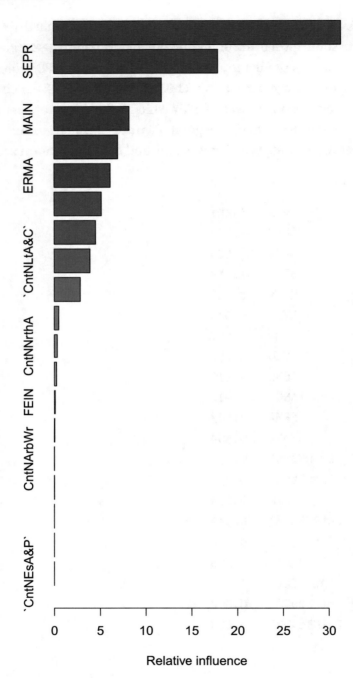

Figure 8-8. *Relative influence visualization*

We next turn our attention to understanding our model's accuracy. The `residuals` function is used to calculate the trained model's MSE on training data (recall the discussion on MSE back in the machine learning introduction chapter). We compare this with the predicted vs. actual values on the validation data—and see it may well be possible the model has overfit.

```
mean(stats::residuals(sgbModel)^2)
```

```
## [1] 3.6
```

```
mean((predict(sgbModel, sgbDataValidation) -
                sgbDataValidation$SPAD)^2)
```

```
## [1] 15
```

A challenge with this method is the precise mechanics of any particular model itself are complex enough that understanding what drives the prediction is difficult. The package `DALEX` has a function aptly named `explain` that seeks to help understand something more of the model and how the predictors interact with the response. While this package is very much of use to compare various models on test data, the syntax may be learned via train vs. validation data.

The `explain` function takes the model used as the first formal, a label to distinguish later graphs (here we distinguish between train and validation), a pointer to our `data`, and an assertion of our response variable SPAD (teen fertility rates).

```
explainSGBt <- explain(sgbModel, label = "sgbt",
                data = sgbDataTrain,
                y = sgbDataTrain$SPAD)

explainSGBv <- explain(sgbModel, label = "sgbv",
                data = sgbDataValidation,
                y = sgbDataValidation$SPAD)
```

The `explain` objects themselves are not of special interest for this text. What is of interest is the information that can be found from them, and the function `model_performance` is the first of these. Reusing the `plot_grid` function from the `cowplot` package, both the distribution and boxplot of the residuals are charted. The left most chart in Figure 8-9 is a plot of the absolute value of the residuals. Keeping in mind the general assumption of normally distributed residuals and then imagining an absolute value transformation would make a normal distribution look like half a distribution at double the height, we see that the validation set mostly still looks normal, although it is more prone to error than the training set. The second chart shows the same truth, with a wider range on the boxplot.

```
performanceSGBt <- model_performance(explainSGBt)
performanceSGBv <- model_performance(explainSGBv)

plot_grid(
  plot(performanceSGBt, performanceSGBv),
  plot(performanceSGBt, performanceSGBv, geom = "boxplot"),
  ncol = 2)
```

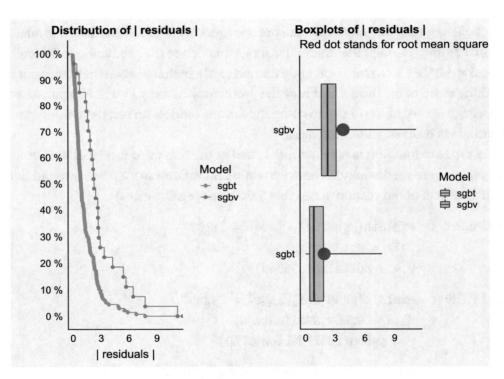

Figure 8-9. *DALEX residual visualizations*

While we have already seen the relative influence by model weight, the function `variable_importance` computes the dropout loss that would be had in a world without information from various variables. It does this per dataset, so notice that values for the training data are different than values for the validation dataset in Figure 8-10. All the same, the topmost key variables do match up (though not in the same order) as the relative influence variables.

```
importanceSGBt <- variable_importance(explainSGBt)
importanceSGBv <- variable_importance(explainSGBv)
plot(importanceSGBt, importanceSGBv)
```

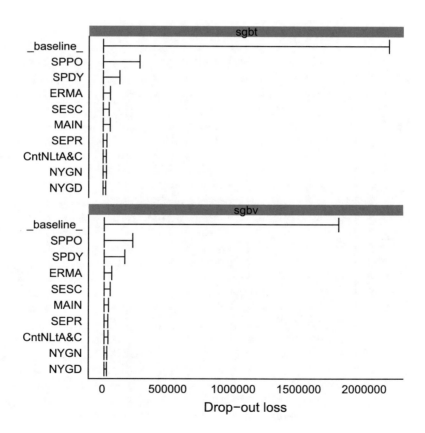

Figure 8-10. *DALEX dropout loss*

The last DALEX function we discuss is the `variable_response` function which takes an additional formal argument of a single variable. In this case, we choose the partial dependence plot type and see the resulting plot in Figure 8-11. The count of female children outside of primary school (SEPR) is plotted against the model outcome. The value of this plot is that, even though the model itself may not be readily understood (unlike a simple regression equation), the relationship between teen fertility and lack of attendance in primary school can be explored.

```
responseSGBprmt <- variable_response(explainSGBt, variable = "SEPR", type = "pdp")
responseSGBprmv <- variable_response(explainSGBv, variable = "SEPR", type = "pdp")
plot(responseSGBprmt, responseSGBprmv)
```

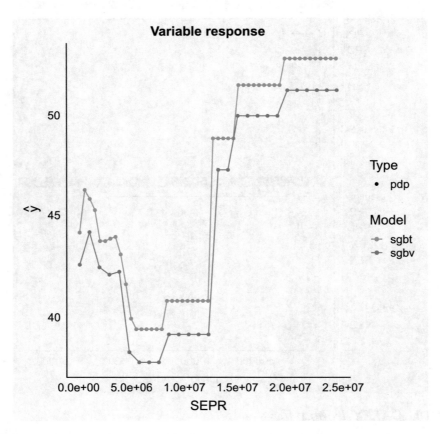

Figure 8-11. *DALEX primary school non-attendance count vs. teen pregnancy.*

We run the code, using the same settings, except this time the death rate per 1,000 (SPDY) is shown in Figure 8-12. Here, the relationship is more complex. This perhaps shows why this type of model may be well suited to addressing data that has a complicated relationship with the predictor. It also shows why this model is at a risk for overfitting.

```
responseSGBdynt <- variable_response(explainSGBt, variable = "SPDY", type = "pdp")
responseSGBdynv <- variable_response(explainSGBv, variable = "SPDY", type = "pdp")
plot(responseSGBdynt, responseSGBdynv)
```

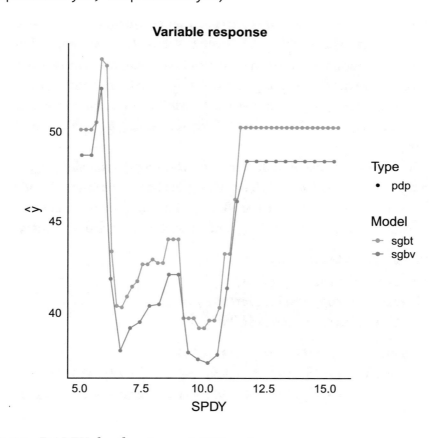

Figure 8-12. *DALEX death rate per 1,000 vs. teen pregnancy*

What is next? Well, this model is not precisely showing a high level of accuracy. Expanding the tuning grid to allow for more trees, deeper trees, and other variable patterns to give the algorithm a wider search grid for the optimal model is in order. We leave such an extension to the interest reader though, as the structure has been built already, and proceed to the next model.

Multilayer Perceptron

A multilayer perceptron (MLP) is a feedforward artificial neural network. This is actually not precisely the latest and greatest of the deep learning or neural network algorithms. All the same, this is the type of neural network one tends to start with learning. At its heart, a neural network is just a lot of linear algebra matrix multiplication that gets optimized by the algorithm. The net effect is that, rather than a standard regression equation that has one weight per predictor variable, there is a rather vast array of weights.

A key feature of neural networks is that rather than attempt to have a single equation create the predictor, multiple equations are used (hence the matrices). Indeed, every portion of the predictor will be treated somewhat separately, weighted, and mapped to possible outputs. Additionally, rather than use a single equation, there are potentially several layers of equations between input and output. These are aptly named "hidden layers," and while they can give the model great flexibility, most multilayer perceptrons are quite difficult to understand as a model. Their redeeming feature is they often make quite accurate predictions.

Indeed, the classic example is of digit or character recognition via image visuals. Our example, being much simpler, may well not be a great fit for this method (indeed it is not). That does not in any way negate the efficacy of the method in other scenarios. As before, we create our last round of training and validation data, using dummy variables.

```
mlpDataTrain <- copy(trainData)
mlpDataTrain[,Year:=as.numeric(Year)]
mlpDataValidation <- copy(validationData)
mlpDataValidation[,Year:=as.numeric(Year)]

ddum <- dummyVars("~.", data = mlpDataTrain)
mlpDataTrain <- data.table(predict(ddum, newdata = mlpDataTrain))
mlpDataValidation <- data.table(predict(ddum, newdata = mlpDataValidation))
rm(ddum)
```

The model and seed stay the same. This time, the method = "mlpML" is used which stands for multilayer perceptron, multilayer (which admittedly sounds redundant, but there it is). We continue to scale and center our data, although we do not use PCA. This first run is done via the default method, which nulls out the second and third layer. This leaves the first hidden layer to be run, and, by default, the model simulates a few different quantities of nodes in that first layer. Due to the way this book was created, warnings are printed (indeed this is a helpful feature often). In the case of this model, many warnings are issued, which in this case only show model nuances beyond the scope of this text. The function supressWarnings is used for its eponymous result.

```
set.seed(12345)
suppressWarnings(
  mlpModel <- train(
    SPAD ~ .,
    data = mlpDataTrain,
    method = "mlpML",
    preProcess = c("scale", "center"),
    metric = "RMSE",
    trControl = trainControl(method = "cv",
                             number = 5)
  )
)
mlpModel

## Multi-Layer Perceptron, with multiple layers
##
## 135 samples
##  20 predictor
##
## Pre-processing: scaled (20), centered (20)
## Resampling: Cross-Validated (5 fold)
## Summary of sample sizes: 108, 108, 108, 107, 109
## Resampling results across tuning parameters:
##
##    layer1  RMSE  Rsquared  MAE
##    1       36    0.31      29
##    3       25    0.49      19
##    5       17    0.76      13
##
## Tuning parameter 'layer2' was held constant at a value of 0
##
## Tuning parameter 'layer3' was held constant at a value of 0
```

```
## RMSE was used to select the optimal model using the smallest value.
## The final values used for the model were layer1 = 5, layer2 = 0
##   and layer3 = 0.
```

As seen from the output, the single hidden layer was tuned by default to five nodes. The summary function provides a bit more detail, showing the activation function weights as well as the bias for each of our 20 input columns, 5 hidden nodes, and 1 output node. This also shows the overall layout of our network, in a 20-5-1 layout.

```
summary(mlpModel)
```

```
## SNNS network definition file V1.4-3D
## generated at Fri Nov 02 19:23:07 2018
##
## network name : RSNNS_untitled
## source files :
## no. of units : 26
## no. of connections : 105
## no. of unit types : 0
## no. of site types : 0
##
##
## learning function : Std_Backpropagation
## update function   : Topological_Order
##
##
## unit default section :
##
## act       | bias      | st | subnet | layer | act func      | out func
## ---------|-----------|----|--------|-------|--------------|-------------
##  0.00000 |  0.00000 | i  |      0 |     1 | Act_Logistic | Out_Identity
## ---------|-----------|----|--------|-------|--------------|-------------
##
##
## unit definition section :
##
```

```
## no. | typeName | unitName        | act       | bias      | st | position |
act func       | out func | sites
## ----|----------|-------------------|----------|----------|----|----------|
--------------|----------|-------
##    1 |          | Input_CntNArbWr   | -0.35224 |  0.25864 | i  | 1, 0, 0 |
Act_Identity |          |
##    2 |          | Input_CntNCEatB   | -0.35224 | -0.07158 | i  | 2, 0, 0 |
Act_Identity |          |
##    3 |          | Input_`CntNEr&CA` | -0.35224 |  0.17340 | i  | 3, 0, 0 |
Act_Identity |          |
##    4 |          | Input_CntNErpnU   | -0.35224 |  0.09913 | i  | 4, 0, 0 |
Act_Identity |          |
##    5 |          | Input_`CntNEsA&P` | -0.35224 |  0.02550 | i  | 5, 0, 0 |
Act_Identity |          |
##    6 |          | Input_`CntNLtA&C` | -0.35224 | -0.07856 | i  | 6, 0, 0 |
Act_Identity |          |
##    7 |          | Input_`CntNME&NA` | -0.35224 |  0.10749 | i  | 7, 0, 0 |
Act_Identity |          |
##    8 |          | Input_CntNNrthA   | -0.35224 | -0.17845 | i  | 8, 0, 0 |
Act_Identity |          |
##    9 |          | Input_`CntNSb-SA` |  2.81793 |  0.20316 | i  | 9, 0, 0 |
Act_Identity |          |
##   10 |          | Input_Year        |  1.51162 |  0.09500 | i  | 10, 0, 0 |
Act_Identity |          |
##   11 |          | Input_NYGD        | -0.84754 | -0.12790 | i  | 11, 0, 0 |
Act_Identity |          |
##   12 |          | Input_NYGN        | -0.83754 | -0.26720 | i  | 12, 0, 0 |
Act_Identity |          |
##   13 |          | Input_SEPR        |  2.37160 |  0.23106 | i  | 13, 0, 0 |
Act_Identity |          |
##   14 |          | Input_ERMA        |  2.28315 | -0.13108 | i  | 14, 0, 0 |
Act_Identity |          |
##   15 |          | Input_SESC        | -1.27208 | -0.13525 | i  | 15, 0, 0 |
Act_Identity |          |
```

```
## 16 |               | Input_FEMA     | -1.28836 |  0.21397 | i | 16, 0, 0 |
Act_Identity |        |
## 17 |               | Input_SPDY     |  0.55710 |  0.19976 | i | 17, 0, 0 |
Act_Identity |        |
## 18 |               | Input_FEIN     | -1.80461 | -0.15568 | i | 18, 0, 0 |
Act_Identity |        |
## 19 |               | Input_MAIN     | -1.77002 |  0.29051 | i | 19, 0, 0 |
Act_Identity |        |
## 20 |               | Input_SPPO     |  2.09958 | -0.29660 | i | 20, 0, 0 |
Act_Identity |        |
## 21 |               | Hidden_2_1     |  1.00000 | 16.57569 | h |  1, 2, 0 |
             |        |
## 22 |               | Hidden_2_2     |  1.00000 | -42.96722| h |  2, 2, 0 |
             |        |
## 23 |               | Hidden_2_3     |  0.00000 | -49.55274| h |  3, 2, 0 |
             |        |
## 24 |               | Hidden_2_4     |  1.00000 | 39.86870 | h |  4, 2, 0 |
             |        |
## 25 |               | Hidden_2_5     |  1.00000 | 21.29272 | h |  5, 2, 0 |
             |        |
## 26 |               | Output_1       | 66.31960 |-1069.78821| o |  1, 4, 0 |
Act_Identity |        |
## ----|----------|-------------------|----------|----------|----|----------|
--------------|----------|-------
##
##
## connection definition section :
##
## target | site | source:weight
## -------|------|--------------------------------------------------------------
##     21 |      | 20:48.96251, 19:-47.13240, 18:-46.89282, 17:29.71989,
                   16:-31.71886, 15:-30.48963, 14:58.16447, 13:58.57929,
                   12:-15.94363,
```

```
##                        11:-15.49402, 10:21.26738,  9:66.89728,  8:-5.93687,
                          7:-24.55808,  6:-8.02125,  5:-5.53956,  4:-5.59685,
                          3:-5.55265,
##                         2:-5.77468,  1:-5.90555
##      22 |            | 20:19.36864, 19:-36.07077, 18:-30.54533, 17:30.76214,
                         16:-19.48156, 15:-10.87526, 14:40.41421, 13:38.18938,
                         12: 4.44169,
##                        11: 0.46633, 10:46.79949,  9:49.01194,  8:15.15641,
                          7:-81.96992,  6:14.43461,  5:-6.70102,  4:-
                         40.07672,  3:12.00658,
##                         2: 7.56544,  1:31.19720
##      23 |            | 20:-12.19030, 19: 2.85565, 18:31.55284, 17:84.06341,
                         16:77.33264, 15:103.85283, 14:-12.62139, 13:-9.16594,
                         12:81.62211,
##                        11:81.21730, 10:-194.05856,  9:49.49640,  8:115.78496,
                          7:-4.44688,  6:179.36331,  5:-26.62095,  4:35.70350,
                          3:-54.66271,
##                         2:33.46564,  1:-328.59045
##      24 |            | 20:22.24673, 19:-17.17476, 18:-17.65513, 17:-30.85148,
                         16:-20.34034, 15:-17.87234, 14:19.58477, 13:16.31513,
                         12:-25.13864,
##                        11:-24.56263, 10:-14.12056,  9:11.75429,  8:-14.03880,
                          7:-0.54804,  6:62.57944,  5:14.06488,  4:-13.82649,
                          3:-14.21823,
##                         2:-31.15477,  1:-13.92656
##      25 |            | 20:35.84281, 19:-31.83327, 18:-33.52740, 17: 7.88547,
                         16:-23.72048, 15:-24.41236, 14:36.01567, 13:37.40243,
                         12:-16.33773,
##                        11:-15.88420, 10:19.37509,  9:42.81319,  8:-7.29130,
                          7:10.05342,  6:-4.69724,  5:-11.28413,  4:-7.28304,
                          3:-7.42553,
##                         2:-7.40186,  1:-7.59370
##      26 |            | 25: 5.30307, 24:31.84159, 23:30.69779, 22:36.21242,
                         21:-7.03748
## -------|------|-------------------------------------------------------
```

As suggested with using an overly complex method for a comparatively simple dataset, the model does not perform all that well. It performs even worse on the validation data. Note that due to the `residuals` function being masked by one of the required packages, we explicitly require the base R version of `residuals`.

```
mean(stats::residuals(mlpModel)^2)
```

```
## [1] 462
```

```
mean((predict(mlpModel, mlpDataValidation) -
                mlpDataValidation$SPAD)^2)
```

```
## [1] 407
```

Again, one challenge with this method is the precise mechanics of any particular model itself are complex enough that understanding what drives the prediction is difficult. We again use the package DALEX's aptly named `explain` function to show how the predictors interact with the response.

Recall that the `explain` function syntax takes the model used as the first formal, a label to distinguish later graphs (here we distinguish between train and validation), a pointer to our `data`, and an assertion of our response variable SPAD.

```
explainMLPt <- explain(mlpModel, label = "mlpt",
               data = mlpDataTrain,
               y = mlpDataTrain$SPAD)
```

```
explainMLPv <- explain(mlpModel, label = "mlpv",
               data = mlpDataValidation,
               y = mlpDataValidation$SPAD)
```

Also as before, the `explain` objects themselves are not of special interest. What is of interest is the information that can be found from them, and the function `model_performance` is the first of these. Reusing the `plot_grid` function from the `cowplot` package, both the distribution and boxplot of the residuals are charted. The left most chart in Figure 8-13 is a plot of the absolute value of the residuals. Keeping in mind the general assumption of normally distributed residuals and then imagining an absolute value transformation would make a normal distribution look like half a distribution at double the height, we see that the validation set mostly still looks normal, although it is more prone to error than the training set. We additionally see that the gbm method from the prior section was less prone to error. The second chart continues to show the same truth, with a wider range on the boxplot. In both cases, the charts also show the perpetual risk of overtraining. The validation data—unseen to the model data—has wider residuals.

```
performanceMLPt <- model_performance(explainMLPt)
performanceMLPv <- model_performance(explainMLPv)

plot_grid(
plot(performanceMLPt, performanceMLPv, performanceSGBt, performanceSGBv),
plot(performanceMLPt, performanceMLPv, performanceSGBt, performanceSGBv,
geom = "boxplot"),
ncol = 2
)
```

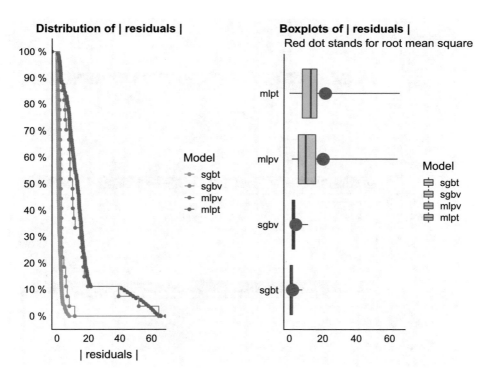

Figure 8-13. *Model performance contrasting SGB vs. MLP methods*

We use the `variable_importance` function to compute the dropout loss had in a world without information from various variables. What is interesting to note in the plot of these—along with the plot from the previous gbm model—is which variables are most key. Recall it is the variables with the most loss after dropout, as seen in Figure 8-14. This shows our multilayer perceptrons model these same data very differently. In this case, that seems to be a weakness. However, with other data, this difference could well be a strength.

```
importanceMLPt <- variable_importance(explainMLPt)
importanceMLPv <- variable_importance(explainMLPv)
plot(importanceMLPt, importanceMLPv, importanceSGBt, importanceSGBv)
```

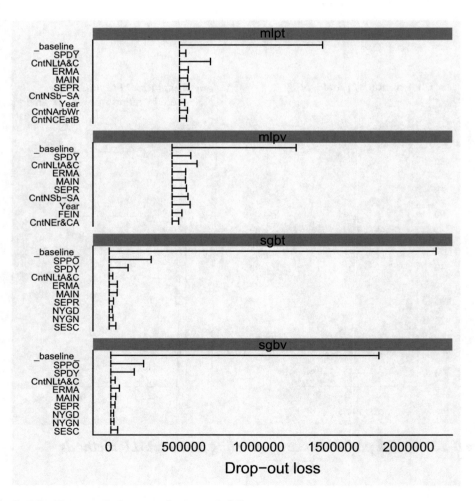

Figure 8-14. *Determining top key variables*

As before, we conclude with the DALEX function variable_response which takes an additional formal argument of a single variable. In this case, we choose the partial dependence plot type and see the resulting plot in Figure 8-15. The count of female children outside of primary school is plotted against the model outcome. The value of this plot remains that, even though the model itself may not be readily understood (unlike a simple regression equation), the relationship between teen fertility and lack of attendance in primary school can be explored.

```
responseMLPprmt <- variable_response(explainMLPt, variable = "SEPR", type = "pdp")
responseMLPprmv <- variable_response(explainMLPv, variable = "SEPR", type = "pdp")
plot(responseMLPprmt, responseMLPprmv, responseSGBprmt, responseSGBprmv)
```

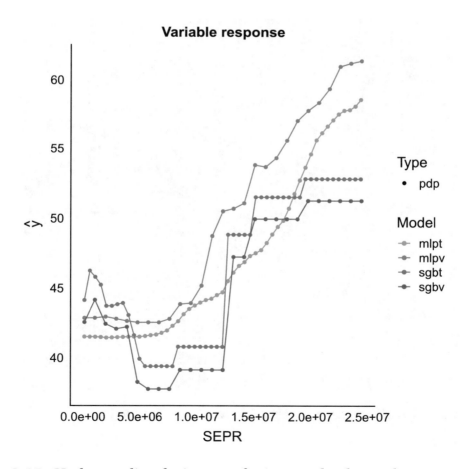

Figure 8-15. *Understanding the impact of primary school attendance on teen pregnancy*

We conclude our DALEX exploration using the death rate per 1,000 as shown in Figure 8-16. Here, the contrast between the two models is starker. Again, this demonstrates that the underlying methodology is quite different, and thus mlpML may be suited to other data where the earlier methods did not achieve effective predictions.

```
responseMLPdynt <- variable_response(explainMLPt, variable = "SPDY", type = "pdp")
responseMLPdynv <- variable_response(explainMLPv, variable = "SPDY", type = "pdp")
plot(responseMLPdynt, responseMLPdynv, responseSGBdynt, responseSGBdynv)
```

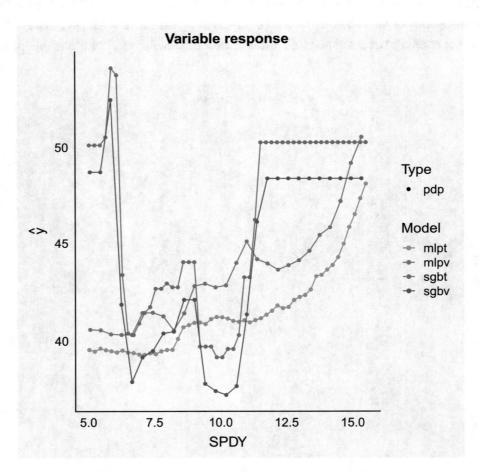

Figure 8-16. *DALEX death rate per 1,000 vs. teen pregnancy*

Before we conclude this section, we consider whether our model was simply not tuned enough. Due to their complexity, multilayer perceptrons tend to have many tuning parameters. For caret and in particular mlpML, the tuning parameters are the number of hidden layers, as well as the number of nodes in each hidden layer. Using our now familiar tuneGride formal and expand.grid function, we have the model explore over a thousand options, to determine an optimal model. For real-life applications where perceptron methods are thought to be the likely optimal model, it is not uncommon to have hundreds of nodes. This creates tens of thousands of variables, each of which is optimized by the model in turn. It is a significant investment in computational resources, which, in this case, will not pay off.

```
set.seed(12345)
suppressWarnings(
  mlpModelb <- train(
    SPAD ~ .,
    data = mlpDataTrain,
    method = "mlpML",
    preProcess = c("scale", "center"),
    metric = "RMSE",
    verbose = FALSE,
    trControl = trainControl(method = "cv",
                             number = 5),
    tuneGrid = expand.grid(
      layer1 = 0:10,
      layer2 = 0:10,
      layer3 = 0:10
    )
  )
)
mlpModelb

## Multi-Layer Perceptron, with multiple layers
##
## 135 samples
##  20 predictor
##
```

```
## Pre-processing: scaled (20), centered (20)
## Resampling: Cross-Validated (5 fold)
## Summary of sample sizes: 108, 108, 108, 107, 109
## Resampling results across tuning parameters:
##
```

##	layer1	layer2	layer3	RMSE	Rsquared	MAE
##	0	0	0	**NaN**	**NaN**	**NaN**
##	0	0	1	41	1.4e-01	30.4
##	0	0	2	36	3.0e-01	27.1
##	0	0	3	26	4.9e-01	17.5
##	0	0	4	18	7.4e-01	14.6
##	0	0	5	18	7.9e-01	14.2
##	0	0	6	16	7.7e-01	11.9
##	0	0	7	14	8.3e-01	10.7
##	0	0	8	16	8.0e-01	12.0
##	0	0	9	15	8.1e-01	9.8
##	0	0	10	11	8.8e-01	8.9
##	0	1	0	37	2.6e-01	29.9
##	0	1	1	33	3.0e-01	25.1
##	0	1	2	36	4.0e-01	28.5
##	0	1	3	38	3.2e-01	28.7
##	0	1	4	35	6.7e-02	24.7
##	0	1	5	33	2.3e-01	24.4
##	0	1	6	35	1.4e-01	26.1
##	0	1	7	34	2.0e-01	28.5
##	0	1	8	56	3.0e-01	50.3
##	0	1	9	41	3.7e-01	36.8
##	0	1	10	41	1.3e-01	32.9
##	0	2	0	28	3.5e-01	18.8
##	0	2	1	42	2.5e-01	32.2
##	0	2	2	40	1.7e-01	33.7
##	0	2	3	36	4.5e-01	27.7
##	0	2	4	34	2.5e-01	24.6
##	0	2	5	34	4.1e-01	28.1
##	0	2	6	32	3.0e-01	23.2

##	0	2	7	35	3.8e-01	27.6
##	0	2	8	39	2.6e-01	29.9
##	0	2	9	40	2.3e-01	31.9
##	0	2	10	249	3.3e-01	243.2
##	0	3	0	24	5.8e-01	19.1
##	0	3	1	40	2.3e-01	30.7
##	0	3	2	40	3.9e-04	33.2
##	0	3	3	38	2.7e-01	30.7
##	0	3	4	35	4.4e-01	26.2
##	0	3	5	36	8.9e-02	30.0
##	0	3	6	32	4.3e-01	21.5
##	0	3	7	32	2.1e-01	25.8
##	0	3	8	46	2.3e-01	37.9
##	0	3	9	30	3.0e-01	21.6
##	0	3	10	48	3.3e-01	42.4
##	0	4	0	22	5.7e-01	17.2
##	0	4	1	51	2.9e-01	42.1
##	0	4	2	38	2.3e-01	30.9
##	0	4	3	35	5.0e-02	27.3
##	0	4	4	36	2.2e-01	24.4
##	0	4	5	35	2.1e-01	29.9
##	0	4	6	31	2.6e-01	23.2
##	0	4	7	38	2.2e-01	28.8
##	0	4	8	31	4.1e-01	23.2
##	0	4	9	33	2.8e-01	27.1
##	0	4	10	35	3.1e-01	27.5
##	0	5	0	24	5.6e-01	16.5
##	0	5	1	37	3.7e-01	29.4
##	0	5	2	32	2.8e-02	25.2
##	0	5	3	34	2.6e-01	25.3
##	0	5	4	31	2.3e-01	22.6
##	0	5	5	40	2.0e-01	31.4
##	0	5	6	33	8.7e-02	25.3
##	0	5	7	30	2.7e-01	22.3
##	0	5	8	35	1.9e-01	25.3

##	0	5	9	28	3.3e-01	21.1
##	0	5	10	34	3.9e-01	27.1
##	0	6	0	18	7.4e-01	14.7
##	0	6	1	34	**NaN**	24.0
##	0	6	2	37	1.3e-01	26.5
##	0	6	3	32	3.9e-01	22.3
##	0	6	4	35	2.5e-01	24.4
##	0	6	5	33	2.1e-01	26.0
##	0	6	6	52	1.9e-01	42.8
##	0	6	7	30	2.7e-01	23.8
##	0	6	8	36	3.3e-01	27.8
##	0	6	9	54	1.5e-01	45.9
##	0	6	10	30	2.8e-01	23.7
##	0	7	0	22	7.0e-01	17.0
##	0	7	1	38	4.4e-02	30.6
##	0	7	2	42	1.5e-01	35.7
##	0	7	3	35	1.2e-01	24.6
##	0	7	4	33	2.4e-01	25.3
##	0	7	5	39	1.5e-01	32.4
##	0	7	6	32	1.7e-01	24.4
##	0	7	7	32	2.7e-01	26.3
##	0	7	8	41	1.3e-01	33.6
##	0	7	9	33	1.4e-01	25.6
##	0	7	10	31	2.5e-01	22.1
##	0	8	0	15	8.3e-01	11.9
##	0	8	1	34	**NaN**	25.8
##	0	8	2	33	6.2e-02	25.1
##	0	8	3	30	3.2e-01	22.5
##	0	8	4	43	1.8e-01	37.1
##	0	8	5	39	3.3e-01	31.3
##	0	8	6	34	1.1e-01	26.0
##	0	8	7	34	1.3e-01	24.7
##	0	8	8	45	1.2e-01	38.6
##	0	8	9	32	3.1e-01	23.3
##	0	8	10	50	4.4e-01	42.3

##	0	9	0	15	8.4e-01	11.9
##	0	9	1	34	**NaN**	24.7
##	0	9	2	34	1.6e-01	26.6
##	0	9	3	34	5.2e-01	25.5
##	0	9	4	43	2.4e-01	33.8
##	0	9	5	44	2.2e-01	37.4
##	0	9	6	37	3.6e-01	30.3
##	0	9	7	39	2.8e-01	27.3
##	0	9	8	35	1.3e-01	27.1
##	0	9	9	31	2.9e-01	22.3
##	0	9	10	51	2.2e-01	41.2
##	0	10	0	19	7.6e-01	15.1
##	0	10	1	44	**NaN**	35.5
##	0	10	2	33	**NaN**	26.3
##	0	10	3	37	4.8e-01	29.8
##	0	10	4	33	9.2e-02	24.2
##	0	10	5	35	1.4e-01	25.9
##	0	10	6	35	1.7e-01	27.7
##	0	10	7	32	1.6e-01	23.0
##	0	10	8	67	5.1e-02	62.2
##	0	10	9	31	1.6e-01	23.7
##	0	10	10	32	3.1e-01	25.1
##	1	0	0	39	4.1e-01	31.2
##	1	0	1	37	5.3e-01	29.1
##	1	0	2	40	3.5e-01	31.7
##	1	0	3	34	7.9e-02	24.5
##	1	0	4	36	3.8e-01	24.6
##	1	0	5	33	3.8e-01	24.7
##	1	0	6	35	3.4e-01	27.3
##	1	0	7	37	2.3e-01	27.4
##	1	0	8	30	3.8e-01	21.4
##	1	0	9	38	2.7e-01	30.3
##	1	0	10	40	2.8e-01	32.9
##	1	1	0	37	2.2e-01	29.0
##	1	1	1	42	7.3e-01	32.5

##	1	1	2	38	6.4e-01	31.6
##	1	1	3	33	**NaN**	25.0
##	1	1	4	36	4.5e-01	24.4
##	1	1	5	32	7.1e-01	26.1
##	1	1	6	34	2.7e-01	25.7
##	1	1	7	39	3.2e-01	31.9
##	1	1	8	37	5.4e-01	30.0
##	1	1	9	37	2.5e-01	28.4
##	1	1	10	35	9.0e-02	27.5
##	1	2	0	40	4.8e-01	34.1
##	1	2	1	34	**NaN**	25.7
##	1	2	2	33	5.0e-01	24.8
##	1	2	3	36	5.1e-01	26.6
##	1	2	4	40	4.6e-01	32.0
##	1	2	5	33	5.1e-01	23.7
##	1	2	6	41	5.5e-01	33.0
##	1	2	7	32	4.7e-01	23.3
##	1	2	8	33	3.5e-01	25.2
##	1	2	9	38	4.5e-01	29.5
##	1	2	10	39	4.5e-01	30.5
##	1	3	0	32	4.1e-01	24.1
##	1	3	1	34	**NaN**	26.3
##	1	3	2	34	2.8e-01	27.0
##	1	3	3	41	**NaN**	31.2
##	1	3	4	33	6.5e-01	23.4
##	1	3	5	31	5.4e-01	23.4
##	1	3	6	41	6.8e-01	33.8
##	1	3	7	44	5.0e-01	32.6
##	1	3	8	57	4.2e-01	47.5
##	1	3	9	41	4.7e-01	34.1
##	1	3	10	42	2.8e-01	32.6
##	1	4	0	38	4.3e-01	28.8

```
##  [ reached getOption("max.print") -- omitted 1165 rows ]
##
## RMSE was used to select the optimal model using the smallest value.
```

```
## The final values used for the model were layer1 = 0, layer2 = 0
##   and layer3 = 10.
```

mean(stats::residuals(mlpModelb)^2)

```
## [1] 668
```

mean((predict(mlpModelb, mlpDataValidation) -
                 mlpDataValidation$SPAD)^2)

```
## [1] 552
```

summary(mlpModelb)

```
## SNNS network definition file V1.4-3D
## generated at Fri Nov 02 19:34:18 2018
##
## network name : RSNNS_untitled
## source files :
## no. of units : 31
## no. of connections : 210
## no. of unit types : 0
## no. of site types : 0
##
##
## learning function : Std_Backpropagation
## update function   : Topological_Order
##
##
## unit default section :
##
## act       | bias      | st | subnet | layer | act func      | out func
## ---------|-----------|----|--------|-------|---------------|-------------
##  0.00000 |  0.00000 | i  |      0 |     1 | Act_Logistic | Out_Identity
## ---------|-----------|----|--------|-------|---------------|-------------
##
##
## unit definition section :
```

```
##
## no. | typeName | unitName            | act       | bias      | st | position |
act func      | out func | sites
## ----|----------|--------------------|----------|----------|----|----------|
--------------|----------|-------
##   1 |          | Input_CntNArbWr     | -0.35224 |  0.27237 | i  |  1, 0, 0 |
Act_Identity |          |
##   2 |          | Input_CntNCEatB     | -0.35224 |  0.12640 | i  |  2, 0, 0 |
Act_Identity |          |
##   3 |          | Input_`CntNEr&CA`   | -0.35224 |  0.14994 | i  |  3, 0, 0 |
Act_Identity |          |
##   4 |          | Input_CntNErpnU     | -0.35224 |  0.05916 | i  |  4, 0, 0 |
Act_Identity |          |
##   5 |          | Input_`CntNEsA&P`   | -0.35224 | -0.03508 | i  |  5, 0, 0 |
Act_Identity |          |
##   6 |          | Input_`CntNLtA&C`   | -0.35224 |  0.20488 | i  |  6, 0, 0 |
Act_Identity |          |
##   7 |          | Input_`CntNME&NA`   | -0.35224 | -0.18422 | i  |  7, 0, 0 |
Act_Identity |          |
##   8 |          | Input_CntNNrthA     | -0.35224 | -0.24506 | i  |  8, 0, 0 |
Act_Identity |          |
##   9 |          | Input_`CntNSb-SA`   |  2.81793 | -0.11938 | i  |  9, 0, 0 |
Act_Identity |          |
##  10 |          | Input_Year          |  1.70822 |  0.19864 | i  | 10, 0, 0 |
Act_Identity |          |
##  11 |          | Input_NYGD          | -0.84501 |  0.20640 | i  | 11, 0, 0 |
Act_Identity |          |
##  12 |          | Input_NYGN          | -0.83288 |  0.20374 | i  | 12, 0, 0 |
Act_Identity |          |
##  13 |          | Input_SEPR          |  2.38957 |  0.21969 | i  | 13, 0, 0 |
Act_Identity |          |
##  14 |          | Input_ERMA          |  2.33470 | -0.14782 | i  | 14, 0, 0 |
Act_Identity |          |
##  15 |          | Input_SESC          | -1.32473 | -0.08605 | i  | 15, 0, 0 |
Act_Identity |          |
##  16 |          | Input_FEMA          | -1.37682 |  0.28955 | i  | 16, 0, 0 |
Act_Identity |          |
```

```
##   17 |              | Input_SPDY        |   0.44706 |   0.01856 | i  | 17, 0, 0 |
Act_Identity |               |
##   18 |              | Input_FEIN        |  -1.72852 |  -0.10129 | i  | 18, 0, 0 |
Act_Identity |               |
##   19 |              | Input_MAIN        |  -1.69223 |  -0.26429 | i  | 19, 0, 0 |
Act_Identity |               |
##   20 |              | Input_SPPO        |   2.06062 |  -0.17451 | i  | 20, 0, 0 |
Act_Identity |               |
##   21 |              | Hidden_2_1        |   1.00000 |  -8.20578 | h  |  1, 2, 0 |
             |               |
##   22 |              | Hidden_2_2        |   0.00000 | -61.29456 | h  |  2, 2, 0 |
             |               |
##   23 |              | Hidden_2_3        |   1.00000 |  -4.93293 | h  |  3, 2, 0 |
             |               |
##   24 |              | Hidden_2_4        |   0.00000 |  -1.26154 | h  |  4, 2, 0 |
             |               |
##   25 |              | Hidden_2_5        |   1.00000 |-136.45082 | h  |  5, 2, 0 |
             |               |
##   26 |              | Hidden_2_6        |   1.00000 |  -3.98742 | h  |  6, 2, 0 |
             |               |
##   27 |              | Hidden_2_7        |   1.00000 | -38.50706 | h  |  7, 2, 0 |
             |               |
##   28 |              | Hidden_2_8        |   1.00000 | -59.23545 | h  |  8, 2, 0 |
             |               |
##   29 |              | Hidden_2_9        |   1.00000 | -13.07257 | h  |  9, 2, 0 |
             |               |
##   30 |              | Hidden_2_10       |   0.00000 |  -3.80823 | h  | 10, 2, 0 |
             |               |
##   31 |              | Output_1          | 80.51889 | 165.28026 | o  |  1, 4, 0 |
Act_Identity |               |
## ----|----------|-------------------|----------|----------|----|----------|
--------------|----------|-------
##
##
## connection definition section :
##
```

```
## target | site | source:weight
## -------|------|-------------------------------------------------------
##     21 |      | 20:14.99080, 19:-4.37922, 18:-5.64107, 17:-14.21278,
                   16:-7.40044, 15:-6.96982, 14: 2.04649, 13: 3.28454,
                   12:-1.09398,
##                 11:-1.05998, 10:-10.35237,  9: 3.07458,  8: 2.81771,
                   7: 7.14501,  6:22.43020,  5:-17.89252,  4: 2.94234,
                   3:-3.41989,
##                 2:-20.86507,  1: 2.99437
##     22 |      | 20:-42.11395, 19:46.16928, 18:46.41003, 17: 5.73816,
                   16:50.23483, 15:47.74510, 14:-48.54859, 13:-49.08794,
                   12:49.66853,
##                 11:51.15318, 10:-5.34949,  9:-55.80005,  8:25.68831,
                   7:-2.84146,  6:-12.38087,  5:-27.17203,  4:23.60743,
                   3:24.90088,
##                 2: 1.37566,  1:22.42043
##     23 |      | 20: 3.55224, 19:-3.82663, 18:-2.81083, 17:-8.62280,
                   16:-6.84271, 15:-6.68200, 14: 2.13178, 13: 1.19267,
                   12:-12.25461,
##                 11:-12.00026, 10: 7.13982,  9:-0.63003,  8:-19.36786,
                   7: 1.56885,  6:18.87510,  5: 1.15033,  4: 1.62183,
                   3: 1.71615,
##                 2:-6.58669,  1: 1.57223
##     24 |      | 20:-3.03815, 19: 2.09218, 18: 5.45140, 17: 8.87992,
                   16: 8.22070, 15:10.87528, 14:-0.88254, 13:-2.19428,
                   12: 8.49697,
##                 11: 8.60248, 10:-5.41809,  9: 0.47726,  8:10.84176,
                   7:-27.30257,  6:14.88415,  5:-4.98278,  4: 0.67941,
                   3: 1.06026,
##                 2: 5.48360,  1:-0.97088
##     25 |      | 0:115.12606, 19:-66.18334, 18:-81.21451, 17:59.74724,
                   16:-65.24387, 15:-91.23738, 14:48.93630, 13:63.63233,
                   12:17.89186,
##                 11:13.69802, 10:-49.68809,  9:48.28717,  8:-4.37396,
                   7:36.57324,  6:-239.25703,  5:-116.16022,  4:45.37873,
                   3:47.11736,
```

```
##                      2:-11.51299,  1:193.68028
##     26 |          | 20: 1.62858, 19:-6.50968, 18:-10.44102, 17:-22.89452,
                       16:-21.50499, 15:-24.28332, 14:12.23182, 13: 8.47606,
                       12:-39.59196,
##                      11:-38.13202, 10:14.99837,  9: 1.17484,  8:-42.70610,
                       7: 0.45940,  6:19.33571,  5:48.32096,  4: 2.15939,
                       3: 1.45809,
##                      2:-34.17036,  1: 4.24288
##     27 |          | 20:24.32948, 19:-21.69843, 18:-29.03406, 17:-39.08598,
                       16:-39.70897, 15:-33.39420, 14:22.44779, 13:21.58632,
                       12:-34.10506,
##                      11:-30.81281, 10: 8.56867,  9:13.75277,  8:15.13938,
                       7: 4.94741,  6:29.65086,  5:20.66949,  4:-30.40738,
                       3:-89.47778,
##                      2: 7.78840,  1:27.43818
##     28 |          | 20:74.58337, 19:-73.31054, 18:-54.84992, 17:38.25743,
                       16:-26.25908, 15:-6.00562, 14:51.08411, 13:57.09294,
                       12:-2.95093,
##                      11: 5.12013, 10:-112.35419,  9:81.15020,  8:53.54121,
                       7:63.02877,  6:169.16539,  5:-121.10607,  4:-34.08242,
                       3:-29.48601,
##                      2:-13.01993,  1:-169.10889
##     29 |          | 20:29.63905, 19:-10.05235, 18:-26.69251, 17:-51.13666,
                       16:-14.93525, 15:-15.09192, 14:20.62878, 13:20.32225,
                       12:-9.60404,
##                      11:-9.00413, 10:72.53555,  9:19.84532,  8: 7.09636,
                       7:60.42725,  6:35.26074,  5:-6.58941,  4: 4.21469,
                       3:-123.86588,
##                      2:-1.78124,  1: 5.16048
##     30 |          | 20:-1.49977, 19: 0.78230, 18: 3.44946, 17: 9.52544,
                       16: 5.21607, 15: 7.46949, 14: 2.12706, 13: 0.74737,
                       12: 7.23410,
##                      11: 7.26661, 10:-4.33287,  9: 1.54910,  8: 7.26920,
                       7:-28.82655,  6:10.88533,  5: 1.27544,  4: 1.88801,
                       3: 1.09380,
##                      2: 3.79257,  1: 1.51486
```

```
##      31 |          | 30: 9.60220, 29:-8.94521, 28:33.58144, 27: 1.55787,
                        26: 8.52473, 25:45.79420, 24:-4.24323, 23:-18.14175,
                        22: 5.18716,
##                      21:18.14762
## -------|------|---------------------------------------------------------
```

Not only does it not pay off, the bold attempt to use several hidden layers of up to ten nodes each yields a best estimated model that includes only one layer of only eight nodes. Despite feeling like a letdown in this case, the field of neural networks is a fascinating and growing area of machine learning.

That concludes the introduction to multilayer perceptrons. It also concludes our introduction to supervised machine learning methods. While in this chapter we have only explored a handful of models, we have explored several different model types. Most importantly, by using caret, rather than practicing a new structure each time, you have the ability to easily try dozens of likely models. Remember, each model type tends to have certain types of data with which it works best, and may well have various criteria in terms of ability to cope or not with categorical or numerical data. Additionally, some models have a requirement for or restriction against principal component analysis. Finally, it is important to keep in mind how a model might cope with missing data. Some methods require complete data—in which case the chapter on missing data may be helpful. Other methods are more robust to missing data.

Lastly, we gently remind our readers that, after determining which method is optimal, the correct final step would be to train the model one last time on all available data—especially in this case where the available data is not extensive. For models tuning, it can still be helpful to use cross-validation to discourage overfitting.

8.3 Summary

This chapter introduced and explored the caret package as a common interface to multiple models. Through this expediency, it becomes much more straightforward to examine multiple algorithms for optimal model fit. Additionally, methods of pre-processing data in preparation for modelling were examined. Finally, while not explicitly coded, the principle of separating raw data into training, validation, and test sets was discussed. Special attention was placed on the concept that all models are trained on the training data, a final, optimal model is selected using validation data, and real-world performance is estimated on that final model via the pristine, unused test data.

A summary of the functions used in this chapter is in Table 8-2.

Table 8-2. *Listing of Key Functions Described in This Chapter and Summary of What They Do*

Function	What It Does
boxplot()	Generates boxplot which allows visual inspection for outliers and other misshapes of data.
cbind()	Binds data together by columns.
colnames()	Lists the names of column headers in a dataset.
confusionMatrix()	Shows instances of actual vs. predicted outputs for categorical predictions.
copy()	Creates a full copy of data rather than using pointers. This prevents edits on the copy from cascading to the original dataset. It does double the memory.
cor()	Computes the correlation.
createDataPartition()	Caret package function that creates train/test data mindful of segmentation of data. In this case, it was important as data was regional.
dummyVars()	One hot encodes categorical data for regression.
expand.grid()	Uses all inputs in all possible combinations to create a data frame.
explain()	DALEX function required prior to using other DALEX methods.
fancyRpartPlot()	More colorful tree plot along with additional useful data.
gbm	Gradient boosting machine formal for caret.
hist()	Plots a histogram.
lapply()	List apply function.
mlpML	Multilayer perceptron (multilayer) formal for caret.
model_performance()	DALEX function that creates plottable residual objects.
par()	Can be used to set graphical parameters (such as columns or rows for plotting graphs) via R's default graphics package.
pca()	Principal component analysis from pcaMethods package.

(continued)

Table 8-2. (*continued*)

Function	What It Does
plot_grid()	Cowplot function used to set columns and rows for plotting multiple graphs—contrast with base R's graphics package function par().
prcomp()	Principal component analysis from base R stats package.
predict()	Given a model, predicts outputs from given inputs.
qqnorm()	Graphs the Q-Q plot.
ranger	Random forest formal for caret.
rpart	Classification and regression tree formal for caret.
sapply()	Simplify apply function.
scale()	Base R function to scale and center data.
shapiro.test()	Calculates the Shapiro-Wilke's normality criterion.
stats::residuals()	Calculates residuals from model (base R stats package version).
summary()	Provides a summary of the given input when possible.
svmLinear	Support vector machines (linear) formal for caret.
svmPoly	Support vector machines (polynomial) formal for caret.
train()	Caret function that trains the specified model via given training data.
trainControl()	Caret formal (and function) that provides input for the method of training provided (e.g., cross-validation and number of folds).
variable_importance()	DALEX function that computes dropout loss.
variable_response()	DALEX function that calculates partial dependence.
View()	Opens data frame/table in new viewing window for an interactive view.

CHAPTER 9

Missing Data

Missing data is common in nearly all real-world analysis. This chapter introduces the concept of missing data formally including common ways of describing missingness. Then we discuss some of the potential ways missing data can be addressed in analysis. The main package we will use in this chapter is the mice package, one package that offers robust features for handling missing data and minimizing the impact of missing data on analysis results [95].

```
library(checkpoint)
checkpoint("2018-09-28", R.version = "3.5.1",
  project = book_directory,
  checkpointLocation = checkpoint_directory,
  scanForPackages = FALSE,
  scan.rnw.with.knitr = TRUE, use.knitr = TRUE)

library(knitr)
library(ggplot2)
library(cowplot)
library(lattice)
library(viridis)
library(VIM)

library(mice)
library(micemd)
library(parallel)

library(data.table)
library(xtable)
library(JWileymisc) # has data

options(width = 70, digits = 2)
```

© Matt Wiley and Joshua F. Wiley 2019
M. Wiley and J. F. Wiley, *Advanced R Statistical Programming and Data Models*,
https://doi.org/10.1007/978-1-4842-2872-2_9

9.1 Conceptual Background

Missing data occurs when one or more variables have missing observations. There are many potential causes of missing data. Missing data also causes problems in data analysis:

- Estimates obtained from the non-missing data only may be biased.

- Missing data results in less efficiency by reducing information.

- Many tools and software are not equipped to handle missing data.

Bias occurs when data are not missing completely at random, so that the estimates obtained from observed data points are systematically different from the estimates that would have been obtained had all cases been observed. Loss of efficiency occurs because with some data missing, there is less information in the data. Further, many simple techniques for addressing missing data, such as complete case analysis (listwise deletion), discard cases where any variable is missing, even if data are available on other variables, again leading to a loss of efficiency. Finally, because many tools cannot handle missing data directly, analysis with missing data is often more time consuming and complex as it may require a first step to address the missing data before using standard tools.

There are multiple methods available to address each of the three problems of missing data discussed. However, before implementing a specific method, there are some general considerations. Fundamentally, we suppose that the goal is to estimate some parameter, θ, and obtain an estimate of the uncertainty of our estimate, $VAR(\theta)$. We assume that the data are sampled from the population we are interested in generalizing to, although this need not be the entire population. The fundamental questions are under what conditions and assumptions does an unbiased estimate of θ exist, and how do we obtain it?

Data are typically classified as

- Missing completely at random (MCAR) when the missingness mechanism is completely independent of the estimate of θ

- Missing at random (MAR) when the missingness mechanism is conditionally independent of the estimate of θ

- Missing not at random (MNAR) when the missingness mechanism is associated with the estimate of θ

If, and only if, the data are MCAR, then a complete case analysis (listwise deletion) will yield an unbiased estimate of θ. In all other cases, we must consider what assumptions we will make about the missingness process and whether, with the data available under our assumed missingness process, we can obtain an unbiased estimate of θ.

One systematic way of representing expected models and explicating the assumed causal and missingness processes is through the use of graphical models such as directed acyclic graphs (DAGs). DAGs have been used to develop ways of identifying under which models unbiased estimates may be recovered by Mohan, Pearl, and Tian [66] and extended to further causal queries in Mohan and Pearl [64], as well as tests for whether data are MCAR or at the variable level MAR+ [65].

There are a few practical implications of this research. First, depending on the causal process and the missingness mechanism, unbiased estimates of θ may or may not be recoverable, even from the same data. It is therefore paramount that model specification be considered. It is not sufficient to simply adjust for all possible variables as other work demonstrates that a previously unbiased estimate of θ may become biased when adding inappropriate variables or paths (the "back-door criterion") to the DAG [73]. Second, rigorously explicating the model and the conditions under which unbiased estimates can or cannot be obtained is both complex and requires knowledge or expectations about both the causal process and the missingness mechanism in greater detail than most researchers and the state of science will have available. The complexity grows exponentially as the number of variables considered increases, particularly when there is missingness on multiple variables.

Although not addressed at length in the theoretical work of Mohan and Pearl, when applied practically it is necessary not only that the correct variables and paths be specified but that the functional form of the relationship be correctly specified. Thus, for $P(Y|f(X))$, $f(X)$ may be a linear or some other non-linear function, and the unbiased estimate of θ will depend on correctly specifying this form. In practice, at the least, the form of the relationship for observed values should be examined and where necessary be allowed to be non-linear.

Although it is beyond the scope of this chapter to provide the necessary background on specific variables to decide on the missing data mechanisms, they are worth considering, and at least if there is no evidence to guide decisions, this should be noted as a limitation in research. The next section discusses multiple imputation as one approach for addressing missing data and provides some options for dealing with unknown functional forms, at least.

Multiple Imputation

General

Many imputation methods exist. Single imputations (e.g., mean and hot deck imputation) as well as conditional mean imputation are discouraged and not discussed further. The problem with single imputations, whatever the imputation model, is that they assume no error in the predictions for missing values. This results in the uncertainty estimates for models based on the imputed data being too low, increasing the type 1 error rate. The degree of impact will depend on the amount of missing data and how accurate (or inaccurate) the imputation model is.

Instead of imputing a dataset a single time, they can be imputed multiple times. The principle behind multiple imputation is that rather than generating a single predicted value for imputation, multiple imputations are drawn randomly from the predicted distribution. The spread of the predicted distribution captures the uncertainty in the predictions, and this results in variability in each imputed dataset.

The general process to generate multiply imputed data is as follows:

- An initial dataset with missing data.

- Fill in the missing data values with initial estimates. Initial estimates are typically easy to generate, such as the mean or median for each variable or random values drawn from the variable.

- For each variable, build a model to predict it from the remaining variables. Use the model to predict the missing data. For parametric models, such as linear regression, draw a random value from the predicted distribution (e.g., normal with mean based on the predicted value and standard deviation capturing uncertainty) or sample from the posterior distribution in Bayesian models. For nonparametric models, such as random forest models [28], uncertainty may be introduced through other means such as bootstrapping to build an empirical distribution around the predicted values and sample from this.

- Repeat the previous step until the predicted complete dataset does not differ substantially from the previous iteration, suggesting the models have converged. This step is needed as the models built may change over time as the initial estimates (often quite bad) become more accurate.

- All these steps are repeated to generate the desired number of multiply imputed datasets (e.g., 5, 100).

Once multiply imputed datasets have been generated, the analysis varies slightly from analyzing a single dataset with no missingness. Multiply imputed data are analyzed separately, and results are pooled using Rubin's rules [58]. Briefly:

- The model is estimated separately in each imputed dataset.

- Many parameters from each model are averaged, including regression coefficients, mean differences, and predicted values. Some values are combined not directly by averaging, such as estimates of the residual variance or correlations, which are typically transformed prior to averaging.

- Uncertainty estimates (e.g., standard errors) for the pooled results include two sources of variability:

  - The average uncertainty in parameters from each model

  - The variability in parameter estimates between models

- Statistical tests based on the average parameters and uncertainty estimates that capture uncertainty from both models and the different datasets.

In general, any analysis should happen on the individual datasets, and then only as the very final step, the results should be pooled across datasets. For instance, if the goal is not just a regression model, but to generate a graph of predicted values, the predicted values should be generated separately for each imputed dataset, and then the predictions combined. Do not first pool the regression model results and then used the pooled model to generate predictions. One common misconception about multiple imputation is that only independent variables should be imputed, and that dependent variables should not be imputed. Simulations have demonstrated that not imputing dependent variables increases bias [67], and it is recommended that all variables be imputed [38]

Approaches to Multiple Imputation

In general, models to multiply impute data take one of two forms:

> JM: A joint model (JM) where the joint, multivariate distribution of all variables is specified. Values are imputed by drawing from the conditional distribution.

> FCS: A fully conditional specification (FCS) where a separate model is built for each missing variable, using all other variables as predictors. Values for each missing variable are imputed by drawing from the conditional densities of the model for that specific variable.

The **JM approach** can be shown to be theoretically valid [85], which is an advantage over FCS methods, which currently lack a strong theoretical justification for why they may produce valid results.

A challenge with the JM approach is that it requires the multivariate distribution to (1) be specified, (2) be a reasonable representation of the data, and (3) be analytically tractable. Often, some but not all of these conditions are readily fulfilled. For example, using a multivariate normal (MVN) model is relatively easy to specify and analytically easy to estimate and draw from. However, only in a subset of research cases can all variables with missing data be reasonably construed as normally distributed. For instance, as soon as a single dichotomous covariate (e.g., sex) is introduced, the MVN model begins to be unrealistic. The MVN model still is used sometimes in cases where the data are unlikely truly MVN as an approximation, but this highlights a drawback of the JM. MVN models are not the only type of multivariate distribution that can be specified. However, multivariate distributions involving both normally distributed and dichotomous or count type variables become extremely difficult to estimate and draw from, often making them analytically intractable.

In our view, the JM approach is particularly difficult in practical settings as many applied research questions with missing data (e.g., epidemiological studies, longitudinal studies) often have models that include many covariates and focal predictors, let alone potential determinants of dropout or missingness that, although not part of the final analytical model, may be critical to include to improve the missing data model. It seems rare that so many different variables will all follow any convenient multivariate distribution, making the JM approach impractical in many circumstances.

Although it lacks the strong theoretical foundations to support it as an appropriate technique, the **FCS approach** is a compelling alternative to the JM approach. The FCS approach is also known as multiple imputation through chained equations (MICE) as there are a series of separate equations. Where there are k variables, rather than define a k dimensional multivariate density, k univariate densities are defined. This both simplifies the computational aspect and makes it relatively easy to use any available distribution, thus accommodating a wide variety of outcomes (e.g., normal for continuous, Poisson for count, binomial for binary, gamma for strictly positive continuous, outcomes, etc.). Another advantage of FCS is that it becomes possible to specify very flexible functional forms for the relations between predictors and each outcome.

A standard way of conducting FCS is by taking Gibbs samples from the (conditional) posterior distribution of each variable, generally of the form $P\left(Y_i | Y_{-i}, \theta_i\right)$ for i in 1,…, k. These can be used to "fill in" the missing data for each k outcome. Once all outcomes have been filled in, the process can be repeated now using both the observed and filled-in information on predictors for each outcome, and this process is iteratively repeated until convergence is achieved. Once convergence is achieved, as many samples as desired can be drawn from the conditional posterior distributions to generate the desired number of multiply imputed datasets.

The FCS (MICE) has been described in detail by Van Buuren, Brand, Groothuis-Oudshoorn, and Rubin [94] and also Van Buuren and Groothuis-Oudshoorn [95]. One issue with FCS is that the conditional distributions may not be compatible. When the conditional distributions are not compatible, the order of variable imputation matters, although these effects may be relatively small. The conditions under which the conditional distributions are compatible and thus yield equivalent draws as from the joint distribution have been discussed by Hughes and colleagues [46], and they also provide a simulation examining the performance of FCS when distributions are and are not compatible. In general, although theoretically less well justified than the JM approach, in simulations and in practice, the FCS appears to perform remarkably well and, given its less restrictive assumptions, can outperform the JM approach when the JM assumptions are not tenable. Because of its increased flexibility, this chapter focuses on the FCS approach to imputation.

Non-linear Effects and Non-normal Outcomes

In many applications, the default assumption is that variables linearly relate to each other. However, this assumption should be checked and non-linear functional forms allowed as dictated by theory or by the data. Even when the data are MAR, for the estimates to be unbiased, the model must be correctly specified including both the correct variables and the correct functional form.

Previously, so-called passive imputation was employed for non-linear terms. For example, for interactions between two variables, X_1, X_2, each variable would be imputed separately, and then after the multiple imputation in the analysis, the product term, X_1X_2, would be formed based on the imputed X_1 and X_2 values, and likewise for squared terms (X_1^2) and other non-linear terms. Simulations have demonstrated that such passive imputation results in biased estimates [101]. Instead, these non-linear transformations should be incorporated into the multiple imputation by first creating the variables and then including them as additional variables in the imputation model. The interactions, squared, or otherwise transformed variables are sometimes called "just another variable" (JAV). However, a limitation of such an approach is that the relationship between X_1 and X_1^2 (or the interaction) is lost. Another approach has been developed by Vink and van Buuren [100] called the polynomial combination approach that both yields unbiased estimates and preserves the relationship between X_1 and X_1^2. However, to our knowledge it is currently only available for squared terms, not for interactions between different variables nor for higher-order polynomials, although theoretically the work may generalize.

A related issue is skewed or non-normally distributed variables. One intuitive approach is to attempt some normalizing transformation; however, von Hippel [102] showed that this results in less than optimal results, and in many cases suggested simply using a normal model. Alternatives are to use the (correct) distribution for each outcome (e.g., gamma, beta, etc.). Currently, most simulations and research have examined only the properties of multiple imputation for normal, binomial outcomes, with somewhat fewer examining ordered outcomes; thus, little guidance is available for variables coming from alternative distributions. Further most software implementations allow a fairly limited set of known distributions.

Just as interactions or other non-linear terms should be imputed where needed, if some other variable, Y, depends not only on X_1, X_2, but also their interaction X_1X_2, in order to properly specify the model, the interaction term must be included as a predictor in the FCS. If Y depends on X_1, X_2, and X_1X_2, but X_1X_2 are omitted from the imputation model, the results will be biased. However, if Y depends on X_1, X_2, but X_1, X_2, and X_1X_2 are all included in the imputation model, the results will not be biased. This is explored in greater detail by Bartlett, Seaman, White, and Carpenter [3] who show that the substantive analytical model can be a restricted version of the imputation model, but not vice versa. Specifically they advise ensuring that the imputation model is compatible with the substantive model or at least is semi-compatible (i.e., the substantive model is a restricted version of the imputation model) .

GLMs for Imputation

The FCS approach does not require any specific model, but the most common class of models used are generalized linear models (GLMs), such as linear regression for continuous, approximately normally distributed variables or logistic regression for dichotomous variables. For interested readers, GLMs are covered in more depth in Chapters 3 and 4. One of the benefits of GLMs is that they are a familiar analytical model for many analysts, making it easier to understand how they are used in multiple imputation. Also, most GLMs are very fast to estimate on modern computers, making this a relatively low-cost computationally type of model to use. Despite their popularity and their speed, there are downsides to GLMs.

First, GLMs default to a linear relationship between all variables and the outcome on the scale of the link function. Although this is always true for any GLM model, the issue is exacerbated by the fact that multiple imputation typically involves many relationships. For example, consider a relatively simple research question: is obesity associated with blood pressure? Blood pressure itself has two components: systolic and diastolic pressure. Further a number of factors may confound this association, and so be relevant covariates, two likely candidates being age and socioeconomic status. Finally, factors related to missing data also should be included. For example, if it is known that employed participants were more likely to have missing data, it would make sense to include that in the imputation model. In this example, barely beginning to consider the myriad factors related to obesity, blood pressure, and potentially missingness, we are dealing with *six* variables. In the final analysis, all we care about are the relations of stress, age, and socioeconomic status (three variables) with blood pressure (two variables), so we only have to worry about six unique pairwise associations. However, for the imputation model, we have *15* unique pairwise associations. If any of those are not linear, either our imputation model will be wrong or we need to identify which is not linear and include the appropriate functional form (e.g., quadratic, etc.). Many applied research questions will include far more variables, either because the fundamental research question involves more variables or because there are more relevant covariates or more factors likely related to missingness. Having the analyst manually decide the functional form of every possible pairwise association quickly becomes impractical (e.g., there are 45 unique pairs for 10 variables; 190 unique pairs for 20 variables, etc.) .

Second, GLMs do not include any interactions between variables, by default. Again as the number of variables in the imputation model increase, the possible number of interactions rapidly becomes impractical for an analyst to manually specify.

Third, especially in smaller datasets or in certain applications, the number of variables to be included may be large relative to the number of observations. In these cases, GLMs will tend to overfit the data and especially for dichotomous outcomes may result in complete separation and estimability challenges, the fundamental issue being that GLMs do not include any method for variable selection by default.

GAMs for Imputation

Faced with the challenge of an unknown functional form, a natural solution is to utilize models that attempt to empirically approximate an unknown form. One such class of models are generalized additive models (GAMs) [40, 122], which use smoothers and penalties to attempt to approximate the functional form without (terribly) overfitting the data. An extension of a "basic" GAM is GAMs for location, scale, and shape parameters (GAMLSS). GAMLSS allow not only the location (e.g., mean in a normal distribution) to be modelled but also allow the scale (variance) of the distribution to be a function of predictors as well as the shape of the distribution [78, 88]. Although GAMs are generally well developed (e.g., [40, 122, 123]), there is little research examining their properties when used in multiple imputation.

One simulation study by de Jong, van Buuren, and Spiess [27] demonstrates good results from using GAMLSS in multiple imputation to allow skewed and non-normally distributed outcomes and non-linear effects. In general, imputation using GAMLSS provided better coverage than using GLMs; however, it is worth noting that the simulation study by [27] only examined a few variables. Non-convergence and other computational challenges may emerge in applied research settings with numerous predictors.

RFs for Imputation

Random forests (RFs) are a type of machine learning and are widely used for predictive modelling as they tend to deliver excellent results [16]. For imputation, RFs offer several advantages over previously discussed alternatives.

One of the challenges we noted in imputation is the need to correctly capture the functional form, which may not be linear, between all variables. GAMs offer one possible solution to this, but RFs can also address this issue as the trees allow splits at different points of a variable allowing non-linear associations.

Unlike GAMs or GLMs, RFs can also identify important interactions between variables. RFs identify interactions efficiently even when systematically screening all possibilities is impractical. This feature of RFs offers a great advantage, as with ten or more variables, it would be virtually impossible for a researcher to specify all interactions between variables that are relevant and even including all the interactions would result in a complex model that may require a very large sample size.

Finally, RFs are advantageous as they essentially have a degree of built-in variable selection. RFs only split on variables that help predict the outcome. Consequently, unlike GLMs or GAMs, RFs can better handle cases where there are a large number of variables in the imputation model relative to the number of cases in the dataset.

Although random forests are very commonly used in prediction and machine learning literatures, there is relatively little research evaluating their performance for imputation. One of the earlier papers examining RFs for imputation was by Stekhoven and Bühlmann [90] who proposed a RF model imputation in the bioinformatics field along with an R package, `missForest`. However, Stekhoven and Bühlmann [90] proposed using RFs for single imputation only, which as noted earlier will result in underestimating uncertainty due to missing data. More recently, two other teams independently developed a method for using RFs for multiple imputation and demonstrated generally good performance of these techniques [28, 86]. In both cases, the models rely on bootstrapping to propagate the uncertainty in imputing missing values to generate multiple imputed datasets.

Other Cases

In addition to the multiple imputation situations discussed thus far, there are a number of more specialized cases that require additional methods. Multiple imputation is complicated in the case of time to event or survival outcomes, and imputation of covariates in these cases is discussed by White and Royston [108]. For multiple imputation with multilevel data, a recent overview is available by van Buuren [93]. Although not directly multiple imputation, a multilevel model that accommodates missing data in both responses and covariates (including non-normal outcomes/covariates and interaction terms) has been described by Goldstein, Carpenter, and Browne [37]. However, the complexity of such an analysis may be prohibitive in many cases.

9.2 R Examples

To apply the missing data approaches discussed, we will use data simulated to closely match a daily study. Most of the imputation models discussed only apply to single-level data, so we will begin by collapsing the daily data into averages across the study.

```
## load example dataset
data("aces_daily")
draw <- as.data.table(aces_daily)[order(UserID)]
davg <- na.omit(draw[, .(
  Female = na.omit(Female)[1],
  Age = na.omit(Age)[1],
  SES_1 = na.omit(SES_1)[1],
  EDU = na.omit(EDU)[1],
  STRESS = mean(STRESS, na.rm = TRUE),
  SUPPORT = mean(SUPPORT, na.rm = TRUE),
  PosAff = mean(PosAff, na.rm = TRUE),
  NegAff = mean(NegAff, na.rm = TRUE)),
  by = UserID])
```

Next, we want to add some missingness into the data. For this example, we will let the probability of being missing depend on a combination of level of perceived support and stress. Then, for a few variables, we generate missingness based on these probabilities. We randomly set stress and support to missing.

```
## missing depending on support and stress
davg[, MissingProb := ifelse(
        SUPPORT < 5,
          ifelse(STRESS > 2.5, .4, .0),
          ifelse(STRESS > 2.5, 0, .4))]

set.seed(1234)
davgmiss <- copy(davg)
davgmiss[, PosAff := ifelse(rbinom(
            .N, size = 1, prob = MissingProb) == 1,
            NA, PosAff)]
davgmiss[, NegAff := ifelse(rbinom(
            .N, size = 1, prob = MissingProb) == 1,
            NA, NegAff)]
## random missingness on stress and support
davgmiss[, STRESS := ifelse(rbinom(
            .N, size = 1, prob = .1) == 1,
            NA, STRESS)]
```

```
davgmiss[, SUPPORT := ifelse(rbinom(
            .N, size = 1, prob = .1) == 1,
            NA, SUPPORT)]
davgmiss[, Age := ifelse(rbinom(
            .N, size = 1, prob = .1) == 1,
            NA, Age)]
davgmiss[, SES_1 := ifelse(rbinom(
            .N, size = 1, prob = .1) == 1,
            NA, SES_1)]
davgmiss[, Female := factor(ifelse(rbinom(
            .N, size = 1, prob = .1) == 1,
            NA, Female), levels = 0:1,
            labels = c("Male", "Female"))]
davgmiss[, EDU := factor(ifelse(rbinom(
            .N, size = 1, prob = .1) == 1,
            NA, EDU), levels = 0:1,
            labels = c("< Uni Graduate", "Uni Graduate +"))]
## drop unneeded variables to make analysis easier
davgmiss[, MissingProb := NULL]
davgmiss[, UserID := NULL]
```

Now to begin, it is helpful to visually examine the missing data patterns in a dataset. This can be done using the VIM package and the aggr() function as shown in Figure 9-1. The left side shows the percentage of missing data on each variable total. The right side shows different patterns of missingness across the study variables, along with their proportion. Critically, we can see that even though the overall rate of missing data is not that high around 15%, only about one third of participants actually have complete data on all variables. This would make a complete case analysis have a large loss of information and power.

```
ggr(davgmiss, prop = TRUE,
    numbers = TRUE)
```

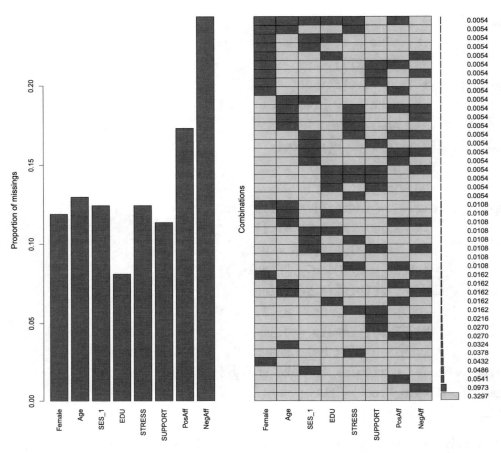

Figure 9-1. *Visual summary of the missingness by variable and missingness patterns*

The VIM package also has a function to help identify whether missingness depends on another variable, using the marginplot() function. These graphs shown in Figure 9-2 include quite a bit of information. The center is a basic scatter plot for the pairwise present data. The margins show the spread of missing data, and the boxplots show the distribution of each variable stratified by whether the other variable is missing or not. These plots make it evident that when affect is missing, stress tends to be higher and support lower.

```
par(mfrow = c(2, 2))
marginplot(davgmiss[,.(STRESS, NegAff)])
marginplot(davgmiss[,.(SUPPORT, NegAff)])
marginplot(davgmiss[,.(STRESS, PosAff)])
marginplot(davgmiss[,.(SUPPORT, PosAff)])
```

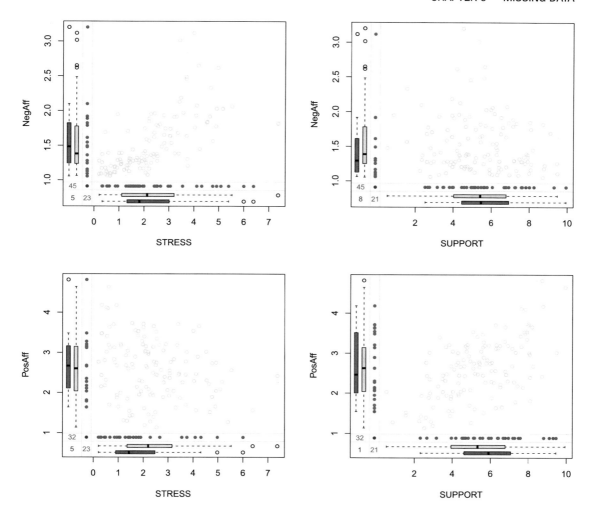

Figure 9-2. *Bivariate plots of missing data. Central dots show the non-missing data. Margin dots show missing data. Boxplots summarize each variable by whether the other variable is missing or present.*

We can apply a statistical test to examine whether missingness on a variable is related to another variable, for example, using t-tests as shown next. We see no significant differences in stress and support by affect. However, in general, relying on statistical significance for determining whether to include a variable in the imputation model is discouraged, as depending on the magnitude of effect and sample size, results may not be statistically significant yet may have an impact on the results of models analyzed on the multiply imputed data.

```
## does age differ by missing on negative affect?
t.test(Age ~ is.na(NegAff), data = davgmiss)$p.value
```

```
## [1] 0.9
```

```
## does age differ by missing on positive affect?
t.test(Age ~ is.na(PosAff), data = davgmiss)$p.value
```

```
## [1] 0.14
```

```
## does stress differ by missing on negative affect?
t.test(STRESS ~ is.na(NegAff), data = davgmiss)$p.value
```

```
## [1] 0.89
```

```
## does stress differ by missing on positive affect?
t.test(STRESS ~ is.na(PosAff), data = davgmiss)$p.value
```

```
## [1] 0.17
```

```
## does social support differ by missing on negative affect?
t.test(SUPPORT ~ is.na(NegAff), data = davgmiss)$p.value
```

```
## [1] 0.49
```

```
## does social support differ by missing on positive affect?
t.test(SUPPORT ~ is.na(PosAff), data = davgmiss)$p.value
```

```
## [1] 0.17
```

Multiple Imputation with Regression

In this example, we will multiply impute the data using regression methods. Because multiple imputation involves a random component (like bootstrapping), the final results will vary randomly each time it is conducted. To make results reproducible, a random seed can be set. For the sake of example and speed, we are only generating six imputed datasets. For actual research, using more (e.g., 50, 100) can be helpful to ensure that even with a different seed, the final results of the pooled models will be more or less identical.

The following code uses the mice() function from the mice package to impute any missing data in the dataset, davgmiss. We ask for six imputations, with m = 6. The default method specifies the type of model to use for continuous (numeric in R) variables, binary variables, nominal (polytomous) variables, and ordered categorical variables. We begin with ten iterations of the algorithm, set the seed for reproducibility, and turn off messages during imputation. The call to `system.time()` will return the number of seconds the imputation took. Note that the results of `mice()` is a `mids` class object, which contains the original data and all the multiple imputations, in one object.

```
system.time(mi.1 <- mice(
  davgmiss,
  m = 6,   maxit = 10,
  defaultMethod = c("norm", "logreg", "polyreg", "polr"),
  seed = 1234, printFlag = FALSE)
)

##    user  system elapsed
##     2.9     0.0     3.0
```

Using a limited number of iterations, it is possible that the model did not converge. Convergence can be checked through plots, similar to Bayesian approaches. Each imputation is plotted with a separate color and the results are converged if they have become stationary, and each of the separate imputations is not systematically different from the other (which may indicate converging to separate local maxima). If there is doubt regarding convergence, we can allow further iterations using `mice.mids()` without having to re-run the initial ten iterations. The results appear reasonably stationary, and there is no clear indication of systematic differences providing reasonable evidence for convergence.

```
## plot convergence diagnostics
plot(mi.1, PosAff + NegAff + SUPPORT ~ .it | .ms)

## run an additional iterations
system.time(mi.1 <- mice.mids(
  mi.1, maxit = 10,
  printFlag = FALSE)
)

##    user  system elapsed
##       3       0       3

## plot convergence diagnostics
plot(mi.1, PosAff + NegAff + SUPPORT ~ .it | .ms)
```

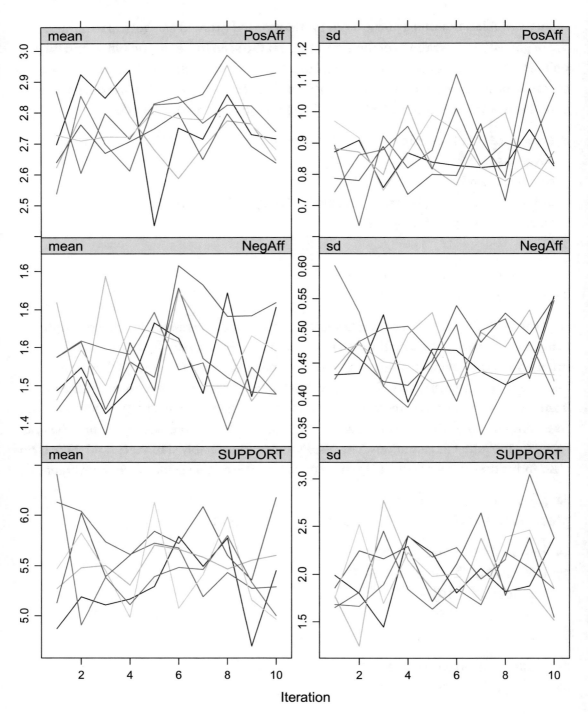

Figure 9-3. *Mice diagnostics for convergence*

Beyond checking if the model converged, it is helpful to evaluate whether the imputed values appear plausible. This can be done by plotting the distribution of the observed and imputed values using the densityplot() from the mice package. Beyond a univariate assessment, bivariate relations can be tested using the xyplot() function also from the mice package. Blue is used for the original observed data and the reddish color for the imputed data. One minor point is that using the regression-based methods, imputations have been generated outside the range of the data, particularly with predicted affect values below 1, which is outside the range of the scale. While this may not seem ideal, it is recommended to keep these values, as excluding them or forcing them to be at 1 reduces the variability in predictions and can make the model appear more certain than it should be.

```
densityplot(mi.1, ~ PosAff + NegAff + SUPPORT + STRESS)

xyplot(mi.1, NegAff + PosAff ~ STRESS + SUPPORT)
```

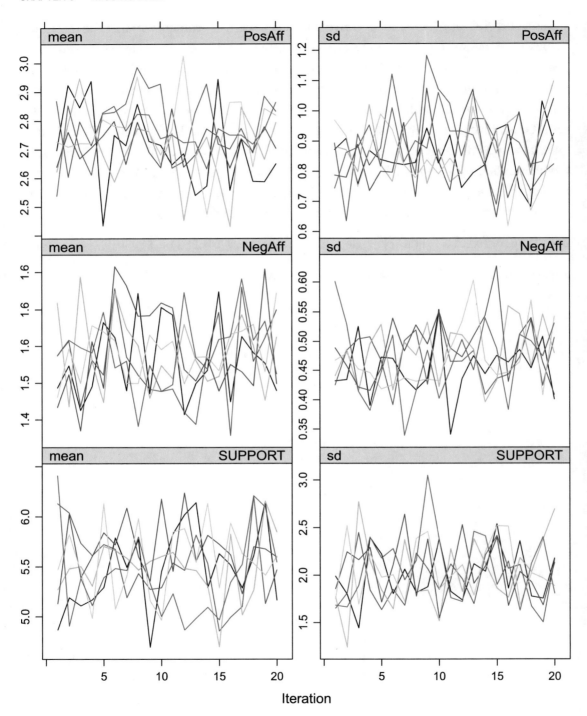

Figure 9-4. *Mice diagnostics for convergence after more iterations*

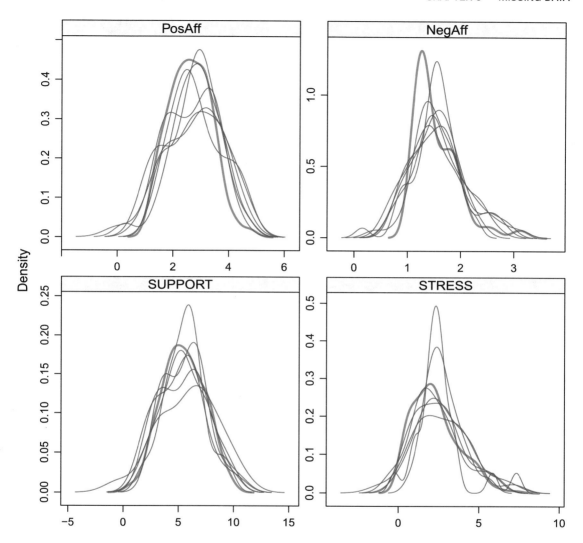

Figure 9-5. *Univariate density plots for observed and imputed data, separated by imputation*

If at this stage, we are comfortable that nothing problematic happened in the imputation model, we can proceed to fit our primary analyses. The mice package includes methods for the with() function to apply a given R expression to each imputed dataset when using it with a mids class object returned from mice(). To begin with, we will just run a linear regression model with positive affect as the outcome and stress as the main explanatory factor, with other factors included as covariates. The results are a mira class object, which is an object class that contains the same analysis repeated on separate multiply imputed datasets. If we print the object, we get the results from each individual regression model.

403

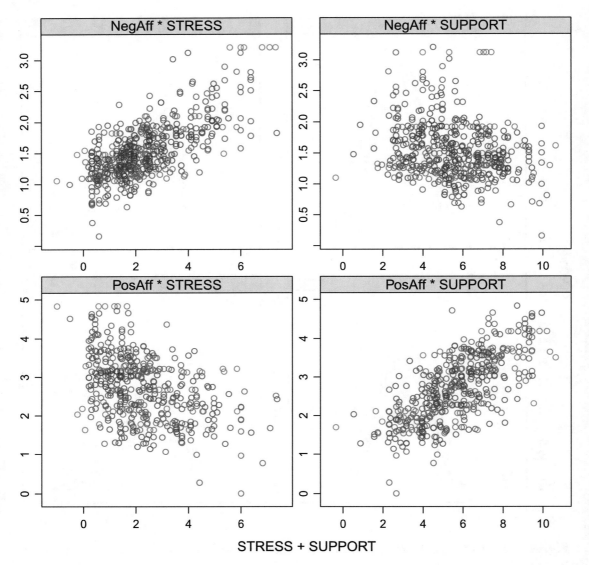

Figure 9-6. *Bivariate scatter plots with imputed data colored separately by observed and imputed data*

```
lm.1 <- with(mi.1, lm(PosAff ~ STRESS + Age + EDU + Female))

lm.1

## call :
## with.mids(data = mi.1, expr = lm(PosAff ~ STRESS + Age + EDU +
##      Female))
##
```

```
## call1 :
## mice.mids(obj = mi.1, maxit = 10, printFlag = FALSE)
##
## nmis :
##  Female      Age    SES_1      EDU  STRESS SUPPORT  PosAff  NegAff
##      22       24       23       15      23      21      32      45
##
## analyses :
## [[1]]
##
## Call:
## lm(formula = PosAff ~ STRESS + Age + EDU + Female)
##
## Coefficients:
##        (Intercept)                STRESS                   Age
##             4.1174               -0.2194               -0.0459
## EDUUni Graduate +        FemaleFemale
##            -0.0715                0.0450
##
##
## [[2]]
##
## Call:
## lm(formula = PosAff ~ STRESS + Age + EDU + Female)
##
## Coefficients:
##        (Intercept)                STRESS                   Age
##             4.0037               -0.1697               -0.0415
## EDUUni Graduate +        FemaleFemale
##            -0.0967               -0.0464
##
##
## [[3]]
##
```

```
## Call:
## lm(formula = PosAff ~ STRESS + Age + EDU + Female)
##
## Coefficients:
##       (Intercept)              STRESS                 Age
##            3.8330             -0.2022             -0.0266
## EDUUni Graduate +       FemaleFemale
##           -0.0267             -0.2389
##
##
## [[4]]
##
## Call:
## lm(formula = PosAff ~ STRESS + Age + EDU + Female)
##
## Coefficients:
##       (Intercept)              STRESS                 Age
##            4.2129             -0.2214             -0.0466
## EDUUni Graduate +       FemaleFemale
##           -0.0529             -0.0799
##
##
## [[5]]
##
## Call:
## lm(formula = PosAff ~ STRESS + Age + EDU + Female)
##
## Coefficients:
##       (Intercept)              STRESS                 Age
##            3.6608             -0.1867             -0.0221
## EDUUni Graduate +       FemaleFemale
##           -0.1658             -0.0691
##
##
## [[6]]
##
```

```
## Call:
## lm(formula = PosAff ~ STRESS + Age + EDU + Female)
##
## Coefficients:
##        (Intercept)              STRESS                 Age
##             3.6097             -0.1872             -0.0232
## EDUUni Graduate +       FemaleFemale
##            -0.0190             -0.0252
```

We could examine the individual models for model diagnostics as shown in Figure 9-7.

```
par(mfcol = c(2,2 ))
plot(lm.1$analyses[[1]])

par(mfcol = c(1,1))
```

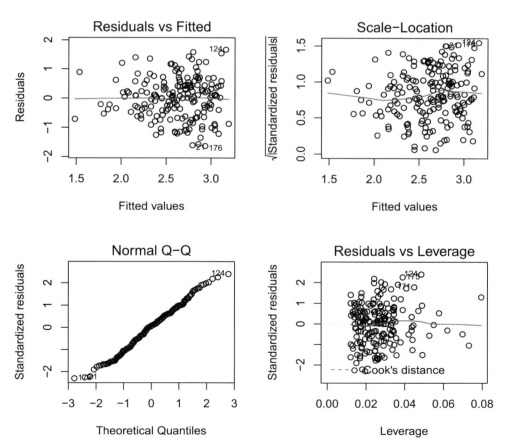

Figure 9-7. *Linear regression model diagnostics from first imputed dataset*

407

However, generally, we are less interested in the individual models and instead want to see what the overall results are pooled across models. The `pool()` function from the `mice` package is designed to do exactly that. Combined with `summary()`, we can get the usual linear regression summary table, but based on the pooled results from the multiply imputed data. The results are the pooled regression coefficients, their standard errors, t-values, estimated degrees of freedom, p-value, and confidence interval. To format it all nicely for the book, we use the `xtable()` function, and the results are in Table 9-1. To do this, not using LaTeX, simply run only `summary(pool(lm.1), conf.int = TRUE)`.

```
xtable(summary(pool(lm.1), conf.int=TRUE),
  digits = 2,
  caption = "Regression results pooled across multiply imputed data",
  label = "tmd-pooledres1")
```

Table 9-1. *Regression Results Pooled Across Multiply Imputed Data*

	Estimate	Std. Error	Statistic	df	p.value	2.5%	97.5%
(Intercept)	3.91	0.68	5.71	89.89	0.00	2.55	5.27
STRESS	−0.20	0.04	−4.48	50.53	0.00	−0.29	−0.11
Age	−0.03	0.03	−1.07	88.88	0.29	−0.10	0.03
EDUUni Graduate +	−0.07	0.15	−0.50	83.78	0.62	−0.36	0.22
FemaleFemale	−0.07	0.15	−0.45	19.98	0.65	−0.39	0.25

To get the pooled R^2 from the model, we can use the `pool.r.squared()` function.

```
pool.r.squared(lm.1)
```

```
##      est lo 95 hi 95 fmi
## R^2 0.14 0.054  0.26 NaN
```

Some additional columns provide information about the amount of missingness and the fraction of missing information, which is an indicator of how much the missing data impacted that specific coefficient, by specifying the argument `type = "all"`. These results are in Table 9-2.

Table 9-2. *Regression Results Pooled Across Multiply Imputed Data with Additional Information*

	Estimate	Std. Error	Statistic	df	p.value	2.5%	97.5%	riv	Lambda	fmi	ubar	b
(Intercept)	3.91	0.68	5.71	89.89	0.00	2.55	5.27	0.18	0.15	0.17	0.40	0.06
STRESS	−0.20	0.04	−4.48	50.53	0.00	−0.29	−0.11	0.33	0.25	0.28	0.00	0.00
Age	−0.03	0.03	−1.07	88.88	0.29	−0.10	0.03	0.18	0.15	0.17	0.00	0.00
EDUUni Graduate +	−0.07	0.15	−0.50	83.78	0.62	−0.36	0.22	0.19	0.16	0.18	0.02	0.00
Female Female	−0.07	0.15	−0.45	19.98	0.65	−0.39	0.25	0.81	0.45	0.49	0.01	0.01

```
xtable(summary(pool(lm.1), type = "all", conf.int=TRUE),
  digits = 2,
  caption = "Regression results pooled across multiply imputed data with
  additional information",
  label = "tmd-pooledres1alt")
```

Finally, suppose that we wanted to get the predicted regression line for the relationship between stress and positive affect. We first set up a dataset and then generate predictions from each separate regression model, and then average the results.

```
newdat <- data.frame(
  STRESS = seq(from = 0, to = 6, length.out = 100),
  Age = mean(davg$Age),
  EDU = factor("< Uni Graduate", levels = levels(davgmiss$EDU)),
  Female = factor("Female", levels = levels(davgmiss$Female)))

newdat$PosAff <- rowMeans(sapply(1:6, function(i) {
  predict(lm.1$analyses[[i]], newdata = newdat)
}))

ggplot(newdat, aes(STRESS, PosAff)) +
  geom_line()
```

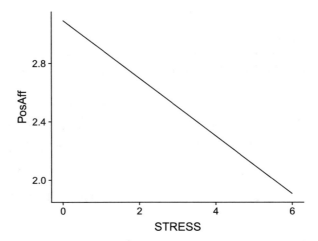

Figure 9-8. *Pooled predictions from linear regression models of the association between stress and positive affect*

Multiple Imputation with Parallel Processing

Although imputation based on regression tends to be fairly fast, with more cases, more variables, more imputations, and especially with more complex imputation methods, suggested as random forests, multiple imputation can be computationally demanding and time consuming. Fortunately, it is straightforward to parallelize to take advantage of multiple cores, if available.

This example shows how to parallelize multiple imputation using a method that will work on both Windows and Linux and Mac operating systems. There are easier strategies on Linux and Mac, but this approach is the most cross-compatible.

First we create a local cluster with two worker processes. If you have more cores, you could set this higher. If you only have two cores, set to two, etc. Next, we make sure that the packages are loaded onto each worker process. Since we will have each worker just work on one imputation, to make the results replicable, we need a separate random seed for each imputation. Finally, we need to export the dataset and the random seeds to each worker for processing.

```
cl <- makeCluster(2)
clusterExport(cl, c("book_directory", "checkpoint_directory" ))

clusterEvalQ(cl, {
  library(checkpoint)
```

```
  checkpoint("2018-09-28", R.version = "3.5.1",
  project = book_directory,
  checkpointLocation = checkpoint_directory,
  scanForPackages = FALSE,
  scan.rnw.with.knitr = TRUE, use.knitr = TRUE)
  library(mice)
  library(randomForest)
  library(data.table)
})
## [[1]]
##  [1] "data.table"    "randomForest"  "mice"         "lattice"
##  [5] "checkpoint"    "RevoUtils"     "stats"        "graphics"
##  [9] "grDevices"     "utils"         "datasets"     "RevoUtilsMath"
## [13] "methods"       "base"
##
## [[2]]
##  [1] "data.table"    "randomForest"  "mice"         "lattice"
##  [5] "checkpoint"    "RevoUtils"     "stats"        "graphics"
##  [9] "grDevices"     "utils"         "datasets"     "RevoUtilsMath"
## [13] "methods"       "base"

imputation_seeds <- c(
  403L, 2L, 2118700268L, 1567504751L,
  -161759579L, -1822093220L)

clusterExport(cl, c("davgmiss", "imputation_seeds"))
```

Now that everything is set up, we can loop from 1 to 6 to get our six multiply imputed datasets. Each are passed to a worker process for the actual running, using the parLapplyLB() function.

```
system.time(mi.par <- parLapplyLB(cl, 1:6, function(i) {
mice(
  davgmiss,
  m = 1,    maxit = 20,
  defaultMethod = c("norm", "logreg", "polyreg", "polr"),
  seed = imputation_seeds[i])
}))
```

```
##     user  system elapsed
##      0.0     0.0     3.5
```

Finally, previously, because we asked `mice()` to do multiple imputations directly, the imputations were already combined into one big object. To split out for parallel processing, we only asked for one imputation each time, so we need to manually combine them, which we do with the `ibind()` function to create a single `mids` class object with the six imputations. When we print the results, we can see the number of multiple imputations is six.

```
## combine the separate imputations into a single object
mi.par2 <- ibind(mi.par[[1]], mi.par[[2]])
for (i in 3:6) {
  mi.par2 <- ibind(mi.par2, mi.par[[i]])
}

mi.par2
```

```
## Class: mids
## Number of multiple imputations:  6
## Imputation methods:
##    Female       Age     SES_1       EDU    STRESS   SUPPORT    PosAff
## "logreg"    "norm"    "norm"  "logreg"    "norm"    "norm"    "norm"
##    NegAff
##    "norm"
## PredictorMatrix:
##          Female Age SES_1 EDU STRESS SUPPORT PosAff NegAff
## Female        0   1     1   1      1       1      1      1
## Age           1   0     1   1      1       1      1      1
## SES_1         1   1     0   1      1       1      1      1
## EDU           1   1     1   0      1       1      1      1
## STRESS        1   1     1   1      0       1      1      1
## SUPPORT       1   1     1   1      1       0      1      1
```

Multiple Imputation Using Random Forests

Imputing data with random forests works basically the same as with the simpler regression methods. Again the basic function call is to `mice()`. The main difference is that a different method is specified, "rf". Since RFs can handle both continuous and categorical variables, we do not need to specify separate methods for different types of variables. Finally, with RFs, there are a few additional options we can set. First is how many trees should be included in each forest. We set this to 100 in this example. Some imputations have suggested fewer trees may be needed, but this will be dependent on the problem, and in some cases to get a good prediction model, more than 100 trees may be needed. We also specify the node size for the RF, in this case 10. The iterations is set at 20, to match the total number of iterations from the regression imputation. In practice it is not always clear how many iterations should be used. Enough iterations are needed that the models converged.

The first thing to notice is that whereas the regression methods took just a few seconds, the RF imputation took substantially longer, even running in parallel. The model appears to have converged. Finally, examining diagnostics, the distributions more closely match the observed data with this RF imputation, and the affect values are not imputed outside the range of the possibility (below 1).

```
system.time(mi.rfpar <- parLapplyLB(cl, 1:6, function(i)  {
  mice(
    davgmiss,
    m = 1, maxit = 30,
    method = "rf",
    seed = imputation_seeds[i],
    ntree = 500, nodesize = 10)
}))

##    user  system elapsed
##    0.16    0.11  850.27

## combine into a single object
mi.rf <- ibind(mi.rfpar[[1]], mi.rfpar[[2]])
for (i in 3:6) {
  mi.rf <- ibind(mi.rf, mi.rfpar[[i]])
}

## plot convergence diagnostics
plot(mi.rf, PosAff + NegAff + SUPPORT ~ .it | .ms)

## model diagnostics
densityplot(mi.rf, ~ PosAff + NegAff + SUPPORT + STRESS)

xyplot(mi.rf, NegAff + PosAff ~ STRESS + SUPPORT)
```

Note that regardless of how the data were multiply imputed (GLMs, GAMs, RFs), once they are imputed, we would analyze them the same way. We can compare the results from the different models to see the impact. Because we generated the missingness, in this example we can not only compare imputation models to complete case results but also compare them to the "truth." The following code runs a GLM with positive affect as the outcome, some sociodemographic covariates in the model and stress included as the focal predictor. The model is repeated using the true data, the complete cases only, imputed data based on chained equations and linear models, and random forest multiple imputations. The estimates and confidence intervals are plotted in Figure 9-12. In this made-up example, we can see that the complete case approach yields larger confidence intervals and generally less accurate results. The different imputation methods have more comparable confidence intervals to the "true" model. Unfortunately, although it is possible to evaluate the performance of different imputation models under known conditions, in practice, it is impossible to know the truth or the exact missing data mechanism, making it difficult to judge which particular approach is most accurate. These results also show that depending on how findings are interpreted, many of the results may be considered quite similar. Given this similarity and the uncertainty of which model is best, it is reasonable to consider computational costs associated with different approaches, which may lead some to choose a simpler imputation model than random forests.

```
m.true <- lm(PosAff ~ STRESS + Age + EDU + Female, data = davg)
m.cc <- lm(PosAff ~ STRESS + Age + EDU + Female, data = davgmiss)
m.mireg <- summary(pool(with(mi.1,
  lm(PosAff ~ STRESS + Age + EDU + Female))),
  conf.int = TRUE)
m.mirf <- summary(pool(with(mi.rf,
  lm(PosAff ~ STRESS + Age + EDU + Female))),
  conf.int = TRUE)

res.true <- as.data.table(cbind(coef(m.true), confint(m.true)))
res.cc <- as.data.table(cbind(coef(m.cc), confint(m.cc)))
res.mireg <- as.data.table(m.mireg[, c("estimate", "2.5 %", "97.5 %")])
res.mirf <- as.data.table(m.mirf[, c("estimate", "2.5 %", "97.5 %")])
```

```
setnames(res.true, c("B", "LL", "UL"))
setnames(res.cc, c("B", "LL", "UL"))
setnames(res.mireg, c("B", "LL", "UL"))
setnames(res.mirf, c("B", "LL", "UL"))

res.compare <- rbind(
  cbind(Type = "Truth", Param = names(coef(m.true)), res.true),
  cbind(Type = "CC", Param = names(coef(m.true)), res.cc),
  cbind(Type = "MI Reg", Param = names(coef(m.true)), res.mireg),
  cbind(Type = "MI RF", Param = names(coef(m.true)), res.mirf))

ggplot(res.compare, aes(factor(""),
   y = B, ymin = LL, ymax = UL, colour = Type)) +
  geom_pointrange(position = position_dodge(.4)) +
  scale_color_viridis(discrete = TRUE) +
  facet_wrap(~Param, scales = "free") +
  theme(
    legend.position = c(1, 0),
    legend.justification = c("right", "bottom"))

## clean up cluster
stopCluster(cl)
rm(cl)
```

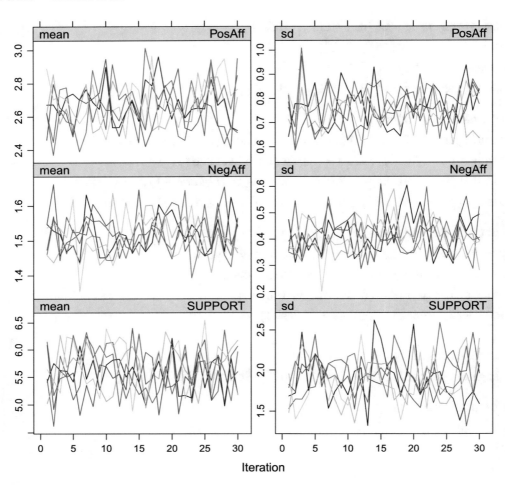

Figure 9-9. *Convergence diagnostics for random forest imputation model*

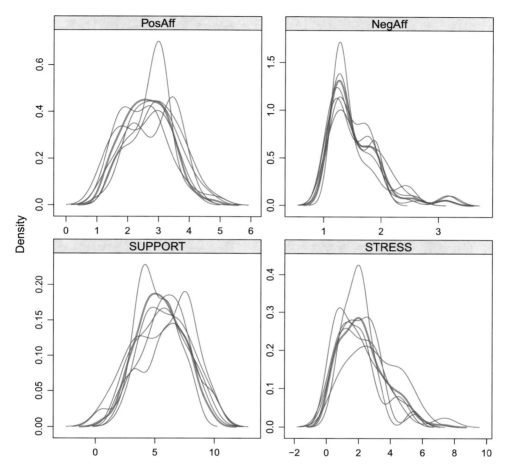

Figure 9-10. *Density plots of observed and imputed values from random forest model*

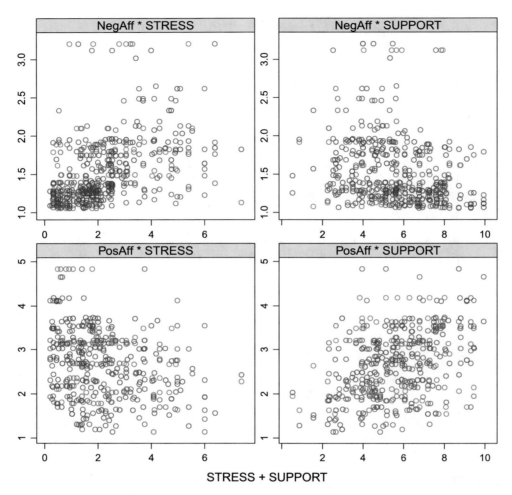

Figure 9-11. *Scatter plots of affect vs. stress and social support for observed and imputed values*

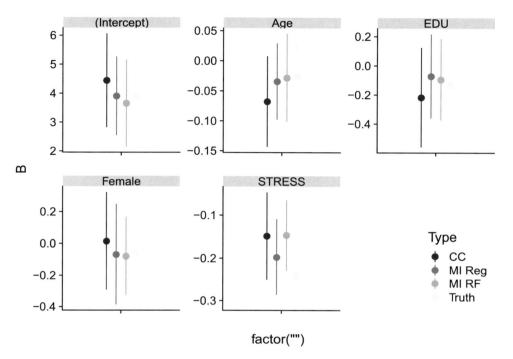

Figure 9-12. *Pooled predictions from linear regression models of the association between stress and positive affect*

9.3 Case Study: Multiple Imputation with RFs

To close out this chapter, we will examine a complete worked example using multiple imputation with random forests. Random forest multiple imputation tends to be time consuming, so in nearly all cases, you should consider using parallel processing. The following code sets up a local cluster and loads all the required packages. As we noted in the introduction, we use the checkpoint package to ensure that we can exactly control and specify which version of R packages to use by specifying the version of R and the date. This helps to make results reproducible and ensure that if you come back to your code at a later time, you know exactly what software you used to run it.

```
cl <- makeCluster(2)
clusterExport(cl, c("book_directory", "checkpoint_directory" ))

clusterEvalQ(cl, {
  library(checkpoint)
  checkpoint("2018-09-28", R.version = "3.5.1",
```

```
  project = book_directory,
  checkpointLocation = checkpoint_directory,
  scanForPackages = FALSE,
  scan.rnw.with.knitr = TRUE, use.knitr = TRUE)
  library(mice)
  library(randomForest)
  library(data.table)
})
```

```
## [[1]]
##  [1] "data.table"    "randomForest"  "mice"          "lattice"
##  [5] "checkpoint"    "RevoUtils"     "stats"         "graphics"
##  [9] "grDevices"     "utils"         "datasets"      "RevoUtilsMath"
## [13] "methods"       "base"
##
## [[2]]
##  [1] "data.table"    "randomForest"  "mice"          "lattice"
##  [5] "checkpoint"    "RevoUtils"     "stats"         "graphics"
##  [9] "grDevices"     "utils"         "datasets"      "RevoUtilsMath"
## [13] "methods"       "base"
```

Because multiple imputation includes a stochastic component, the random seeds also need to be set, if you want to get the same results if you were to re-run your multiple imputation model. One easy way to get many random seeds is to use the built-in .Random.seed variable in R. Because this can change over time, rather than rely directly on the results, use the dput() function to get them exported as copy-and-pasteable R code. In the following code, we show a brief example and then show reusing seeds so that we can be certain which seeds to use. To use in the parallel processing, we have to export the results to our local cluster, using clusterExport().

```
## example of how to have R return some seed values
dput(.Random.seed[1:5])

## c(403L, 148L, -1767993668L, 1417792552L, 298386660L)

## random seeds
imputation_seeds <- c(403L, 148L, -1767993668L,
  1417792552L, 298386660L, 1360311820L,
1356573822L, -1472988872L, 1215046494L, 759520201L,
1399305648L, -455288776L, 969619279L, 518793662L,
-383967014L, -1983801345L, -698559309L, 1957301883L,
-1457959076L, 1321574932L, -537238757L,
11573466L, 1466816383L, -2113923363L, 1663041018L)

clusterExport(cl, c("davgmiss", "imputation_seeds"))
```

The code to actually run the random forest multiple imputation is fairly brief. Little setup is needed for predictors in random forest models, as the models readily accommodate both categorical and continuous predictors and do not impose assumptions regarding the functional form of the association. Thus concerns about transforming data are not particularly relevant. Although outliers can have some influence, because random forests rely on splits in the data, outliers or extreme values in predictors also tend to have less impact. Outliers on continuous outcomes may pose some challenges, so it is a good idea to check for those. An example of quick diagnostics for the data is shown in the following code and Figure 9-13. The graphs suggest the data are probably good enough to proceed.

```
ggplot(melt(davgmiss[, sapply(davgmiss, is.numeric),
            with = FALSE], measure.vars = 1:6), aes(value)) +
        geom_density() + geom_rug() +
        facet_wrap(~variable, scales = "free")

## Warning: Removed 168 rows containing non-finite values (stat_density) .
```

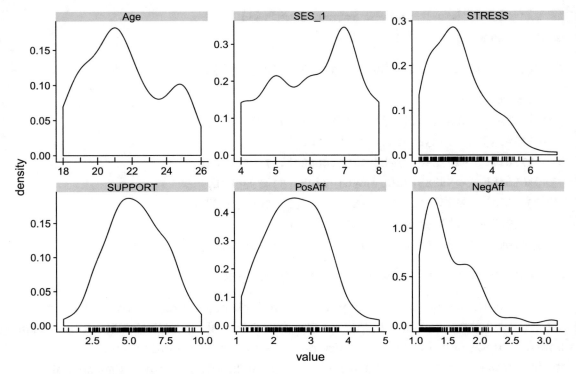

Figure 9-13. *Density plots for continuous variables to be included in the imputation model*

To begin with, it often is a good idea to run a relatively simple model and generate few multiply imputed datasets. Together, this means it takes less time to get initial results, which will help to identify any errors in your setup, ensure the results from the imputation model seem reasonable, and can form a basis to estimate how long it will take to complete the full imputation. The following code is one such simpler example. Since we set up a local cluster with just two processors, we are making four imputed datasets, so each processor only has to generate two datasets. We also limit the number of iterations to five, which improves speed considerably. We use the proc.time() function at the beginning and end so that we can get a log of how long it took to run, in total.

```
start.time <- proc.time()
mi.rfpar1 <- parLapplyLB(cl, 1:4, function(i) {
  mice(
    davgmiss,
    m = 1, maxit = 5,
    method = "rf",
    seed = imputation_seeds[i],
    ntree = 100, nodesize = 10)
})
stop.time <- proc.time()

## estimate of how long it took
stop.time - start.time

##    user  system elapsed
##       0       0      15

## combine into a single object
mi.rf1 <- ibind(mi.rfpar1[[1]], mi.rfpar1[[2]])
for (i in 3:4) {
  mi.rf1 <- ibind(mi.rf1, mi.rfpar1[[i]])
}
```

We can see that about 15.5 seconds elapsed. That will increase roughly linearly as imputations per core increase, and also with more iterations. We should also check diagnostics from this simpler model and see whether anything seems unusual or may suggest a problem in the models. We look first at some convergence diagnostics in Figure 9-14.

```
## plot convergence diagnostics
plot(mi.rf1, NegAff + STRESS + Age ~ .it | .ms)
```

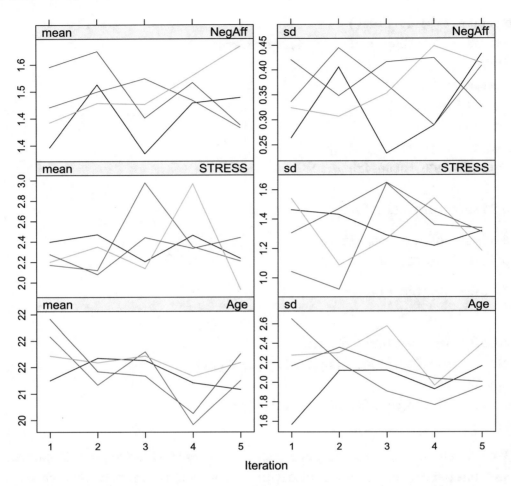

Figure 9-14. *Convergence diagnostics for random forest imputation model*

Next we compare the imputations to observed distributions in Figure 9-15.

```
## model diagnostics for continuous study variables
densityplot(mi.rf1, ~ NegAff + STRESS + Age)
```

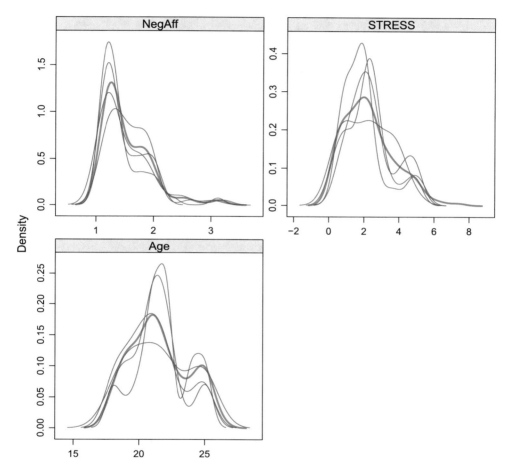

Figure 9-15. *Density plots of observed and imputed values from random forest model*

Now we can fit our target model, a linear regression, extract the residuals from each imputed dataset, and check these for assumptions, outliers, etc. Because there are multiple regression models (one for each imputed dataset), we use `lapply()` to loop through these, combine into a single vector of standardized residuals, and plot. The results are in Figure 9-16. While the distribution is somewhat skewed and there are a few relatively extreme values, they are not too terrible, and the residual distribution is not terribly skewed.

```
## fit the models
fit.mirf1 <- with(mi.rf1,
  lm(NegAff ~ STRESS + Age + EDU + Female + SES_1))

testdistr(unlist(lapply(fit.mirf1$analyses, rstandard)))
```

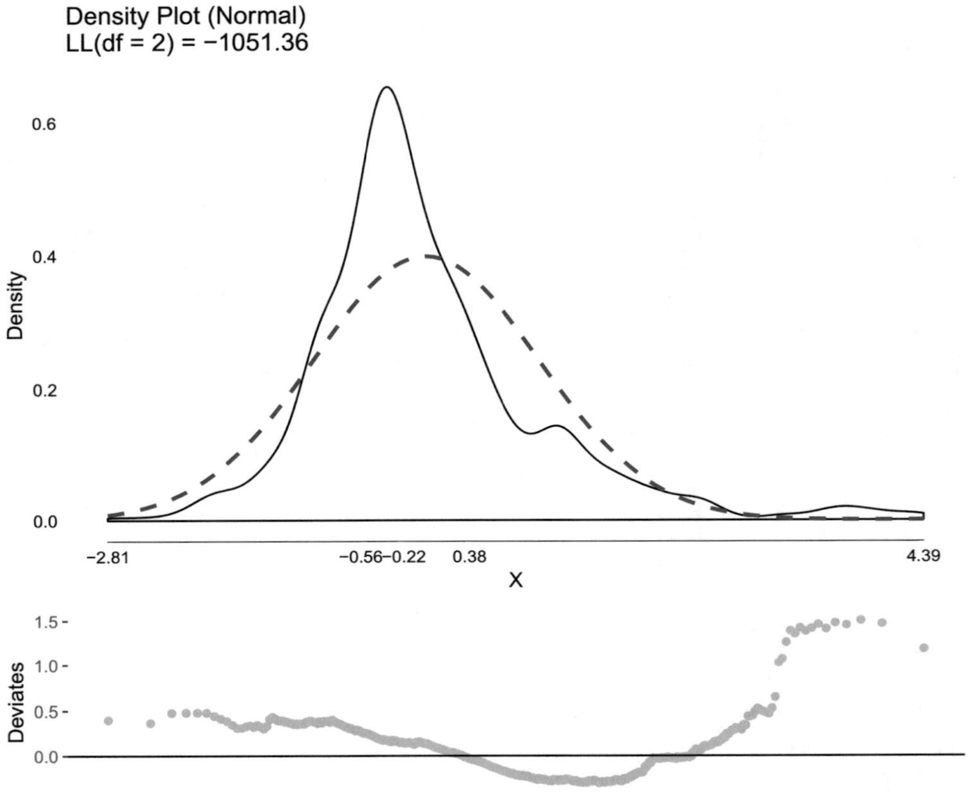

Figure 9-16. *Distribution plot (density and Q-Q deviates) for model residuals*

Finally, we can pool and summarize results and then view them in Table 9-3. This is helpful to identify early any apparent issues in the data, coding, or final analysis model. If there were issues, they can be addressed faster this way, instead of running the full imputation model which may be time consuming and only then realizing there are issues. Anecdotally, our experience is that often there are some data issues the first few times that are not caught until later, and because the imputation relies on the data being correct, any issues in the data require re-running the imputation model and all subsequent analyses.

```
## pool results and summarize
m.mirf1 <- summary(pool(fit.mirf1), conf.int = TRUE)

xtable(m.mirf1,
  digits = 2,
  caption = "Regression results pooled across multiply imputed data test run",
  label = "tmd-pooledres2")
```

Table 9-3. *Regression Results Pooled Across Multiply Imputed Data Test Run*

	Estimate	Std. Error	Statistic	df	p.value	2.5%	97.5%
(Intercept)	1.05	0.32	3.30	75.58	0.00	0.42	1.69
STRESS	0.19	0.02	10.41	118.97	0.00	0.15	0.22
Age	0.01	0.01	0.60	103.88	0.55	−0.02	0.03
EDUUni Graduate +	0.02	0.07	0.34	25.31	0.73	−0.12	0.16
FemaleFemale	−0.01	0.06	−0.19	24.86	0.85	−0.14	0.11
SES_1	−0.02	0.02	−0.97	60.39	0.33	−0.06	0.02

Once we are reasonably satisfied that the imputation model worked, there are no data issues to be addressed, and our final analytic model is likely to work, then we proceed to our final imputation. In this example, we increase the maximum iterations from 5 to 30, to help ensure convergence. We also increase the number of imputed datasets from 4 to 10. In practice, it is more common to use 25–100 imputed datasets, but we keep it briefer so that the examples do not take too long to run.

Previously, we saw that it took 15.5 seconds for 2 imputations per core with 5 iterations. We expect that to take roughly 6 times longer when using 30 instead of 5 maximum iterations, and if we are doing 10 imputations (5 per core), about 2.5 times longer for the number of imputations. In summary, we would estimate it takes about 232.5 seconds for this longer imputation to complete. If we were planning 50 imputed datasets, it would be about 5 times longer still.

```
start.time2 <- proc.time()
mi.rfpar2 <- parLapplyLB(cl, 1:10, function(i) {
  mice(
    davgmiss,
```

```
    m = 1, maxit = 30,
    method = "rf",
    seed = imputation_seeds[i],
    ntree = 100, nodesize = 10)
})
stop.time2 <- proc.time()

## time taken
stop.time2 - start.time2

##    user  system elapsed
##    0.04    0.02  274.58

## combine into a single object
mi.rf2 <- ibind(mi.rfpar2[[1]], mi.rfpar2[[2]])
for (i in 3:10) {
  mi.rf2 <- ibind(mi.rf2, mi.rfpar2[[i]])
}
```

We can see that about 274.6 seconds elapsed, compared to our predicted 232.5 seconds. As with the simpler model, we should check diagnostics. These are in Figure 9-17.

```
## plot convergence diagnostics
plot(mi.rf2, NegAff + STRESS + Age ~ .it | .ms)
```

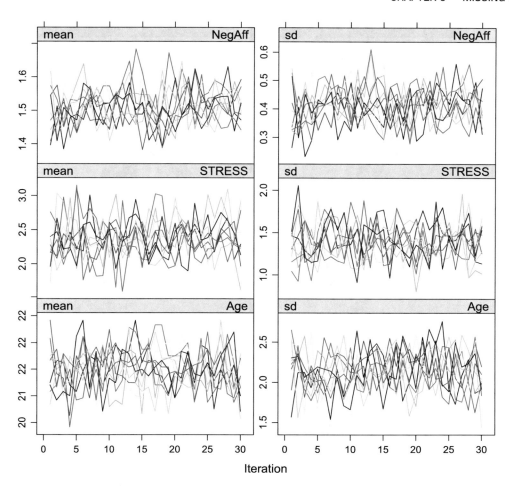

Figure 9-17. *Convergence diagnostics for random forest imputation model*

Next we compare the imputations to observed distributions in Figure 9-18.

```
## model diagnostics for continuous study variables
densityplot(mi.rf2, ~ NegAff + STRESS + Age)
```

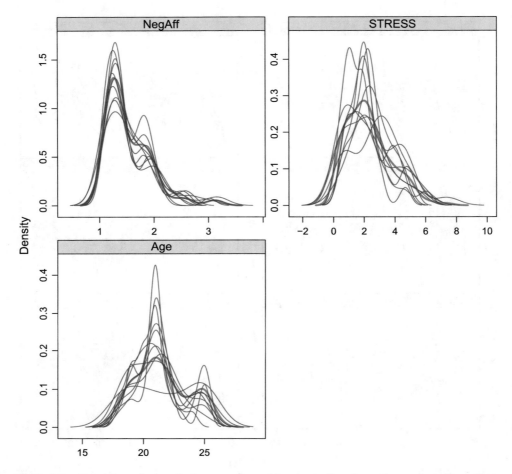

Figure 9-18. *Density plots of observed and imputed values from random forest model*

Now we fit our target model, a linear regression, and check the standardized residuals in Figure 9-19.

```
## fit the models
fit.mirf2 <- with(mi.rf2,
  lm(NegAff ~ STRESS + Age + EDU + Female + SES_1))

testdistr(unlist(lapply(fit.mirf2$analyses, rstandard)))
```

430

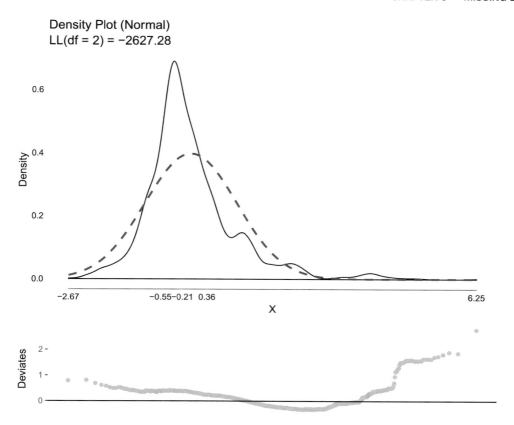

Figure 9-19. *Distribution plot (density and Q-Q deviates) for model residuals*

Finally, we can pool and summarize results and then view them in Table 9-4.

```
## pool results and summarize
m.mirf2 <- summary(pool(fit.mirf2), conf.int = TRUE)

xtable(m.mirf2,
  digits = 2,
  caption = "Regression results pooled across multiply imputed data final run",
  label = "tmd-pooledres3")
```

If you compare Table 9-3 with Table 9-4, you can see there are some differences. If we had run even more imputations, there would likely still be more differences. One reason for choosing 50 to 100 imputations is that with higher numbers, differences from one random set of imputations to another tend to be smaller. With only five to ten imputations, there may be relatively large variations due to random chance.

Table 9-4. *Regression Results Pooled Across Multiply Imputed Data Final Run*

	Estimate	Std. Error	Statistic	df	p.value	2.5%	97.5%
(Intercept)	1.11	0.33	3.39	95.52	0.00	0.46	1.76
STRESS	0.18	0.02	8.94	69.74	0.00	0.14	0.22
Age	0.01	0.02	0.46	44.03	0.65	−0.02	0.04
EDUUni Graduate +	0.03	0.08	0.44	31.10	0.66	−0.12	0.19
FemaleFemale	−0.01	0.07	−0.12	27.86	0.91	−0.15	0.14
SES_1	−0.02	0.02	−1.03	62.65	0.31	−0.07	0.02

9.4 Summary

This chapter introduced multiple imputation through chained equations (MICE) as a flexible technique to address missing data. MICE can improve efficiency by using all available data, compared to easier but often suboptimal complete case analysis. MICE also can reduce bias in estimates due to missing data, if the necessary variables related to missingness are available to include in the MICE model. Addressing missing data often is a key first step in many analyses. Also, even though we only showed a few follow-up analyses in this chapter, MICE could be used as a first step to almost any subsequent analysis, including all of those discussed in other chapters in this book. Finally, a brief summary of some of the key functions used in this chapter to address missing data is shown in Table 9-5.

Table 9-5. *Listing of Key Functions Described in This Chapter and Summary of What They Do*

Function	What It Does
aggr()	Visualizes the missing data in a dataset by variable and examines different patterns of missing data
marginplot()	Visualizes whether missingness on one variable depends on a second variable
mice()	Runs multiple imputation using the fully conditional specification approach
mice.mids()	Runs additional iterations on a mids class object from a previous run of mice to check for convergence
densityplot()	Method for a mids class object to plot the density of the observed and imputed data; can help to see if the distribution of imputed data is similar or different from the observed data
xyplot()	Method for a mids class object to create a scatter plot of two variables showing the observed and imputed values separately
with()	Method for a mids class object to run the specified analysis on each imputed dataset separately
pool()	Pools the results of analyses run separately on different imputed datasets
ibind()	Combines separate mids objects into a single object with all the multiple imputations
complete()	Extracts complete datasets from a mids class object

CHAPTER 10

GLMMs: Introduction

Generalized linear *mixed* models (GLMMs) extend the generalized linear models (GLMs), introduced in previous chapters, to statistically account for data that are clustered (e.g., children within schools, individuals within a particular hospital clinic, repeated measures on the same person) and render these non-independent observations *conditionally* independent.

Accounting for the relations between observations, whether due to clustering within a hierarchy like classes, schools, workplaces, or due to similarities within families, or repeated measures, is critical because many statistical models assume that observations are independent or at least that observations are conditionally independent. This may be a tenable assumption in many cases, particularly when the fundamental unit of observation (e.g., individual people, etc.) each contributes a single observation to our data. The current chapter introduces situations where observations are likely *not* independent. When working with data where observations may not be independent, not only does the final model and statistical inference need to be adapted, data visualization, exploration, and descriptive statistics also must be adapted. Therefore, this chapter covers aspects of working with non-independent data, and the subsequent chapters cover statistical modelling for such data.

```
library(checkpoint)
checkpoint("2018-09-28", R.version = "3.5.1",
  project = book_directory,
  checkpointLocation = checkpoint_directory,
  scanForPackages = FALSE,
  scan.rnw.with.knitr = TRUE, use.knitr = TRUE)

library(knitr)
library(ggplot2)
library(cowplot)
```

© Matt Wiley and Joshua F. Wiley 2019
M. Wiley and J. F. Wiley, *Advanced R Statistical Programming and Data Models*,
https://doi.org/10.1007/978-1-4842-2872-2_10

```
library(viridis)
library(JWileymisc)
library(data.table)
library(lme4)
library(lmerTest)
library(chron)
library(zoo)
library(pander)
library(texreg)

options(width = 70, digits = 2)
```

10.1 Multilevel Data

Two common contexts where observations are not independent are when repeated measures are taken on each individual, such as in longitudinal studies or repeated measures experiments, and when individuals are clustered or grouped, such as if families, schools, companies, etc., were recruited and then individual participants are clustered or nested within these higher-order units. For instance, if families are recruited, observations from siblings belonging to the same family may be more similar than they are to observations from siblings from other families, both due to shared genetic factors and due to shared environmental influences. Likewise, if individuals are followed over time and assessed every day for a week, observations from the same person are naturally more likely to be related to each other than to observations from other people.

In this chapter we will focus on an example of repeated measures data. Although repeated measures data may appear quite different from data on siblings within families or multiple workers within companies, they pose many of the same challenges and have common solutions. The primary difference is that repeated measures data are naturally ordered by time, whereas clustered data are most often unordered (e.g., if 10 employees are sampled from 100 companies, there often is not a natural way to sort or order the employees).

One of the first differences with multilevel data is that there are two common ways to structure data. One approach, sometimes called a wide dataset, is similar to single-level data. Each row represents a single unit of observation (e.g., one person, one school), and additional variables (columns) are added for repeated measures within units. For example, suppose that people with hypertension had their blood pressure measured longitudinally at the start of a study (T1), 6 months later (T2), and at a 1 year follow-up (T3). An example of how these data may look is in Table 10-1. In this example, ID 3 missed the final time point.

Table 10-1. *Example Wide Dataset of Systolic Blood Pressure (SBP) Measured at Three Time Points*

ID	SBPT1	SBPT2	SBPT3
1	135	130	125
2	120	125	121
3	121	125	.

Although wide data can be convenient for some circumstances, it often is not the ideal structure for multilevel data. First, for longitudinal data, such as the example in Table 10-1, part of the information is actually encoded in the variable name: T1, T2, T3. Generally, it is preferable for variable names only to describe the variable or measure and additional information, such as time, to be captured and coded in another variable, such as time point. Second, if there are unequal numbers of assessments within a unit, such as different lengths of longitudinal follow-up or different numbers of students within schools, the wide format becomes highly inefficient. For example, imagine a dataset where 600 students are recruited from a large school and just 15 are recruited from a smaller school. To place these data in wide format would require 600 variables (one for each student in the largest school) for each measure. In the small school, all but 15 of those 600 variables would be missing.

Another way to structure multilevel datasets is sometimes called in long format. In long format, multiple rows may belong to any single unit, and this is indicated by an identification (ID) variable. Table 10-2 shows an example of a long data using the same hypothetical longitudinal study of blood pressure. In this case, if a particular unit missed an observation or has fewer observations than other units (e.g., a small vs. large school), those rows can be omitted. For example, in Table 10-2, ID 3 does not have a Time 3 blood pressure reading, so that row is entirely absent from the data.

Table 10-2. *Example Long Dataset of Systolic Blood Pressure (SBP) Measured at Three Time Points*

ID	SBP	Time
1	135	1
1	130	2
1	125	3
2	120	1
2	125	2
2	121	3
3	121	1
3	125	2

Reshaping Data

If data are stored in one format, it is possible and sometimes perhaps necessary to convert it to the other format. The code that follows creates a wide dataset and then uses the reshape() function to convert it from wide to long format. The varying argument indicates variables in the wide data that vary by time. The v.names argument indicates what each set of varying variables should be named in the long dataset. In this example, we only have SBP, but if there were more variables, those could be added. The direction argument indicates whether the data should be reshaped from wide to long, or from long to wide.

```
ex.wide <- data.table(
  ID = c(1, 2, 3),
  SBPT1 = c(135, 120, 121),
  SBPT2 = c(130, 125, 125),
  SBPT3 = c(125, 121, NA))

print(ex.wide)

##    ID SBPT1 SBPT2 SBPT3
## 1:  1   135   130   125
## 2:  2   120   125   121
## 3:  3   121   125    NA
```

```
reshape(
  data = ex.wide,
  varying = list(paste0("SBPT", 1:3)),
  v.names = c("SBP"),
  idvar = "ID",
  direction = "long")
```

```
##      ID time SBP
## 1:   1    1 135
## 2:   2    1 120
## 3:   3    1 121
## 4:   1    2 130
## 5:   2    2 125
## 6:   3    2 125
## 7:   1    3 125
## 8:   2    3 121
## 9:   3    3  NA
```

Conversely, if data already come in long format, it is possible to convert them to wide format. The code that follows creates a long data and converts it to wide, again using the reshape() function. The sep argument indicates how to make the wide variable names. In this case the base, SBP, then the separator, T, followed by the time: 1, 2, 3.

```
ex.long <- data.table(
  ID = c(1, 1, 1, 2, 2, 2, 3, 3),
  SBP = c(135, 130, 125, 120, 125, 121, 121, 125),
  Time = c(1, 2, 3, 1, 2, 3, 1, 2))
```

print(ex.long)

```
##      ID SBP Time
## 1:   1 135    1
## 2:   1 130    2
## 3:   1 125    3
## 4:   2 120    1
## 5:   2 125    2
## 6:   2 121    3
## 7:   3 121    1
## 8:   3 125    2
```

```
reshape(
  data = ex.long,
  v.names = "SBP",
  timevar = "Time",
  sep = "T",
  idvar = "ID",
  direction = "wide")
```

```
##      ID SBPT1 SBPT2 SBPT3
## 1:   1   135   130   125
## 2:   2   120   125   121
## 3:   3   121   125    NA
```

Daily Dataset

Before we begin, we will introduce a new dataset. These data are drawn from a daily diary study conducted at Monash University in 2017, where young adults completed measures up to three times per day (morning, afternoon, and evening) for about 12 days. Thus each participant contributed about 36 observations to the dataset. To protect participant confidentiality and anonymity, the data used here were simulated from the original data, but in such a way as to preserve the relations among variables and most features of the raw data.

The variable names and a brief description of each are in Table 10-3.

The simulated data are included as part of the JWileymisc package and can be loaded using the data() function.

```
data(aces_daily)
str(aces_daily)
```

```
## 'data.frame':        6599 obs. of  19 variables:
## $ UserID           : int  1 1 1 1 1 1 1 1 1 1 ...
## $ SurveyDay        : Date, format: "2017-02-24" ...
## $ SurveyInteger    : int  2 3 1 2 3 1 2 3 1 2 ...
## $ SurveyStartTimec11: num  1.93e-01 4.86e-01 1.16e-05 1.93e-01 4.06e-01 ...
## $ Female           : int  0 0 0 0 0 0 0 0 0 0 ...
## $ Age              : num  21 21 21 21 21 21 21 21 21 21 ...
## $ BornAUS          : int  0 0 0 0 0 0 0 0 0 0 ...
## $ SES_1            : num  5 5 5 5 5 5 5 5 5 5 ...
```

```
##  $ EDU            : int  0 0 0 0 0 0 0 0 0 0 ...
##  $ SOLs           : num  NA 0 NA NA 6.92 ...
##  $ WASONs         : num  NA 0 NA NA 0 NA NA 1 NA NA ...
##  $ STRESS         : num  5 1 1 2 0 0 3 1 0 3 ...
##  $ SUPPORT        : num  NA 7.02 NA NA 6.15 ...
##  $ PosAff         : num  1.52 1.51 1.56 1.56 1.13 ...
##  $ NegAff         : num  1.67 1 NA 1.36 1 ...
##  $ COPEPrb        : num  NA 2.26 NA NA NA ...
##  $ COPEPrc        : num  NA 2.38 NA NA NA ...
##  $ COPEExp        : num  NA 2.41 NA NA 2.03 ...
##  $ COPEDis        : num  NA 2.18 NA NA NA ...
```

Table 10-3. *Listing of Variable Names in the Daily Diary Study Data*

Variable Name	Description
UserID	A unique identifier for each individual
SurveyDay	The date each observation occurred on
SurveyInteger	The survey coded as an integer (1 = morning, 2 = afternoon, 3 = evening)
SurveyStartTimec11	Survey start time, centered at hours since 11:00 a.m.
Female	A 0 or 1 variable, where 1 = female and 0 = male
Age	Participant age in years, top coded at 25
BornAUS	A 0 or 1 variable where 1 = born in Australia and 0 = born outside of Australia
SES_1	Participants' subjective SES, bottom coded at 4 and top coded at 8
EDU	Education level where 1 = university graduate or above and 0 = less than university graduate
SOLs	Self-reported sleep onset latency in minutes, morning survey only
WASONs	Self-reported number of awakenings after sleep onset, top coded at 4, morning survey only
STRESS	Overall stress ratings on a 0–10 scale, repeated 3x daily
SUPPORT	Overall social support ratings on a 0–10 scale, repeated 3x daily
PosAff	Positive affect ratings on a 1–5 scale, repeated 3x daily

(continued)

Table 10-3. (*continued*)

Variable Name	Description
NegAff	Negative affect ratings on a 1–5 scale, repeated 3x daily
COPEPrb	Problem focused coping on a 1–4 scale, repeated 1x daily at the evening survey
COPEPrc	Emotional processing coping on a 1–4 scale, repeated 1x daily at the evening survey
COPEExp	Emotional expression coping on a 1–4 scale, repeated 1x daily at the evening survey
COPEDis	Mental disengagement coping on a 1–4 scale, repeated 1x daily at the evening survey

The data are in a long format, so that each row represents observations from one survey on one day for one individual. In long format, each individual contributes about 36 rows of data. Although long format is an efficient way to store data with many repeated measures, it can make it difficult to identify missing data, as a missed survey is not registered as a row of missing data—rather the entire row is absent. A first step that can be helpful to identify rates of missing data and that will be useful for several future analyses is to add any missed surveys with the observations simply set to missing. To accomplish this, we can create a temporary dataset that has all surveys and dates from an individual's first survey/date to their last survey/date and then perform a full join (or merge, keeping all rows) with the original data.

The following code makes such a temporary dataset, by finding the minimum and maximum dates, and the earliest and latest survey on each of those days. Then using this information, we can create the "complete" dataset with all surveys and all days between the first survey on the first day and the last survey on the last day. Finally we can merge the two datasets together, keeping all rows which will fill in the missing values as needed.

```
draw <- as.data.table(aces_daily)
draw <- draw[order(UserID, SurveyDay, SurveyInteger)]
draw[, UserID := factor(UserID)]
```

```
tmpdata <- draw[!is.na(SurveyDay) & !is.na(SurveyInteger)][, .(
  MinD = min(SurveyDay),
  MinS = min(SurveyInteger[SurveyDay == min(SurveyDay)]),
  MaxD = max(SurveyDay),
  MaxS = max(SurveyInteger[SurveyDay == max(SurveyDay)])),
  by = UserID]

tmpdata <- tmpdata[, .(
  SurveyInteger = c(
    MinS:3L, #first day
    rep(1L:3L, times = MaxD - MinD - 1), #all days between first/last
    1L:MaxS), #last day
  SurveyDay = as.Date(rep(MinD:MaxD, c(
      4L - MinS, #first day
      rep(3, MaxD - MinD - 1), #all days between first/last
      MaxS)), origin = "1970-01-01")), #lastday
  by = UserID]

d <- merge(draw, tmpdata, by = c("UserID", "SurveyDay", "SurveyInteger"),
           all = TRUE)

nrow(draw)

## [1] 6599

nrow(d)

## [1] 6927

nrow(draw)/nrow(d)

## [1] 0.95
```

The number of rows in the data increases from the original data after adding in the missing rows, reflecting the missing data.

10.2 Descriptive Statistics

With non-independent data, basic descriptive statistics also can be calculated in different ways. To begin to understand these differences, and also what a multilevel structure implies, examine the two plots in Figure 10-1. This plot shows made-up data with ten assessments on four different people under some different conditions. The solid lines indicate the average for each individual, and the dots indicate observed data.

With multilevel data in long format, if we calculate the mean and variance of a variable, y, that would average across people and time for the mean. The variance will incorporate both differences between people (the distance between lines) and variance within people (how much the data points vary around each individual's own mean). Conversely, if we first average observations across time within a person, then the mean will be the average of the four lines and the variance will only be the variability between individual means, the lines in Figure 10-1.

In addition, calculating descriptives on all data points versus calculating descriptives on data averaged by person results in weighting participants differently. When calculating summaries on all observations, a participant with ten observations would receive ten times the weight of a participant who only had one observation (e.g., due to missing data). The issue of weighting tends to be less important to the extent that clusters are all about the same size and makes no difference if all clusters are identical (e.g., everyone has exactly ten observations).

```
set.seed(1234)
ex.data.1 <- data.table(
  ID = factor(rep(1:4, each = 10)),
  time = rep(1:10, times = 4),
  y = rnorm(40, rep(1:4, each = 10), .2))

ex.data.2 <- data.table(
  ID = factor(rep(1:4, each = 10)),
  time = rep(1:10, times = 4),
  y = rnorm(40, 2.5, 1))

plot_grid(
 ggplot(ex.data.1,
        aes(time, y, colour = ID, shape = ID)) +
  stat_smooth(method = "lm", formula = y ~ 1, se=FALSE) +
```

```
 geom_point() +
 scale_color_viridis(discrete = TRUE),
ggplot(ex.data.2,
        aes(time, y, colour = ID, shape = ID)) +
 stat_smooth(method = "lm", formula = y ~ 1, se=FALSE) +
 geom_point() +
 scale_color_viridis(discrete = TRUE),
 ncol = 1,
 labels = c(
   "High Between Variance",
   "Low Between Variance"),
 align = "hv")
```

With multilevel data, there is not a single right or wrong way to calculate descriptive statistics, but it is important that one understand the differences and accurately describe whatever approach was used. Generally, three common approaches are as follows:

- Ignore structure and calculate descriptives on all observations. This may weight units differently, if they have varying numbers of observations. It also will provide a variance estimate that combines both variations between people and within people. That is, the variability is the total variability in a variable.

- First average (or otherwise combine) observations within a unit, and then calculate descriptives. This gives each unit equal weight. The variance estimate will only capture the variance of a variable between people, so it may be more helpful for describing the sample characteristics, rather than the total variability of a variable.

- Calculate descriptive statistics for the first time point only. This will still tend to include some between- and within-person variability, as the within-person variability is not averaged out. If there are meaningful changes over time, the mean may not represent the overall mean of the study. This approach only makes sense for longitudinal data. For other multilevel structures (e.g., students nested within classrooms), there is no sensible way to pick which one student to use.

Basic Descriptives

Examples of each of these approaches to calculating descriptive statistics are shown in the example daily data we loaded earlier for positive affect. The data management necessary is performed using data.table() directly in the dataset on the fly, rather than creating a new variable or new dataset. To get the first observation by ID, we sort by ID, day of the week, and survey (morning, afternoon, evening), and then pick the first observation by ID.

```
## mean and SD on all observations
egltable("PosAff", data = d)
```

```
##                M (SD)
## 1: PosAff 2.68 (1.07)
```

```
## mean and SD first averaging within ID
egltable("PosAff",
  data = d[, .(
    PosAff = mean(PosAff, na.rm = TRUE)) ,
    by = UserID])
```

```
##                M (SD)
## 1: PosAff 2.68 (0.80)
```

```
## mean and SD on first observations
egltable("PosAff", data = d[
  order(UserID, SurveyDay, SurveyInteger)][,
    .(PosAff = PosAff[1]), by = UserID])
```

```
##                M (SD)
## 1: PosAff 2.71 (1.02)
```

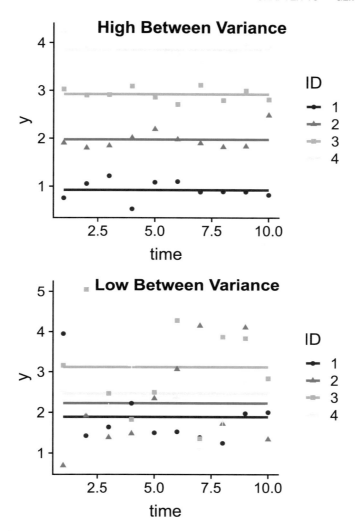

Figure 10-1. *Plot showing hypothetical data with high between variance and low between variance. In the high between variance, observations within a person vary little, but there are large individual differences. In low between variance, there are not many individual differences, but large variability within each person.*

With a long dataset, to calculate summary measures on time-invariant variables, rather than averaging within participants, we must first subset the data to remove duplicate rows, so that it is back to a single-level structure. This can be done by asking for the data dropping any duplicated IDs.

```
tab <- egltable(c("Female", "Age", "BornAUS", "SES_1", "EDU"),
                data = d[!duplicated(UserID)],
                strict = FALSE)
tab
```

```
##                M (SD)/N (%)
##  1:   Female
##  2:        0    28 (40.6)
##  3:        1    41 (59.4)
##  4:      Age 21.91 (2.38)
##  5: BornAUS
##  6:        0    41 (60.3)
##  7:        1    27 (39.7)
##  8:    SES_1  6.05 (1.21)
##  9:      EDU
## 10:        0    45 (66.2)
## 11:        1    23 (33.8)
```

Plots can be a helpful way to present descriptive statistics, such as by different groups. Figure 10-2 shows the average of individual mean coping for women and men. The means are plotted as points with the anchors from the scale added to facilitate interpretation.

```
## create a dataset of the means and labels by gender
copeplotdata <- d[!is.na(Female), .(
  M = c(
    mean(COPEPrb, na.rm = TRUE),
    mean(COPEPrc, na.rm = TRUE),
    mean(COPEExp, na.rm = TRUE),
    mean(COPEDis, na.rm = TRUE)),
  Var = 1:4,
  Low = sprintf("I usually don't do this at all\n[%s]",
                c("Problem Focused", "Emotional Processing",
                  "Emotional Expression", "Disengagement")),
  High = sprintf("I usually do this a lot\n[%s]",
                 c("Problem Focused", "Emotional Processing",
                   "Emotional Expression", "Disengagement"))),
  by = Female]
```

```
## coded 0/1 but for plotting, R needs to know
## it is discrete not a continuous number
copeplotdata[, Female := factor(Female)]

## create a plot
gglikert(x = "M", y = "Var", leftLab = "Low", rightLab = "High",
        data = copeplotdata, colour = "Female",
  xlim = c(1, 4), title = "Average Coping") +
  scale_colour_manual(values =
    c("1" = "grey70", "0" = "grey30"))
```

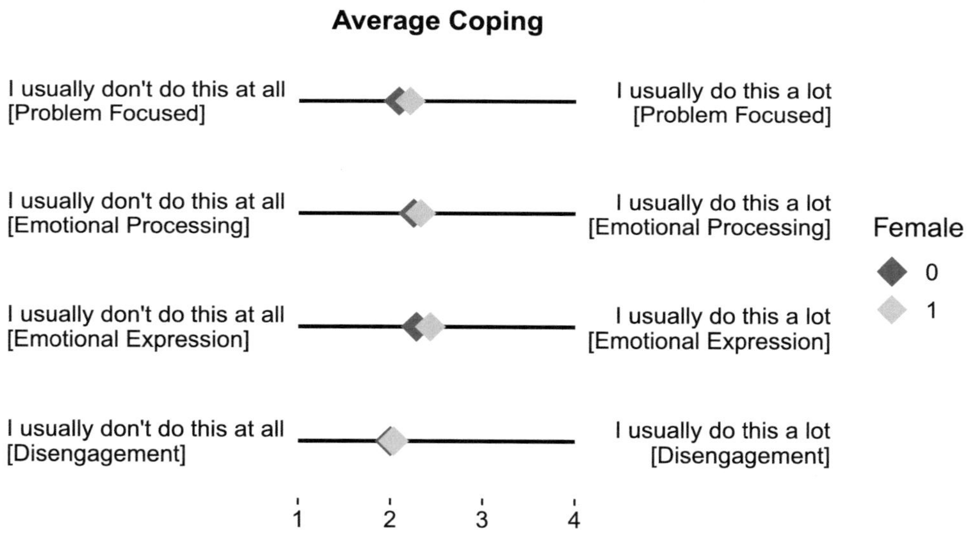

Figure 10-2. *Plot showing average coping ratings for women and men*

Descriptive statistics also can be broken down by other variables. For example the following code calculates and plots in Figure 10-3 the average level of positive and negative affect based on the level of reported stress at that time. Note that because this is by survey, it would be interpreted at the survey, not person, level. That is, on surveys where people rated their stress above 5, how did they rate their affect on average? It does not tell us the average affect of people with high or low stress, on average. Indeed, the same person may contribute some surveys to the average for high stress and some surveys to the average for low stress, if they sometimes reported stress above 5 and other times stress below 5.

```
## create a dataset of the means and labels by stress
afplotdata <- d[!is.na(STRESS), .(
  M = c(
    mean(PosAff, na.rm = TRUE),
    mean(NegAff, na.rm = TRUE)),
  Var = 1:2,
  Low = sprintf("Very Slightly or\nNot at all\n[%s]",
                c("Positive Affect", "Negative Affect")),
  High = sprintf("Extremely\n\n[%s]",
                 c("Positive Affect", "Negative Affect"))),
  by = .(Stress = STRESS > 5)]

## add labels to understand stress
afplotdata[, Stress := factor(Stress, levels = c(FALSE, TRUE),
                              labels = c("<= 5", "> 5"))]

## create a plot
gglikert(x = "M", y = "Var", leftLab = "Low", rightLab = "High",
         data = afplotdata, colour = "Stress",
  xlim = c(1, 5), title = "Affect by Stress") +
  scale_colour_manual(values =
    c("<= 5" = "grey70", "> 5" = "grey30"))
```

When observations can be meaningfully ordered, such as by time of day, we also may wish to calculate descriptive statistics separately by time point. In the code that follows, we create a new variable for survey with nicer labels, then average responses within participants but separately by survey, and finally calculate the descriptive statistics. To get a nice grouped summary, we can pass Survey as a grouping variable. However, note that the statistical test of group differences will be inaccurate in this case as it assumes independent groups. We ignore the test and focus on the descriptive statistics, which are still accurate. This example also shows getting descriptive stats on multiple variables at once.

```
d[, Survey := factor(SurveyInteger, levels = 1:3,
    labels = c("Morning", "Afternoon", "Evening"))]

egltable(c("PosAff", "NegAff", "STRESS"), g = "Survey",
  data = d[, .(
```

```
  PosAff = mean(PosAff, na.rm = TRUE),
  NegAff = mean(NegAff, na.rm = TRUE),
  STRESS = mean(STRESS, na.rm = TRUE)
  ), by = .(UserID, Survey)])
```

```
##            Morning M (SD) Afternoon M (SD) Evening M (SD)
## 1: PosAff     2.67 (0.84)      2.69 (0.81)    2.67 (0.81)
## 2: NegAff     1.53 (0.46)      1.57 (0.49)    1.56 (0.49)
## 3: STRESS     2.14 (1.47)      2.52 (1.60)    2.39 (1.56)
##                           Test
## 1: F(2, 570) = 0.05, p = .947
## 2: F(2, 570) = 0.33, p = .720
## 3: F(2, 570) = 3.01, p = .050
```

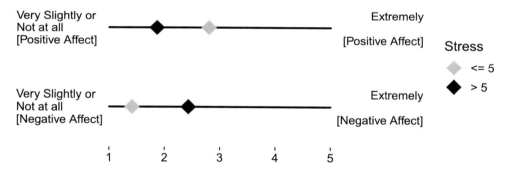

Figure 10-3. *Plot showing average coping ratings for women and men*

The additional choices (average within participants or not, report separately by time point, etc.) arise because with repeated measures or non-independent data, observations can be decomposed into different levels. Concretely, we can imagine that specific positive affect observations are a combination of a participant's typical or average positive affect plus an effect of that particular day or time.

We can decompose a variable into "between"- and "within"-person aspects by calculating the mean for each participant (i.e., the between component) and then taking the difference between observations and each participant's own mean (i.e., the within component). The code that follows shows an example with positive affect. After calculating the two components, we can also get descriptive statistics on each separately.

Note that for the average for each participant, we should drop duplicated values first; otherwise it will be weighted more toward participants with more repeated measures, which is not appropriate for between-level or time-invariance variables.

```
d[, BPosAff := mean(PosAff, na.rm = TRUE), by = UserID]
d[, WPosAff := PosAff - BPosAff]

egltable("BPosAff", data = d[!duplicated(UserID)])

##                   M (SD)
## 1: BPosAff 2.68 (0.80)

egltable("WPosAff", data = d)

##                   M (SD)
## 1: WPosAff 0.00 (0.72)
```

Decomposing variables often is useful not only for descriptive statistics but also for analyses where one may want to look at how variables relate between people and within people. To facilitate such work, we can make between and within variables for all the time-varying variables. To reduce the amount of code we write, we can define a function, bwmean(), that calculates the average and deviations from the average, and then use data.table() to apply this function by ID.

```
## define a new function
bwmean <- function(x, na.rm = TRUE) {
  m <- mean(x, na.rm = na.rm)
  list(m, x - m)
}
```

```
## apply it to affect, support, and stress, by ID
d[, c("BNegAff", "WNegAff") := bwmean(NegAff), by = UserID]
d[, c("BSUPPORT", "WSUPPORT") := bwmean(SUPPORT), by = UserID]
d[, c("BSTRESS", "WSTRESS") := bwmean(STRESS), by = UserID]
```

We also can make a between and within variable for the sleep and coping measures. However, these work a little differently. Sleep was only measured in the morning, and coping was only measured in the evening as overall coping that day. This is easily seen if we count the number of non-missing observations by survey, accomplished by first converting each value into a 0 or 1 based on whether it is missing or not, using the is.na()

function, and then summing using `sum()`, all broken down by survey. Note that `is.na()` return TRUE or 1 for missing and FALSE or 0 for not missing. To reverse this, we add the exclamation mark, which would mean the code is read, and sum the non-missing values by survey.

```
d[, .(
  NCope = sum(!is.na(COPEPrb)),
  NSOLs = sum(!is.na(SOLs))),
  by = Survey]
```

```
##         Survey NCope NSOLs
## 1: Afternoon     0     0
## 2:   Evening  2090  2097
## 3:   Morning     0     0
```

Even though sleep and coping are only assessed at a particular survey, they apply to the whole day so we could fill them in at the other surveys. It is easy to do this for the between-participant variables: we simply fill in the mean for all surveys. To fill in the other survey time points is more complicated. The key is to pass a single value, which R will recycle as needed. We can accomplish this by having `data.table()` apply the operation by both ID and survey day, then omit any missing coping values (leaving at most one per day), and subtract the between-person coping variable. Since we already filled that in, the between-person coping variable will have values at every survey time and so the within coping variable will too.

```
d[, BCOPEPrb := mean(COPEPrb, na.rm = TRUE), by = UserID]
d[, WCOPEPrb := na.omit(COPEPrb) - BCOPEPrb,
  by = .(UserID, SurveyDay)]
d[, BCOPEPrc := mean(COPEPrc, na.rm = TRUE), by = UserID]
d[, WCOPEPrc := na.omit(COPEPrc) - BCOPEPrc,
  by = .(UserID, SurveyDay)]
d[, BCOPEExp := mean(COPEExp, na.rm = TRUE), by = UserID]
d[, WCOPEExp := na.omit(COPEExp) - BCOPEExp,
  by = .(UserID, SurveyDay)]
d[, BCOPEDis := mean(COPEDis, na.rm = TRUE), by = UserID]
d[, WCOPEDis := na.omit(COPEDis) - BCOPEDis,
  by = .(UserID, SurveyDay)]
```

```
d[, BSOLs := mean(SOLs, na.rm = TRUE), by = UserID]
d[, WSOLs := na.omit(SOLs) - BSOLs,
  by = .(UserID, SurveyDay)]
d[, BWASONs := mean(WASONs, na.rm = TRUE), by = UserID]
d[, WWASONs := na.omit(WASONs) - BWASONs,
  by = .(UserID, SurveyDay)]
```

Intraclass Correlation Coefficient (ICC)

Another descriptive statistic that is useful for multilevel models is called the intraclass correlation coefficient or ICC. In a multilevel context, the ICC is based on decomposing variability into two sources: variability between individuals and variability within individuals. A robust way to calculate the ICC for a variable is by using the simplest type of multilevel model: a model with only a random intercept, that is, a model that only includes an intercept but allows this to vary randomly by ID. The variance of the random intercept is the between variance, as it is essentially the variance in the individual means, and the residual variance is the within-person variance, what is not explained by individuals' own means alone. Together, these two sources of variance form the total variance.

$$TotalVariance = \sigma^2_{between} + \sigma^2_{within} = \sigma^2_{randomintercept} + \sigma^2_{residual} \qquad (10.1)$$

Using these two sources of variance, we can calculate the ratio of the variability between individuals to the total variability. This ratio varies between 0 and 1 and provides information regarding how much of the total variability occurs between individuals. A value of 0 indicates that all individual averages are equal to each other so that all variability occurs within individuals. Conversely, a value of 1 indicates that within individuals all values are the same with all variability occurring between individuals. This ratio is referred to as the intraclass correlation coefficient (ICC) and can be calculated as shown in the following equation:

$$ICC = \frac{\sigma^2_{between}}{\sigma^2_{between} + \sigma^2_{within}} = \frac{\sigma^2_{randomintercept}}{\sigma^2_{randomintercept} + \sigma^2_{residual}} \qquad (10.2)$$

The ICC can be calculated by hand by fitting a random intercept-only model, or more conveniently by using the iccMixed() function. iccMixed() requires the variable name, the name of the ID variable (or variables if multiple IDs), and the dataset. It returns the variance estimated at each level, called Sigma, and the ratio of the variance at each level to the total variance, which is the ICC. While ICCs are commonly used for two-level structures, the function can generalize to higher-order structures, such as observations within students and students with classes. If student and class IDs were available, the variance at each level could be calculated, and the ICC would be the ratio of variance at each level to total variance.

```
iccMixed("NegAff", "UserID", d)

##             Var Sigma   ICC
## 1:      UserID  0.21 0.44
## 2: Residual  0.27 0.56

iccMixed("PosAff", "UserID", d)

##             Var Sigma   ICC
## 1:      UserID  0.63 0.54
## 2: Residual  0.53 0.46
```

Beyond providing an index of how much variability occurs between or within individuals, the ICC also is used in calculating the "effective" sample size. The "effective" sample size is an approximate estimate of how many independent samples the data provide. For example, if data were collected on 10 people every day for 10 days, there are a total of 100 observations, but the data is unlikely to provide the same effective information as 100 people measured once (independent samples) .

To better understand the effective sample size, often referred to as NEffective, consider two extremes. First, suppose that within an individual every observation is identical. For example, imagine measuring an adult's height. Once height was measured on day 1, another 9 days of assessment are unlikely to provide any further useful information. In this example, the ICC would be 1: all variability occurs between individuals, with no variability within individuals. That is, adults have many different heights (between variability), but the same adult has essentially the same height every day (no within variability).

On the other extreme, some variables may change as much day to day as between people. Imagine assessing commute times for different adults in the same city. Ignoring different routes, for the moment, variation may exist only because of different days and traffic conditions. Thus, on average, all their commute times may be identical (no between variability) with all variability occurring on a day-to-day basis (within variability).

In this latter example, 10 people measured for 10 days provides 100 observations and equivalent information as 100 people assessed on one day.

Calculating the NEffective attempts to provide an estimate as to what the equivalent independent sample size would be. The formula depends on the number of participants or truly independent units, N; the number of assessments per individual (unit), k; and the ICC, and is shown as follows:

$$N\,Effective = \frac{N^* k}{\left(\left(1+\left(k-1\right)^* ICC\right)\right)} \qquad (10.3)$$

The effective sample size can also be calculated in R using the nEffective() function. The NEffective also helps highlight how large an impact the ICC can have. The higher the ICC, the lower the NEffective. The following R code shows the NEffective for negative and positive affect. Despite a similar number of observations, the NEffective for negative and positive affect is quite different due to differences in their ICCs, which are calculated automatically by the nEffective() function.

```
## number of units
n <- length(unique(d$UserID))

## average observations per unit
k <- nrow(d[!is.na(NegAff)])/n

## effective sample size
nEffective(n, k, dv = "NegAff", id = "UserID", data = d)

##                       Type     N
## 1: Effective Sample Size   420
## 2:      Independent Units   191
## 3:     Total Observations 6389

k <- nrow(d[!is.na(PosAff)])/n
nEffective(n, k, dv = "PosAff", id = "UserID", data = d)

##                       Type     N
## 1: Effective Sample Size   343
## 2:      Independent Units   191
## 3:     Total Observations 6399
```

10.3 Exploration and Assumptions

Distribution and Outliers

At the beginning of the book, we examined a variety of ways to visualize univariate and multivariate data, in preparation for analyses. For non-independent or repeated measures data, similar approaches apply, except they can be applied at different levels or to different units (e.g., observations, between-person averages, etc.). To facilitate decompositions and graphical examination, we can use the meanDecompose() function. It uses a formula interface, with the primary variable on the left-hand side and the IDs or other variables to decompose by on the right-hand side. It works similarly to how we created between and within variables, except it creates separate datasets so that at the between level, there is no duplication.

```
tmp <- meanDecompose(PosAff ~ UserID, data = d)
str(tmp, max.level = 1)

## List of 2
## $ PosAff by UserID  :Classes 'data.table' and 'data.frame': 191 obs. of
2 variables:
##   ..- attr(*, "sorted")= chr "UserID"
##   ..- attr(*, ".internal.selfref")=<externalptr>
## $ PosAff by residual:Classes 'data.table' and 'data.frame': 6927 obs.
of  1 variable:
##   ..- attr(*, ".internal.selfref")=<externalptr>
```

The separate datasets are stored in a list, and they are named by the variable and the level. Here "UserID" and "residual" correspond to the between and within levels. As in previous chapters, we can examine each variable against a normal (or other) distribution. In the code that follows, we look at just the between-person positive affect values, which are plotted in Figure 10-4.

```
testdistr(tmp[[1]]$X, varlab = names(tmp)[1],
          extremevalues = "theoretical", robust=TRUE)
```

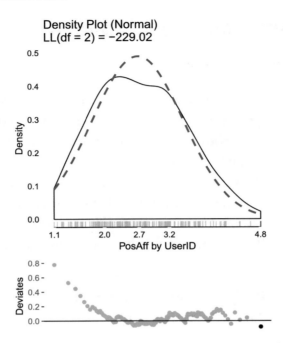

Figure 10-4. *Between-person positive affect against a normal distribution*

Rather than write out code for each level, we can have R loop through all the levels of data and plot them. We do not plot them separately, instead we combine them at the end using plot_grid() as shown in Figure 10-5.

```
plots <- lapply(names(tmp), function(x) {
  testdistr(tmp[[x]]$X, plot = FALSE, varlab = x,
          extremevalues = "theoretical", robust=TRUE)[1:2]
})

do.call(plot_grid, c(unlist(plots, FALSE), ncol = 2))
```

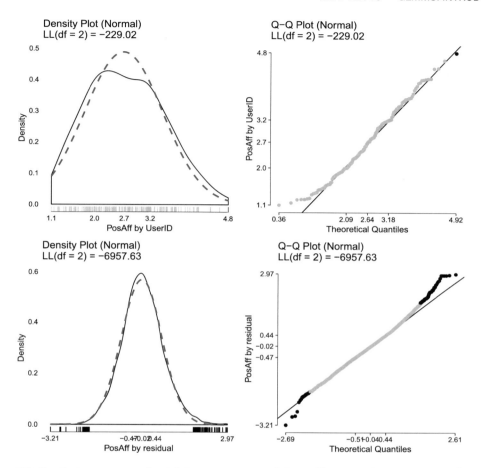

Figure 10-5. *Between- and within-person positive affect against a normal distribution*

Another feature of the meanDecompose() function is that we can add more levels. For instance, we can look at between participants, daily differences within participants, and finally residuals from participants and days. An example of this is shown in Figure 10-6.

```
tmp <- meanDecompose(NegAff ~ UserID + SurveyDay, data = d)
do.call(plot_grid, c(unlist(lapply(names(tmp), function(x) {
  testdistr(tmp[[x]]$X, plot = FALSE, varlab = x,
          extremevalues = "theoretical", robust=TRUE)[1:2]
}), FALSE), ncol = 2))
```

459

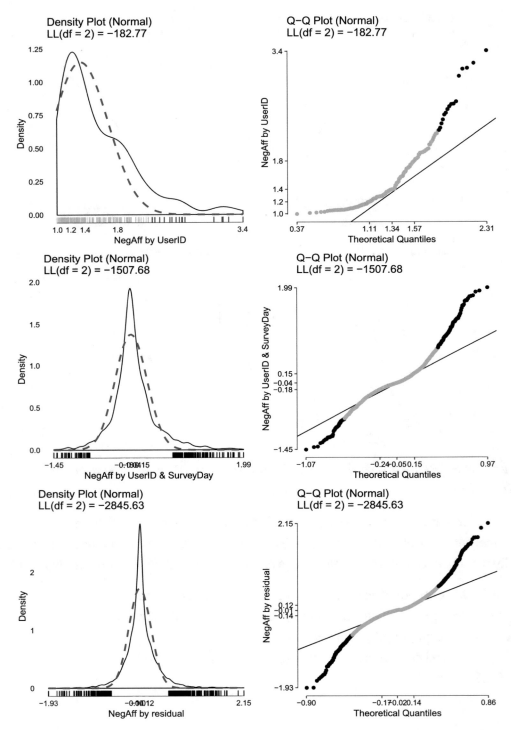

Figure 10-6. *Different levels of negative affect against a normal distribution*

These graphs (Figure 10-6) show that although negative affect by participant and day or residuals are symmetric, between-person negative affect is quite skewed. We could try a logarithm transformation. The results are in Figure 10-7. Although the skew of negative affect between participants is improved, it is still present. Levels by ID and day are approximately normally distributed. However, distribution of the residuals, while symmetric, is leptokurtic. These results suggest we should be cautious in assuming normality for negative affect.

```
d[, logNegAff := log(NegAff)]
tmp <- meanDecompose(logNegAff ~ UserID + SurveyDay, data = d)
do.call(plot_grid, c(unlist(lapply(names(tmp), function(x) {
  testdistr(tmp[[x]]$X, plot = FALSE, varlab = x,
            extremevalues = "theoretical", robust=TRUE)[1:2]
}), FALSE), ncol = 2))
```

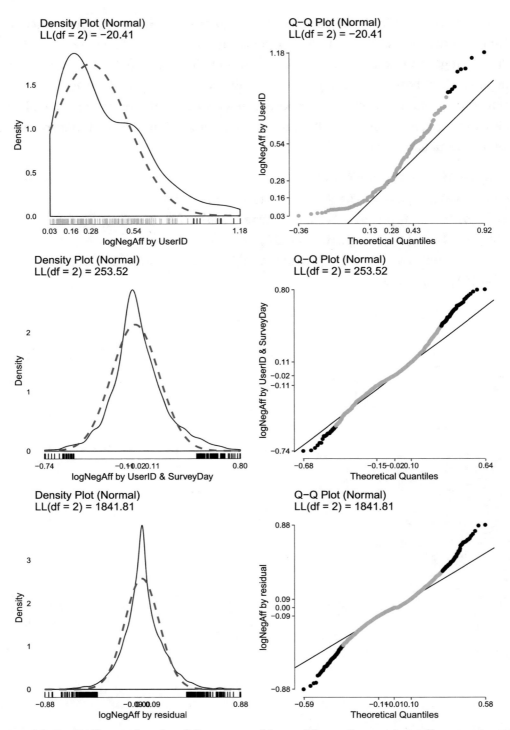

Figure 10-7. *Different levels of the natural logarithm of negative affect against a normal distribution*

Time Trends

In addition to exploring distributions and outliers, longitudinal repeated measures data have additional diagnostics that are helpful. If time trends are not expected in the data or are not the focus of the study, it is useful to empirically demonstrate that there are no time trends. This is important as many analyses assume stationary processes (i.e., a process that does not change substantially over time).

A simple way to begin is by plotting the means across time. We will use our within-participant variables for this, to omit any potential differences driven only by different participants. To plot many variables at once, we can melt the data to a long dataset by variable. The results are graphed in Figure 10-8.

```
dt <- d[, .(
  WPosAff = mean(WPosAff, na.rm = TRUE),
  WNegAff = mean(WNegAff, na.rm = TRUE),
  WSTRESS = mean(WSTRESS, na.rm = TRUE),
  WSUPPORT = mean(WSUPPORT, na.rm = TRUE),
  WSOLs = mean(WSOLs, na.rm = TRUE),
  WWASONs = mean(WWASONs, na.rm = TRUE)) , by = SurveyDay]
dt <- melt(dt, id.var = "SurveyDay")

ggplot(dt, aes(SurveyDay, value)) +
  geom_point() +
  stat_smooth(method = "gam", formula = y ~ s(x, k = 10)) +
  facet_wrap(~ variable, scales = "free")
```

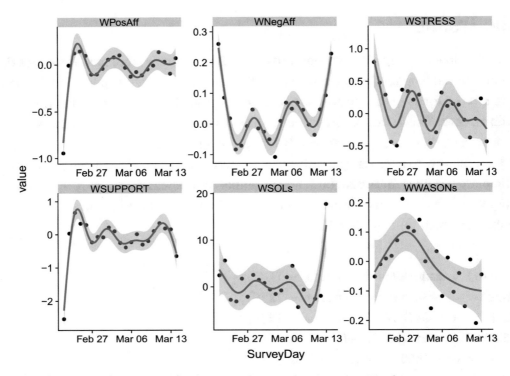

Figure 10-8. *Trends in variables over time with a gam smooth*

In addition to systematic time trends, we could look at day of the week differences, or perhaps most commonly weekday vs. weekend differences. We can use the weekdays() function to convert dates into days of the week and then test whether those match Saturday or Sunday to get a logical comparison that shows us if it's a weekend or not. Results are graphed in Figure 10-9.

```
dt[, Weekend := weekdays(SurveyDay) %in% c("Saturday", "Sunday")]
ggplot(dt, aes(Weekend, value)) +
  stat_summary(fun.data = mean_cl_boot) +
  facet_wrap(~ variable, scales = "free")
```

```
## Warning: Removed 2 rows containing non-finite values (stat_summary).
```

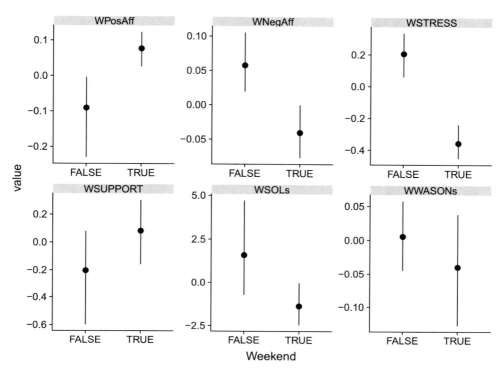

Figure 10-9. *Trends in variables over time with a gam smooth*

Overall, both time trends and some weekday vs. weekend differences emerged. These preliminary results suggest future analyses should adjust for such differences or we could calculate new variables that are residuals after removing any time trends.

Autocorrelation

Beyond time trends, it is helpful to look at how correlated variables are to themselves over time. This is referred to as autocorrelation. Using default autocorrelation tools, it is commonly assumed that observations are equally spaced in time, and there cannot be any missing values. For now as a quick and dirty exploratory approach, we will fill in missing data in two steps. First, we will estimate any missing survey start times as the mean time for each individual for each survey time (morning, afternoon, evening). Then we can combine the dates with the times to get a date and time variable, as shown in the code that follows.

```
d[, StartTimec11Alt := ifelse(is.na(SurveyStartTimec11),
                    mean(SurveyStartTimec11, na.rm = TRUE),
                    SurveyStartTimec11),
```

```
  by = .(UserID, Survey)]
d[, StartDayTimec11Alt := chron(
      dates. = format(SurveyDay, "%m/%d/%Y"),
    times. = StartTimec11Alt)]
```

Most autocorrelation functions are designed for individual time series, so we will operate one participant at a time. To see an example, we begin with a plot for a single participant. We first make the data a time series object, using the zoo() function (the "zoo" stands for Z's ordered observations, with "Z" being the first letter of the author's last name). Next, we fill in missing values by interpolation, using the na.approx() function. Finally we are ready to calculate the autocorrelation using the acf() function. The results are in Figure 10-10 and show that positive affect is perfectly correlated at lag 0 (i.e., the same time point), and it shows that the autocorrelation diminishes at later lags. Since we have three surveys per day, it would also be possible that observations at lag 3 are more highly autocorrelated as these would represent the same time on a different day. The data do not support that, though, suggesting instead that absolute time past is perhaps the most salient factor.

```
tmpd <- d[UserID == 1]
acf(na.approx(zoo(tmpd$PosAff,
    order.by = tmpd$StartDayTimec11Alt)),
    lag.max = 10)
```

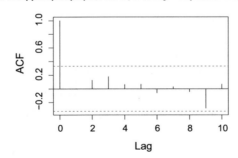

Figure 10-10. *Autocorrelation for one participant*

The preceding code showed the results for one participant, but we have many participants. The code that follows uses an extension, acfByID() designed to calculate the autocorrelation, by ID. We can use this to generate a dataset for the autocorrelations for lag 0 to 10 for positive affect and each ID. We can repeat the process for negative affect and stress and then visualize the results using boxplots, shown in Figure 10-11. The boxplots show the distribution across IDs of autocorrelations at each lag. Lines are added at 0 (no autocorrelation) and at ±0.5 as a rough indication of relatively strong autocorrelation.

```
acf.posaff <- acfByID("PosAff", "StartDayTimec11Alt",
                      "UserID", d)
```

```
print(acf.posaff)
```

```
##         UserID Variable Lag AutoCorrelation
##    1:       1   PosAff   0          1.0000
##    2:       1   PosAff   1          0.0016
##    3:       1   PosAff   2          0.1249
##    ---
## 2099:     191   PosAff   8         -0.0199
## 2100:     191   PosAff   9          0.1337
## 2101:     191   PosAff  10          0.1828
```

```
## make for other measures
acf.negaff <- acfByID("NegAff", "StartDayTimec11Alt",
                      "UserID", d)
acf.stress <- acfByID("STRESS", "StartDayTimec11Alt",
                      "UserID", d)
```

```
## put into one dataset for plotting a panel
acf.all <- rbind(
  acf.posaff, acf.negaff,
  acf.stress)
```

```
ggplot(acf.all,
    aes(factor(Lag), y = AutoCorrelation)) +
  geom_hline(yintercept = 0, colour = "grey50", size = 1) +
  geom_hline(yintercept = c(-.5, .5),
             linetype = 2, colour = "grey50", size = 1) +
  geom_boxplot() + ylab("Auto Correlation") +
  facet_wrap(~ Variable, ncol = 1)
```

The results from Figure 10-11 suggest that the autocorrelations are fairly small after a lag of 1. So for analyses, we will explore lag1 predictions. To prepare for this, we need to create variables that contain the lag 1 values, which we do in the code that follows. First, because lag 1 is one survey difference, we need a new measure that orders surveys from first to last for each participant. Note that the code that follows is only appropriate if there are no missing surveys or the missing surveys have already been added, as we did toward the start of the chapter. Then, we can use it to calculate the lagged value. For coping and sleep, we can only lag by one day, since they were measured daily.

Finally, we can compress and save the processed data as an RDS file, for ease of use in later chapters.

```r
## ensure data ordered by ID, date, and time
d <- d[order(UserID, SurveyDay, SurveyInteger)]
## calculate a number for the survey from 1 to total
d[, USURVEYID := 1:.N, by = .(UserID)]

d[,
  c("NegAffLag1", "WNegAffLag1",
    "PosAffLag1", "WPosAffLag1",
    "STRESSLag1", "WSTRESSLag1") :=
    .SD[.(UserID = UserID, USURVEYID = USURVEYID - 1),
    .(NegAff, WNegAff,
      PosAff, WPosAff,
      STRESS, WSTRESS),
      on = c("UserID", "USURVEYID")]]

d[,
  c("WCOPEPrbLag1", "WCOPEPrcLag1",
    "WCOPEExpLag1", "WCOPEDisLag1",
    "WSOLsLag1", "WWASONsLag1") :=
  .SD[.(UserID = UserID, Survey = Survey, SurveyDay = SurveyDay - 1),
      .(WCOPEPrb, WCOPEPrc, WCOPEExp, WCOPEDis,
        WSOLs, WWASONs),
      on = c("UserID", "Survey", "SurveyDay")]]

## save data after processing, with compression
## for use in subsequent chapters
saveRDS(d, file = "aces_daily_sim_processed.RDS",
        compress = "xz")
```

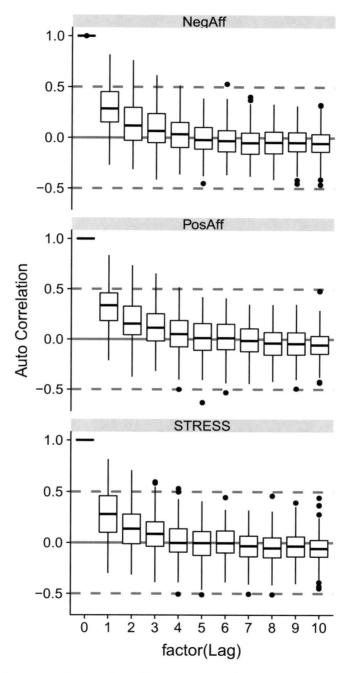

Figure 10-11. *Autocorrelation for all participant for positive and negative affect and stress*

Assumptions

GLMMs have similar assumptions to other regression models. They assume that

- The units used for random effects are independent

- There is a linear relationship between the predictors and the outcome on the link scale

- For normally distributed outcomes, the residual variance is homogenous

- The random effects follow a (multivariate) normal distribution

- The outcome comes from the expected distribution (normal, Poisson, etc.)

Although there are not good tests of independence, the remaining assumptions can be assessed visually. For GLMMs with non-normal outcomes, assessing the residuals and distribution of the outcome can be trickier. For GLMMs with (at least assumed) normally distributed outcomes, standard plots of residuals and fitted vs. residual values can be examined to examine whether the residual variance is homogenous and whether all the distributional assumptions are met. A number of these diagnostic plots are bundled together in the plotDiagnosticsLMER() function. To show these, we fit a model predicting negative affect from between- and within-person stress including both random intercept and slope. We will wait on interpreting the model itself until later. For now, the focus is on checking the assumptions. The plots in Figure 10-12 reveal that the residuals are too clustered together for a normal distribution.

```
m.negaff <- lmer(NegAff ~ 1 + BSTRESS + WSTRESS +
          (1 + WSTRESS | UserID), data = d)
assumptiontests <- plotDiagnosticsLMER(m.negaff, plot = FALSE)
do.call(plot_grid, c(
  assumptiontests[c("ResPlot", "ResFittedPlot")],
  assumptiontests$RanefPlot, ncol = 2))
```

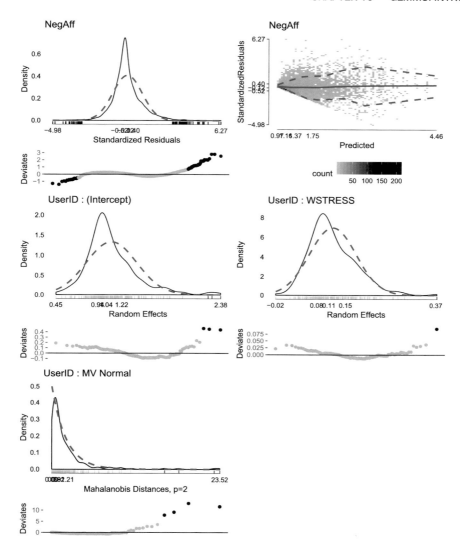

Figure 10-12. *Negative affect mixed effects model diagnostic plots showing the distribution of residuals (top left), the residuals vs. fitted values to assess homogeneity of variance (top right), the distribution of the random intercept (middle left), the distribution of the random slope (middle right), and whether the random effects are multivariate normal (bottom left).*

The plots in Figure 10-13 reveal that overall, the normality assumption is approximately met. There does appear to be a clear multivariate outlier.

```
m.posaff <- lmer(PosAff ~ 1 + BSTRESS + WSTRESS +
        (1 + WSTRESS | UserID), data = d)
```

```
assumptiontests <- plotDiagnosticsLMER(m.posaff, plot = FALSE)
do.call(plot_grid, c(
  assumptiontests[c("ResPlot", "ResFittedPlot")],
  assumptiontests$RanefPlot, ncol = 2))
```

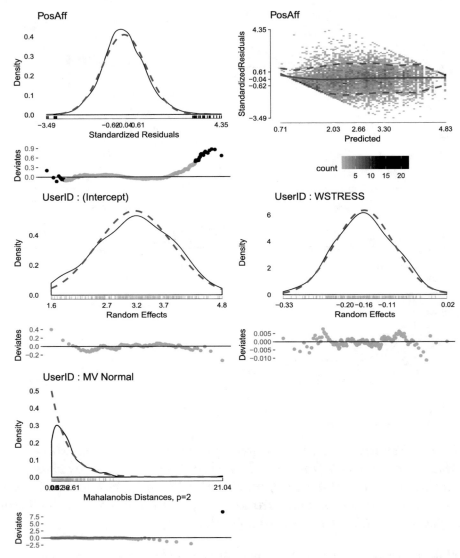

Figure 10-13. *Positive affect mixed effects model diagnostic plots showing the distribution of residuals (top left), the residuals vs. fitted values to assess homogeneity of variance (top right), the distribution of the random intercept (middle left), the distribution of the random slope (middle right), and whether the random effects are multivariate normal (bottom left).*

To further explore the multivariate outlier, we can examine the extreme values. These are separated by type: Residuals, Random Effect UserID : WSTRESS, and Multivariate Random Effect UserID. Based on the plots, we may not be that worried about residuals, since we have such large sample size. We may check out the multivariate outlier. In this case, we see that IDs 57 and 123 appear to be the culprits. To see the results without the influence of these multivariate outliers, we can remove them and re-estimate the model. The plots in Figure 10-14 reveal that overall, the normality assumption is approximately met and there are no longer any obvious outliers on the random effects individually nor multivariate outliers.

```
assumptiontests$ExtremeValues[
  EffectType == "Multivariate Random Effect UserID"]
```

```
##      PosAff UserID                       EffectType
## 1:     4.7    123 Multivariate Random Effect UserID
## 2:     3.9    123 Multivariate Random Effect UserID
## 3:     3.8    123 Multivariate Random Effect UserID
## ---
## 20:    3.7    123 Multivariate Random Effect UserID
## 21:    4.9    123 Multivariate Random Effect UserID
## 22:    4.6    123 Multivariate Random Effect UserID
```

```
m.posaff <- lmer(PosAff ~ 1 + BSTRESS + WSTRESS +
             (1 + WSTRESS | UserID),
             data = d[!UserID %in% c(57, 123)])

assumptiontests <- plotDiagnosticsLMER(m.posaff, plot = FALSE)
do.call(plot_grid, c(
  assumptiontests[c("ResPlot", "ResFittedPlot")],
  assumptiontests$RanefPlot, ncol = 2))
```

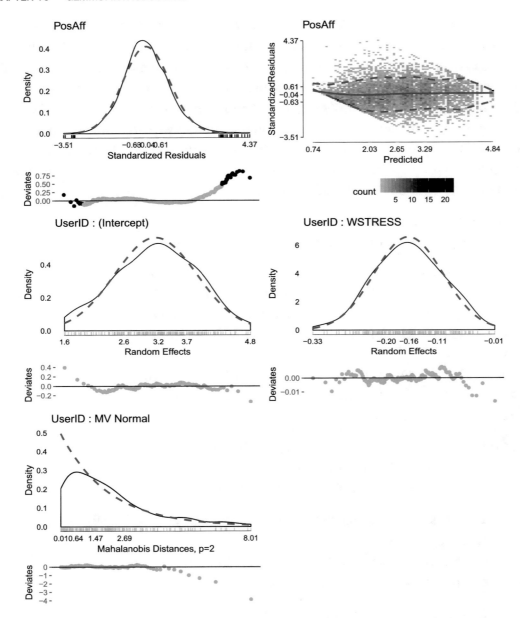

Figure 10-14. *Positive affect mixed effects model diagnostic plots showing the distribution of residuals (top left), the residuals vs. fitted values to assess homogeneity of variance (top right), the distribution of the random intercept (middle left), the distribution of the random slope (middle right), and whether the random effects are multivariate normal (bottom left). Results after removing two multivariate outliers, IDs 57 and 123.*

To better understand what is unusual about these multivariate outliers, we could also examine their data compared to everyone else on positive affect and stress. One way to do this is to plot the slopes for the stress and positive affect relationship for everyone in the dataset, and then separately highlight the relationship for an extreme case. We might also plot data points, just for the extreme IDs to ensure that a single extreme observation is not driving the effect. This is done in Figure 10-15, which shows that IDs 57 and 123 are indeed extreme, although the observed values seem consistent with the estimated slopes.

```
ggplot() +
  stat_smooth(aes(WSTRESS, PosAff, group = UserID),
    data = d[!UserID %in% c(123)], method = "lm",
  se = FALSE, colour = "grey50") +
  stat_smooth(aes(WSTRESS, PosAff, group = UserID),
    data = d[UserID %in% c(123)], method = "lm",
  se = FALSE, colour = "blue", size = 2) +
  geom_point(aes(WSTRESS, PosAff),
    data = d[UserID %in% c(123)], colour = "blue", size = 2) +
  stat_smooth(aes(WSTRESS, PosAff, group = UserID),
  data = d[UserID %in% c(57)], method = "lm",
 se = FALSE, colour = "orange", size = 2) +
geom_point(aes(WSTRESS, PosAff),
  data = d[UserID %in% c(57)], colour = "orange", size = 2)
```

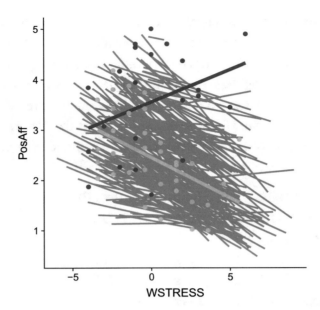

Figure 10-15. *Positive affect and stress associations highlight extreme cases.*

10.4 Summary

This chapter introduced multilevel data structures and how they are commonly formatted or stored (wide or long). It showed how basic data exploration and reporting of descriptive statistics can change when working with multilevel vs. single-level data. It also showed how to visualize and assess some common assumptions of generalized linear mixed models (GLMMs) by plotting one level of data at a time. Finally, it covered the specific case of longitudinal data, where additional diagnostics, such as whether there are consistent time trends and the autocorrelation of a variable with itself over time, are important. A summary of the functions used is in Table 10-4.

Table 10-4. *Listing of Key Functions Described in This Chapter and Summary of What They Do*

Function	What It Does
`[]`	Performs data management in `data.table` objects to calculate missing data and performs operations for all data or to first combine across IDs and then calculate results; operations commonly used when working with multilevel data
`acfByID()`	Calculates the autocorrelation coefficient for various lags by ID and returns a dataset suitable for summaries or plotting
`egltable()`	Calculates descriptive statistics, optionally by group levels of another variable
`gglikert()`	Presents descriptive statistics, such as means, with the response anchors at the left and right side of the graph
`iccMixed()`	Calculates the intraclass correlation coefficient for a variable
`meanDecompose()`	Takes a variable and decomposes it into means and residuals across different levels for fast plotting at each level in multilevel data
`nEffective()`	Calculates the effective sample size for a variable in multilevel data
`plotDiagnosticsLMER()`	Creates plots of a variety of diagnostics for a linear mixed model
`reshape()`	Reshapes data from wide to long or long to wide format

CHAPTER 11

GLMMs: Linear

This chapter builds on the foundation of working with multilevel data and introduces a class of statistical models—generalized linear *mixed* models (GLMMs)—that are appropriate for such data.

```
library(checkpoint)
checkpoint("2018-09-28", R.version = "3.5.1",
  project = book_directory,
  checkpointLocation = checkpoint_directory,
  scanForPackages = FALSE,
  scan.rnw.with.knitr = TRUE, use.knitr = TRUE)

library(knitr)
library(ggplot2)
library(cowplot)
library(viridis)
library(JWileymisc)
library(data.table)
library(lme4)
library(lmerTest)
library(chron)
library(zoo)
library(pander)
library(texreg)
library(xtable)
library(splines)
library(parallel)
library(boot)

options(width = 70, digits = 2)
```

© Matt Wiley and Joshua F. Wiley 2019
M. Wiley and J. F. Wiley, *Advanced R Statistical Programming and Data Models*,
https://doi.org/10.1007/978-1-4842-2872-2_11

11.1 Theory

This section introduces GLMMs more formally. GLMMs extend the fixed effects–only GLMs that were discussed in previous chapters. As a reminder, for GLMs, we defined the expected linear outcome, η, as

$$\eta = X\beta \tag{11.1}$$

The expected linear outcome, η, was mapped to the raw outcome, y, via a link function, $g(\cdot)$.

$$\eta = g(\mu) = g(E(y)) \tag{11.2}$$

The inverse link function, $g^{-1}(\cdot)$, backtransforms the scale of η to the scale of y.

$$E(y) = \mu = g^{-1}(\eta) \tag{11.3}$$

GLMMs build on this structure with a few additional components that are not necessary in GLMs that include only fixed effects.

Generalized Linear Mixed Models

For GLMs, the expected value was a function of the predictors weighted (multiplied) by the parameter estimates, $X\beta$. These are called the fixed effects, although that is often not stated explicitly in GLMs because there are only fixed effects. They are fixed effects in the sense that the parameter estimates, β, do not vary; they are not random variables.

With repeated measures or non-independent data, we need some way to capture the dependency in the observations. Another way of thinking about it is that non-independent data implies that there are systematic differences between units (people, schools, hospitals, etc.). GLMMs address this by adding another component to the model that explicitly captures these differences between units. As for the fixed effects, this component has two parts, a data matrix, by convention referred to as Z, and a paramater component, by convention referred to as γ. The most basic way of capturing systematic differences between units is to allow each unit to have its own intercept. In longitudinal studies of people, that equates to each participant having her or his own intercept. In that case, Z will be a block diagonal matrix of 0 and 1 values (imagine dummy coding the participant ID variable in a longitudinal dataset). This is easier to vizualize than it is to describe. First we can create a matrix of dummy codes for each ID, stored as mat. Then we can plot the matrix for the first 10 participants and the first 300 values, Figure 11-1. In the image, black regions indicate 1s and white space indicates 0s. Note we load the raw data from the JWileymisc package using the data() function and then read in the processed data we made and saved in the previous chapter introducing some basics of LMMs.

```
data(aces_daily)
draw <- as.data.table(aces_daily)
d <- readRDS("aces_daily_sim_processed.RDS")

mat <- model.matrix(~ 0 + factor(UserID), data = d)

image(t(mat[1:300, 1:10]), col = c("white", "black"),
      xlab = "Participants", ylab = "Observation",
      xaxt = "n", yaxt = "n")
```

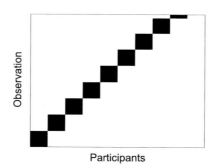

Figure 11-1. *Graph of block diagonal dummy code matrix of UserID. Black values indicate rows of the data that belong to a particular participant. Different columns represent different participants.*

In this simple example, the γ vector has one element for each participant, which is the estimated intercept for that particular participant. With the current dataset, there are 191 participants, so γ will contain 191 parameter estimates.

Putting this additional piece together with the GLM model we already learned, the overall GLMM is defined as

$$\eta = \boldsymbol{X}\beta + Z\gamma \tag{11.4}$$

481

For these specific data, with an intercept-only model, 191 participants, and a total of 6,599 observations, the dimensions of each vector/matrix would be as follows:

$$\underset{6599\times1}{\eta} = \underset{\substack{6599\times1 \\ 6599\times1_{1\times1}}}{X\ \beta} + \underset{\substack{6599\times1 \\ 6599\times191_{191\times1}}}{Z\ \gamma}$$

One aspect that becomes clear under this approach is that if we estimated each parameter in γ individually, we will end up with at least as many parameters as participants, plus whatever other parameters are needed for the rest of the model. In fact, if we estimated each parameter in γ individually, what we have currently described could be estimated as a GLM. Although we conceptually separated $Z\gamma$ from $X\beta$, a GLM could estimate both. What makes GLMMs *mixed* models rather than GLMs with lots of dummy codes is that we do not directly estimate γ. In GLMMs, rather than estimating every parameter in γ, we treat γ as a random variable or a random effect (hence *mixed* effects as there are both fixed and random effects). We assume that the random variable γ comes from some distribution. Specifically, with few exceptions, we assume that γ comes from a *normal* distribution.

For ease, we use N to indicate a normal or Gaussian distribution. The normal distribution is governed by two parameters, the mean (or location), μ, and the standard deviation (or scale), σ. Formally then, we can say that γ is distributed as a normal distribution with a mean and standard deviation, or as an equation:

$$\gamma \sim N(\mu, \sigma) \tag{11.5}$$

The benefit of this approach is that rather than having to estimate each parameter of γ individually, no matter how many participants we have, the intercept-only GLMM only needs to estimate two extra parameters, μ and σ, the parameters of the normal distribution. In fact, we do not even need two extra parameters. Because the fixed effects portion of the model, $X\beta$, includes an estimate of the intercept, we already know what the mean (intercept) will be. Conventionally, γ is defined as the *deviations* from the overall mean, captured by the fixed effects. These deviations will always be zero, on average, and so in fact there is just one extra parameter to estimate, σ, and we write

$$\gamma \sim N(0, \sigma) \tag{11.6}$$

We noted that in almost all cases, γ is assumed to follow a normal distribution. Intuitively, it might seem that for GLMMs, other distributions would be used. However, even though in GLMMs we may assume that the *outcome* follows a different distribution (Normal, Bernoulli, Poisson, etc.), the random effects are still typically assumed to be normally distributed. Although not strictly a requirement of GLMMs, most software only implements normally distributed random effects, the only common exception being software for Bayesian GLMMs, where more flexibility is commonly allowed.

One final important point. There may be more than one random effect. If there are multiple random effects, then γ will be assumed distributed as a multivariate normal distribution. This is indicated by using a bold 0 to indicate it is a vector of means at 0 and an uppercase sigma, Σ, to indicate it is a variance-covariance matrix rather than a single variance or standard deviation.

$$\gamma \sim N\left(0, \sum_{\theta}\right) \tag{11.7}$$

To see how these models translate in terms of a concrete model, we can estimate an intercept-only GLM and an intercept-only GLMM. For this example, we look at positive affect in the daily diary study. We will focus on how to actually code the analyses in R later, for now we just show the model results to focus on the conceptual aspects of the models. Table 11-1 compares a GLM and GLMM intercept-only model for positive affect.

Table 11-1. *Statistical Models*

	GLM	GLMM
(Intercept)	2.68***	2.68***
	(0.01)	(0.06)
R^2	0.00	
Adj. R^2	0.00	
Num. obs.	6399	6399
RMSE	1.07	
AIC		14800.56
BIC		14820.85
Log Likelihood		−7397.28
Num. groups: UserID		191
Var: UserID (Intercept)		0.63
Var: Residual		0.53

$***p < 0.001, **p < 0.01, *p < 0.05$

The first row shows the fixed effect intercept. We can see that the estimates are comparable, but the standard error (in parentheses) is larger in the GLMM than in the GLM. In the GLM, the only other parameter is the residual standard deviation, labelled "RMSE". In the GLMM, there is a residual variance, labelled "Var: Residual"; there is also the variance of the random intercepts, labelled "Var: UserID (Intercept)".

The GLM has a familiar interpretation. In the GLMM, the intercept estimate can be interpreted as the average intercept across participants. The variance of the intercept gives us a sense of how spread out or dispersed individual participants' intercepts are. Since we assume that the intercepts come from a normal distribution, standard rules of thumb apply. That is, about two thirds of participants will fall within one standard deviation of the mean. Taking the square root of the intercept variance, we can find the intercept standard deviation. With that we can calculate that about two thirds of participants should have an intercept between

```
## 2.68 - 0.63 = 2.05
```

```
## 2.68 + 0.63 = 3.31
```

Mixed Effects vs. Multilevel Model Terminology

Before continuing, it is worth noting a common distinction in terminology. This book presents GLMMs from the perspective and using the terminology of mixed effects models. However, the same models also are commonly called and presented from the perspective of multilevel models. Rather than using matrices, multilevel models use subscripts to indicate which parameters vary by participant. First, going back to the GLM, we could write the model algebraically, rather than with matrices as follows:

$$\eta_i = b_0 \tag{11.8}$$

The subscript indicates the expected value for the ith participant. For GLMMs or multilevel models, we need at least two subscripts: one for participants and one for observations within participants. Conventionally, we talk about the ith observation from the jth participant. We can write the GLMM model then as follows:

$$\eta_{ij} = b_0 + \gamma_{0j} \tag{11.9}$$

This equation highlights that any particular observation is a combination of the average intercept, b_0, plus a particular participant's deviation from the average, γ_{0j}. In multilevel models, the nesting of observations is referred to as different levels. So in our daily study, observations within participants would be level 1 and effects at the participant level would be level 2. When there are only two levels, people often refer to level 1 as "within" and level 2 as "between." This makes sense with our daily data as level 1 observations are observations or differences within an individual participant and level 2 or participant level captures differences between participants.

Both mixed effects and multilevel terminology and notation are common, so it is helpful to know that they are the same underlying model. Familiarity with both notations will make it easier to work with people used to either framework.

Statistical Inference

In linear regression, statistical inferences (p-values, confidence intervals) are straightforward to calculate. For the regression coefficients divided by their standard error can be shown to follow a t-distribution with degrees of freedom equal to the number of observations minus the number of estimated parameters. In linear mixed effects, there is no formula to calculate the correct degrees of freedom.

As a consequence, what the "correct" p-values and confidence intervals should be is unknown. Because of this, the lme4 package in R does not print p-values by default nor provide any degrees of freedom. However, there are several strategies to estimate confidence intervals and p-values.

The simplest approach is to assume that the sample size is large enough that the t-distribution approximates the normal distribution so that rather than needing the degrees of freedom for a t-distribution, the normal distribution can be used as a "close enough" proxy. This can be done to calculate both p-values and confidence intervals.

A more accurate way to calculate confidence intervals is to profile the likelihood function. Indeed, this is the default way that is used when asking for confidence intervals from a model in lme4. While profile confidence intervals are more accurate, they are computationally demanding and in some cases may not converge or be estimable.

Another approach is to try to estimate the approximate degrees of freedom. Degrees of freedom based on Satterthwaite's approximation are available in the lmerTest package by Kuznetsova and colleagues [54]. Once the lmerTest package is loaded, it actually masks the lmer function we use so that approximate degrees of freedom are calculated and (approximate) p-values are reported by default. Compared to profiling the likelihood function, approximating the degrees of freedom for statistical inference has low computational cost. For linear mixed effects models, that is, GLMMs where the outcome is continuous and assumed normally distributed, this approach is easy and may represent reasonable "default" choice.

Two final possibilities that we will not go over in depth in this book are to use bootstrapping or Bayesian estimation. Although we will show some examples, we leave coverage of the theoretical justification of these methods to other, more detailed texts. For excellent, advanced coverage of Bayesian methods, see Gelman and colleagues [36]. For in-depth coverage of the theory to bootstrapping, see [26].

Bootstrapping involves drawing a random sample of data points, estimating the model, storing the results, and repeating the process many times to build up an empirical distribution of the parameter values. The benefit of such an approach is that it empirically estimates the distribution, so that it does not require making assumptions about the shape of the parameter sampling distribution. Although bootstrapped confidence intervals have many desirable properties, they are very computationally demanding and can take a long time to complete for all but the most basic models. This makes bootstrapped inference a difficult choice in iterative model building processes, such as when trying various covariates, functional forms of the relationship (linear, non-linear, etc.). However, bootstrapping may be an excellent choice to verify the results from a final model or in "high stakes" cases, such as randomized controlled trials. Bayesian estimation relies on statistical inference through an entirely different framework. Bayesian approaches are powerful and provide a very useful alternative to classical frequentist statistics described so far in this book.

One special consideration when choosing the approach to statistical inference is whether you want to draw inference regarding the random effect variance components. Relying on an approximately normal distribution or approximated degrees of freedom is inappropriate for variance components as they cannot be less than zero, so symmetrical confidence intervals are not sensible. For variance components, confidence intervals may be obtained through profiling the likelihood function, bootstrapping, or Bayesian methods.

Finally, although we have focused on continuous, normally distributed outcomes in this chapter, for other types of GLMMs, such as for binary or count outcomes, the degrees of freedom cannot be approximated, so statistical inference must proceed through assuming a normal distribution, profiling the likelihood, bootstrapping, or Bayesian methods.

Effect Sizes

In linear regression, a common effect size is the proportion of variance accounted for by the model, R^2. In linear regression, R^2 is readily calculated. The total variance is the combination of the variance explained by the model and the residual variance.

$$R^2 = 1 - \frac{\sigma^2_{residual}}{\sigma^2_{total}} \tag{11.10}$$

In mixed effects models, calculating R^2 is not as simple. Variance can be accounted for by the fixed effects and by the random effects. One approach to calculating a pseudo-R^2 is to calculate the squared correlations between predicted and actual values.

```
m <- lmer(NegAff ~ 1 + (1 | UserID), data = d)
cor(na.omit(d$NegAff), fitted(m))^2

## [1] 0.45
```

More recently, Nakagawa and Schielzeth [69] suggested two versions, called the marginal and conditional R^2 for random intercept models. The marginal R^2 is the ratio of the variance explained by the fixed effects to the total variance. The total variance is defined as the sum of the variance explained by the fixed effects, the variance explained by the random effects (may be one or multiple random intercepts), and the residual variance. The marginal R^2 represents the percent of variance explained by the fixed (marginal) effects. In their original formulation, Nakagawa and Schielzeth [69] had a slightly different equation as they had separate terms for error and dispersion variance, but these are the same for continuous normally distributed outcomes. The simplified equation for normally distributed variables is as follows:

$$Marginal\,R^2 = \frac{\sigma^2_{fixedeffects}}{\sigma^2_{fixedeffects} + \sum_{i=1}^{k}\sigma^2_{random_i} + \sigma^2_{residual}} \tag{11.11}$$

The conditional R^2 is defined similarly, but it is the percentage of variance accounted for by both the fixed and random effects, and is given as follows:

$$Conditional\,R^2 = \frac{\sigma^2_{fixedeffects} + \sum_{i=1}^{k}\sigma^2_{random_i}}{\sigma^2_{fixedeffects} + \sum_{i=1}^{k}\sigma^2_{random_i} + \sigma^2_{residual}} \tag{11.12}$$

In a random intercept-only model where the only fixed effect is the intercept, the conditional R^2 as defined in the preceding equation will be the same as the ICC. The R2LMER function calculates both marginal and conditional R^2 from linear mixed models using these formulae. The following example also shows that in the case of model with intercepts only, the conditional R^2 is the same as the ICC.

```
m <- lmer(NegAff ~ 1 + (1 | UserID), data = d)
R2LMER(m, summary(m))

##     MarginalR2 ConditionalR2
##           0.00          0.44

iccMixed("NegAff", "UserID", d)

##             Var Sigma   ICC
## 1:    UserID  0.21  0.44
## 2: Residual  0.27  0.56
```

Since Nakagawa and Schielzeth [69] derived equations for random intercept models, Johnson [48] extended their approach to incorporate models with both random intercepts and slopes. The R2LMER function incorporates these updates to accommodate random intercept-only and random intercept and slope models.

Random Intercept Model

The simplest mixed effects model is a random intercept-only model. Random intercept models may include any number of fixed effects, but by convention the name refers to models where the only random effect is a random intercept.

The random intercept captures the dependency in observations by allowing differences in the intercept for each participant. That way, after accounting for individual differences in the intercept, the residuals will be (conditionally) independent. Beyond the random intercept, any number of fixed effects can be added.

Beyond the theoretical aspects and equations underpinning mixed effects models, visualizing different models can help in understanding what a "random" effect really means.

Visualizing Random Effects

We assume readers are familiar with standard linear regression models (those with only fixed effects). In the daily diary study data we have been using, each participant reports how many minutes it took to fall asleep for up to 12 days. If we wanted to examine the relationship between whether it was someone's first, second, etc., day in the study and how many minutes it took them to fall asleep (sleep onset latency; SOL), we could use a linear regression model, as follows.

```
## data setup
d[,
  SurveyDayCount := as.integer(SurveyDay - min(SurveyDay)),
  by = UserID]

## setup mini dataset
tmpd <- d[!is.na(SOLs) & !is.na(SurveyDayCount),
  .(SOLs, SurveyDayCount, UserID)]
```

```
## fixed effects, all people
mreg <- lm(SOLs ~ 1 + SurveyDayCount, data = tmpd)
## add predictions to the dataset
tmpd[, Fixed := predict(mreg, newdata = tmpd)]
```

This linear regression provides two average (fixed) effects. The **intercept** is the expected SOLs on participants' first night in the study, day 0. The **slope** is the expected change in SOLs for a one day change in study day. These two numbers are the average across every participant in the study. They do not capture any individual differences between participants. A benefit of a model like this is that it combines all participant data, so it will be relatively robust to outliers. It also does not matter if a particular participant has only few data points, because all participants' data are combined. A final point worth noting is that while the average intercept and slope are accurate estimates, their associated p-values will be downwardly biased because the data violate the assumption that observations are independent of each other.

Another simple approach would be to run a fixed effects linear regression model, but run separate models for each individual participant, shown as follows. Because we want to focus on differences in the intercept only, we use a fixed offset to force the slope for SurveyDay to be the same as the overall population average fixed effects model we fit earlier.

```
## fixed effects, individual models
tmpd[, Individual := fitted(lm(SOLs ~ 1 +
  offset(coef(mreg)[2] * SurveyDayCount))),
  by = UserID]
```

These individual models estimate a different intercept for each individual participant, but use the average slope. Because each model is fit on one participant's data, they are more sensitive to outliers, and these models may become very unstable if a particular participant only has few data points (e.g., only 2 or 3 days). A benefit of this individual approach is that the statistical tests may be accurate as within a participant days may be independent.

Finally, we can run a random intercept model with a fixed slope for study day, shown in the following code.

```
## random intercept model, all people
m <- lmer(SOLs ~ 1 + + SurveyDayCount + (1 | UserID), data = tmpd)

## add predictions to the dataset
tmpd[, Random := predict(m, newdata = tmpd)]
```

The random intercept model allows each individual to have a different intercept, but rather than estimating each intercept individually, they are assumed to come from a normal distribution, and the mean and variance of that distribution are estimated. All of the participants' data are included in one model, so again this model would be relatively robust to outliers or extreme values. At the same time, the random intercept ensures that a single average value is not used for every participant.

Two other ways to think about the random intercept model is that the intercept estimates from the random model are a weighted combination of the intercept for an individual and the population average intercept. The more data available for a particular participant, the closer the random intercept estimate will be to their individually estimated intercept. Conversely, the less data a participant has, the closer the random intercept will be to the population average. At the extreme (no data for a participant), the best the model can do is estimate that participant would have the population average value. This approach has the effect of pulling individual estimates toward the population mean, which is called *shrinkage*: extreme estimates are shrunken toward the mean. If you come from a machine learning background, this is also a form of model regularization as constraints are placed on the individual estimates to approximate a normal distribution.

To visualize the differences between these models, we can graph the predicted relations between study day and SOL for a few participants. Figure 11-2 shows the estimated trajectories from the fixed individual model and the random model with the bold blue lines showing the intercept and slope from the linear regression model on all participants.

```
## select a few example IDs to plot
tmpdselect <- melt(tmpd[UserID %in% unique(UserID)[107:115]],
     id.vars = c("UserID", "SurveyDayCount", "SOLs"))

ggplot(tmpdselect[variable != "Fixed"],
       aes(SurveyDayCount, value, group = UserID)) +
  geom_abline(intercept = coef(mreg)[1], slope = coef(mreg)[2],
              size = 2, colour = "blue") +
  geom_line() +
  facet_wrap(~ variable)
```

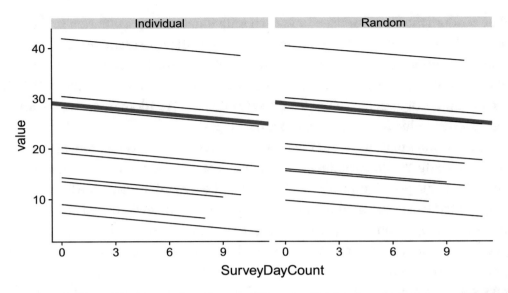

Figure 11-2. *Plot of estimated trajectory from individual regression models and a random intercept model. Population average shown in blue. While all lines have the same slope, the intercepts of each line are closer to the population average for the Random model than the Individual models, showing the effect of shrinkage.*

Another way to view the differences between models is to plot the estimated trajectories against the raw data. This is shown in Figure 11-3. These figures highlight that the individual fixed effects models or random intercept models are as or more accurate than the population average. The figures also highlight that in every case, the random intercept line is the same as the fixed effect or closer to the population average, never more extreme than the individual fixed effects model.

```
## plots against individual data
ggplot(tmpdselect, aes(SurveyDayCount)) +
  geom_point(aes(y = SOLs), size = 1) +
  geom_line(aes(y = value,
                colour = variable,
                linetype = variable), size = 1.5) +
  facet_wrap(~UserID, scales = "free_y") +
  scale_color_viridis(discrete = TRUE) +
  theme(legend.position = "bottom",
        legend.title = element_blank(),
        legend.key.width = unit(2, "cm"))
```

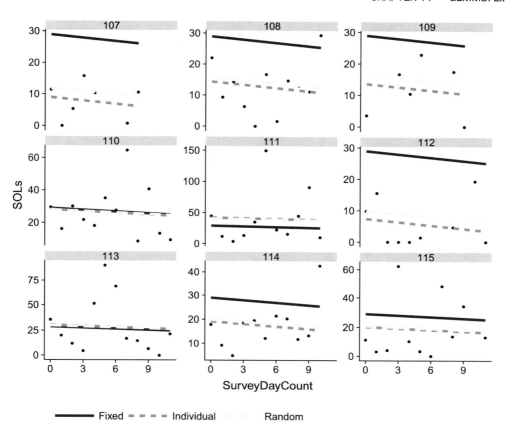

Figure 11-3. *Plots of different model estimated lines against raw data values for nine participants*

The shrinkage can more explicitly be viewed by plotting the intercept from the individual models with the intercept from the random effects models and showing the change. That is done in Figure 11-4. For this, we plot all participants' data and order it from those with the highest to lower mean value in the individual models. This highlights that the greatest shrinkage toward the sample mean (the vertical line) occurs for people whose individual estimates are the furthest away.

```
tmpd <- tmpd[SurveyDayCount==0][order(Individual)]
tmpd[, UserID := factor(UserID, levels = UserID)]

ggplot(tmpd, aes(x = Individual, xend = Random,
                 y = UserID, yend = UserID)) +
  geom_segment(
```

```
    arrow = arrow(length = unit(0.01, "npc"))) +
  geom_vline(xintercept = tmpd[SurveyDayCount==0][1, Fixed]) +
  xlab("Estimated Intercept") +
  theme(axis.text.y = element_blank())
```

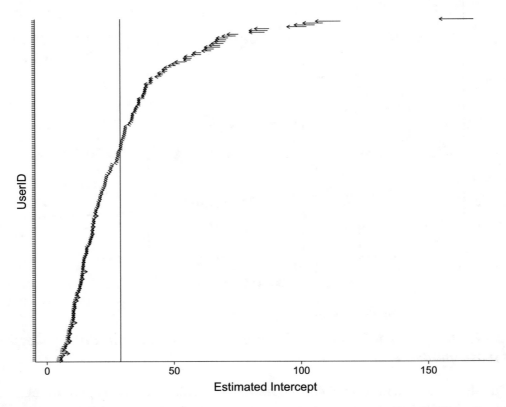

Figure 11-4. *Plot of estimated intercept in the individual models and random effects model for each participant, with arrows showing the shrinkage toward the population mean intercept.*

Interpreting Random Intercept Models

Typically the first step in interpreting a model is evaluating diagnostics to ensure that the model is reasonable. Some basic diagnostics are plotted in Figure 11-5. They show that the residuals are about symmetrically distributed, although not exactly normally distributed with some potential outliers. They also show that the residual variance increases with higher predicted values. There appear to be some fairly extreme random intercepts, and the distribution of random intercepts is generally positively skewed, suggesting that a transformation may be helpful here. A generalized additive model fit to the survey day and sleep onset latency did not select any non-linearity, suggesting that assuming a linear association may be a reasonable approximation in these data.

```
assumptiontests <- plotDiagnosticsLMER(m, plot = FALSE)
do.call(plot_grid, c(
  assumptiontests[c("ResPlot", "ResFittedPlot")],
  assumptiontests$RanefPlot,
  list(ggplot(d, aes(SurveyDayCount, SOLs)) +
      stat_smooth()),
  ncol = 2))

## `geom_smooth()` using method = 'gam' and formula 'y_~_s(x,_bs_=_"cs")'
```

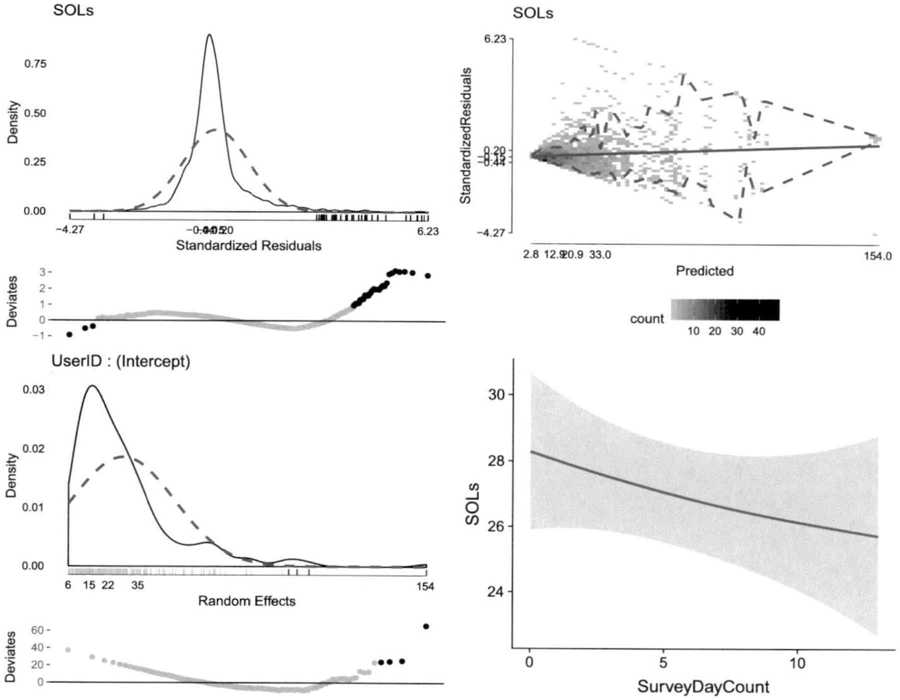

Figure 11-5. *Mixed effects model diagnostic plots showing the distribution of residuals (top left), the residuals vs. fitted values to assess homogeneity of variance (top right), the distribution of the random intercept (bottom left), and a simple single-level generalized additive model smooth of the association between survey day and sleep onset latency to assess for non-linearity (bottom right).*

Because there are some zero values in the outcome, a log transformation will not work well, so we might try a square root transformation. This requires refitting the model and checking the diagnostics again. This is accomplished in the following code and graphed in Figure 11-6. The results show improvements in several areas. The residuals have fewer extreme values. The residual variance is more homogenous across the range of predictions, and the random intercept distribution is closer to normal. It appears there also is an approximately linear association between survey day and sleep onset latency on the square root scale, so this appears to be a reasonable model to present and interpret.

```
d[, sqrtSOLs := sqrt(SOLs)]
m2 <- lmer(sqrtSOLs ~ SurveyDayCount + (1 | UserID),
           data = d)

assumptiontests <- plotDiagnosticsLMER(m2, plot = FALSE)
do.call(plot_grid, c(
  assumptiontests[c("ResPlot", "ResFittedPlot")],
  assumptiontests$RanefPlot,
  list(ggplot(d, aes(SurveyDayCount, sqrtSOLs)) +
       stat_smooth()),
  ncol = 2))

## `geom_smooth()` using method = 'gam' and formula 'y_~_s(x,_bs_=_"cs")'
```

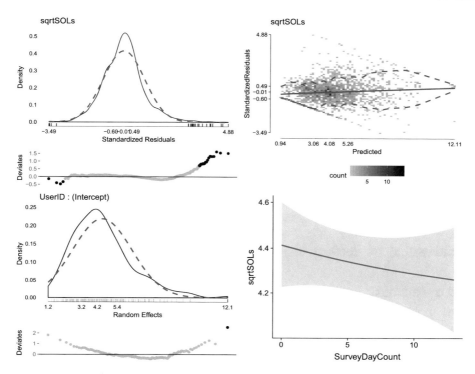

Figure 11-6. *Mixed effects model diagnostic plots showing the distribution of residuals (top left), the residuals vs. fitted values to assess homogeneity of variance (top right), the distribution of the random intercept (bottom left), and a simple single-level generalized additive model smooth of the association between survey day and sleep onset latency to assess for non-linearity (bottom right).*

A good beginning point for interpretation and presentation of models is the `summary()` function, shown in the following code. Because we had the `lmerTest` package loaded, the results include approximate degrees of freedom and p-values based on t-tests from these. Under the random effects heading, we can see that the estimated standard deviation of the random intercept by IT and residuals were about equal, suggesting approximately equal levels of variance at each level. The number of observations included in the model and the number of units also are displayed. Under the fixed effects heading, the overall average intercept is shown as is the average association between survey day and square root transformed sleep onset latency. The statistical tests test whether each of these is statistically significantly different from zero. Perhaps a reasonable enough question for the association with survey day, but not especially interesting for the intercept, as it is implausible that the average sleep onset latency would be zero. In general, the fixed effects can be interpreted similarly to how they would be interpreted in a single-level generalized linear model. In this case, there is a statistically significant association between survey day and square root sleep onset latency, such that each additional day in the study was associated with -0.02 less square root minutes until sleep onset.

summary(m2)

```
## Linear mixed model fit by REML. t-tests use Satterthwaite's method [
## lmerModLmerTest]
## Formula: sqrtSOLs ~ SurveyDayCount + (1 | UserID)
##     Data: d
##
## REML criterion at convergence: 9552
##
## Scaled residuals:
##     Min      1Q Median      3Q     Max
## -3.486  -0.601 -0.012   0.490   4.878
##
## Random effects:
##  Groups    Name         Variance Std.Dev.
##  UserID    (Intercept)  3.72     1.93
##  Residual               4.50     2.12
## Number of obs: 2097, groups:  UserID, 191
##
## Fixed effects:
##                 Estimate Std. Error       df t value Pr(>|t|)
## (Intercept)       4.4578     0.1639  289.5176   27.20   <2e-16 ***
## SurveyDayCount   -0.0223     0.0135 1914.7904   -1.65      0.1 .
## ---
## Signif. codes:  0 '***' 0.001 '**' 0.01 '*' 0.05 '.' 0.1 '_' 1
##
## Correlation of Fixed Effects:
##             (Intr)
## SurveyDyCnt -0.438
```

Recently, many journals want confidence intervals presented along with estimates, rather than estimates and standard errors. As noted earlier in this chapter when discussing statistical inference, there are several ways confidence intervals may be calculated. In general, confidence intervals are obtained using the confint function, but the methods can vary from the simple Wald method, which uses the standard error and essentially assumes large enough degrees of freedom that the t-distribution approximates a normal distribution, to the more accurate but time-consuming profile and bootstrap methods. Each of these is timed to provide some indication of the relative intensity of each approach.

```
system.time(
  ci.wald <- confint(m2,
    method = "Wald", oldNames = FALSE))

##    user  system elapsed
##    0.02    0.00    0.01

system.time(
  ci.profile <- confint(m2,
    method = "profile", oldNames = FALSE))

## Computing profile confidence intervals ...

##    user  system elapsed
##    0.98    0.00    0.99

system.time(
  ci.boot <- confint(m2,
    method = "boot", oldNames = FALSE,
    nsim = 200, seed = 1234))

## Computing bootstrap confidence intervals ...

##    user  system elapsed
##     4.3     0.0     4.4

ci.compare <- data.table(
  Param = rownames(ci.wald),
  Wald = sprintf("%0.2f, %0.2f",
    ci.wald[,1], ci.wald[,2]),
```

```
   Profile = sprintf("%0.2f, %0.2f",
     ci.profile[,1], ci.profile[,2]),
   Boot = sprintf("%0.2f, %0.2f",
     ci.boot[,1], ci.boot[,2]))
```

```
print(ci.compare)
```

```
##                          Param      Wald      Profile        Boot
## 1: sd_(Intercept)|UserID      NA, NA  1.72, 2.16  1.70, 2.15
## 2:                 sigma      NA, NA  2.06, 2.19  2.05, 2.18
## 3:           (Intercept)  4.14, 4.78  4.14, 4.78  4.12, 4.80
## 4:        SurveyDayCount -0.05, 0.00 -0.05, 0.00 -0.05, 0.01
```

Although there are minor differences, in general the three methods have a high degree of agreement in this instance, but the Wald method is virtually instant, the profile method takes a brief period, and the bootstrap takes long enough that it is noticeable, especially during interactive model building.

Here are the full results based on the Wald method.

```
testm2 <- detailedTests(m2, method = "Wald")
```

```
## Parameters and CIs are based on REML,
## but detailedTests requires ML not REML fit for comparisons,
## and these are used in effect sizes. Refitting.
```

```
formatLMER(list(testm2))
```

```
##                          Term             Model 1
##   1:           Fixed Effects
##   2:             (Intercept) 4.46*** [ 4.14, 4.78]
##   3:          SurveyDayCount   -0.02 [-0.05, 0.00]
##   4:          Random Effects
##   5:  sd_(Intercept)|UserID                   1.93
##   6:                 sigma                    2.12
##   7:          Overall Model
##   8:              Model DF                       4
##   9:            N (UserID)                     191
## 10:       N (Observations)                    2097
## 11:                logLik                -4771.82
```

```
## 12:                       AIC          9551.65
## 13:                       BIC          9574.24
## 14:             Marginal R2               0.00
## 15:          Conditional R2               0.45
## 16:            Effect Sizes
## 17: SurveyDayCount (Fixed)    0.00/0.00, p = .099
```

Here are the full results based on the profile likelihood method. The fixed effects are not changed. However, the profile likelihood is able to estimate confidence intervals for random effects.

```
testm2b <- detailedTests(m2, method = "profile")

## Computing profile confidence intervals ...

## Parameters and CIs are based on REML,
## but detailedTests requires ML not REML fit for comparisons,
## and these are used in effect sizes. Refitting.

formatLMER(list(testm2b))

##                       Term              Model 1
##  1:           Fixed Effects
##  2:             (Intercept)  4.46*** [ 4.14, 4.78]
##  3:          SurveyDayCount    -0.02 [-0.05, 0.00]
##  4:          Random Effects
##  5:  sd_(Intercept)|UserID     1.93 [1.72, 2.16]
##  6:                  sigma     2.12 [2.06, 2.19]
##  7:          Overall Model
##  8:               Model DF                     4
##  9:             N (UserID)                   191
## 10:        N (Observations)                  2097
## 11:                 logLik              -4771.82
## 12:                    AIC               9551.65
## 13:                    BIC               9574.24
## 14:            Marginal R2                  0.00
## 15:         Conditional R2                  0.45
## 16:            Effect Sizes
## 17: SurveyDayCount (Fixed)    0.00/0.00, p = .099
```

Random Intercept and Slope Model

Previously we have only introduced random intercept models. However, mixed effects models can allow include one or more random slope parameters. The only requirement for a predictor to be included as a random slope is that it must vary within participants (or whatever higher-order clustering unit is being used for random effects).

Using the data we have been working with in this chapter, study day, stress, and sleep could all be random slopes. Age, education, and whether participants were born in or out of Australia could not be random slopes. Put differently, to be included as a random slope, a variable must have at least some variability within units (participants here). Variables that only vary between (but not within) units cannot be random slopes.

Random slopes work the same way that random intercepts do. That is, we can imagine estimating separate slopes between a predictor and outcome for each participant (or any other higher-order unit). However, rather than estimate individual slopes, the random slope model assumes that the slopes come from a distribution and estimates the parameters of that distribution instead. Almost always the distribution is a normal distribution and so the mean and variance are estimated.

In terms of equations, we previously defined GLMMs as

$$\eta = X\beta + Z\gamma \tag{11.13}$$

For these specific data, with a random intercept and slope, 191 participants, and a total of 6,599 observations, the dimensions of each vector/matrix would be as follows:

$$\underset{6599\times1}{\eta} = \underset{\substack{6599\times1 \\ 6599\times1}}{X} \underset{1\times1}{\beta} + \underset{\substack{6599\times1 \\ 6599\times382}}{Z} \underset{382\times1}{\gamma}$$

In this case, Z has twice as many columns as participants, because there is one column for each participant for the random intercept and one for each participant for the random slope.

Another change in this random intercept and slope model compared to a model with only one random effect, like the random intercept-only model, is that the random effects now include both variances and covariances. The covariances represent how related the random effects are to each other. For example, if participants who started higher (more positive random intercept) tended to have a more negative slope, there would be a negative relationship between the intercept and slope. To give a practical example, if a participant joins the study and sleeps 12 hours on the first night, it is unlikely that sleep will increase on the next day as that would require sleeping even more than 12 hours on a night. Conversely, if a participant stayed up all night on the first day, sleeping 0 hours, it is almost certain that sleep would increase on subsequent days. The key point is that in many situations, it makes sense that random intercepts and slopes may be correlated with each other, and in the default way of modelling random effects in R, the full variance-covariance matrix of random effects is estimated. It is possible to fix covariances to zero, forcing the random effects to be independent, but generally it is better to avoid this if possible.

Theoretically, mixed effects models could include a random slope without including a random intercept; however, in practice this is almost never done. A model with random slopes but not random intercepts only makes sense if it is plausible that all participants start at the same point but have different slopes. If participants do have different intercepts in reality, forcing them to be identical will distort the random slopes as their slopes must all pass through the same average intercept. Conversely, including an unnecessary random intercept will not meaningfully bias the model or results.

To explore the differences between a random intercept and slope model vs. standard linear regression, we will follow similar steps as we did with the random intercept. In the daily diary study data we have been using, each participant reports how many minutes it took to fall asleep for up to 12 days. If we wanted to examine the relationship between whether it was someone's first, second, etc., day in the study and how many minutes it took them to fall asleep (sleep onset latency; SOL), to begin with, we estimate a linear regression model, as follows.

```
## setup dataset
tmpd <- d[!is.na(sqrtSOLs) & !is.na(SurveyDayCount),
  .(sqrtSOLs, SurveyDayCount, UserID)]

## fixed effects, all people
mreg <- lm(sqrtSOLs ~ 1 + SurveyDayCount, data = tmpd)
## add predictions to the dataset
tmpd[, Fixed := predict(mreg, newdata = tmpd)]
```

This linear regression provides two average (fixed) effects. The **intercept** is the expected SOLs on participants' first night in the study, day 0. The **slope** is the expected change in SOLs for a one day change in study day. These two numbers are the average across every participant in the study. They do not capture any individual differences between participants. A benefit of a model like this is that it combines all participant data so it will be relatively robust to outliers. It also does not matter if a particular participant has only few data points, because all participants' data are combined. As with the random intercept models, while the linear regression accurately estimates the intercept and slope, the standard errors, p-values, and any confidence intervals will be biased.

We can also run separate linear regression models for each individual participant, as follows.

```
## fixed effects, individual models
tmpd[, Individual := fitted(lm(sqrtSOLs ~ 1 + SurveyDayCount)),
  by = UserID]
```

These individual models estimate a different intercept and different slopes for each individual participant. With both the intercept and slope estimated separately for individual participants, both intercepts and slopes will be sensitive to outliers and unstable in participants with few observations.

Finally, we can run a random intercept and slope model, shown in the following code. To include the random slope, we add SurveyDayCount to the random section of the model (inside the parentheses, before the vertical bar indicating these parameters should vary randomly by UserID). Note that even though we want a random slope for SurveyDayCount, we also include a fixed effect for SurveyDayCount.

```
## random intercept model, all people
m <- lmer(sqrtSOLs ~ 1 + SurveyDayCount +
          (1 + SurveyDayCount | UserID), data = tmpd)
## add predictions to the dataset
tmpd[, Random := predict(m, newdata = tmpd)]
```

The random model allows each individual to have a different intercept and slope, but these are assumed to come from a multivariate normal distribution, and the parameters of the (multivariate) normal distribution are estimated. As with the random intercept models we examined earlier, we can visualize the differences between random and fixed effects models by graphing the predicted relations between study day and SOL for a few participants. Figure 11-7 shows the estimated trajectories from the fixed individual model and the random model with and slope from the linear regression model on all participants.

```
## select a few example IDs to plot
tmpdselect <- melt(tmpd[UserID %in% unique(UserID)[107:115]],
    id.vars = c("UserID", "SurveyDayCount", "sqrtSOLs"))

ggplot(tmpdselect[variable != "Fixed"],
      aes(SurveyDayCount, value, group = UserID)) +
  geom_abline(intercept = coef(mreg)[1], slope = coef(mreg)[2],
            size = 2, colour = "blue") +
  geom_line() +
  facet_wrap(~ variable)
```

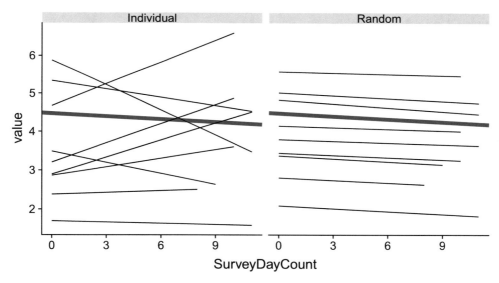

Figure 11-7. *Plot of estimated trajectory from individual regression models and a random intercept model. Population average shown in blue. The random model pulls both the intercepts and slopes closer to the population average intercept and slope, showing the effect of shrinkage.*

Another way to view the differences between models is to plot the estimated trajectories against the raw data. This is shown in Figure 11-8. These figures highlight that the individual fixed effects models or random effects models are as or more accurate than the population average. The figures also highlight that in every case, the random intercept line is the same as the fixed effect or closer to the population average, never more extreme than the individual fixed effects model. Indeed, for participant 114, the random effects model pulls the slope back toward the population average substantially, minimizing the impact of the extreme SOL at study entry.

```
## plots against individual data
ggplot(tmpdselect, aes(SurveyDayCount)) +
  geom_point(aes(y = sqrtSOLs), size = 1) +
  geom_line(aes(y = value,
                colour = variable,
                linetype = variable), size = 1.5) +
  facet_wrap(~UserID, scales = "free_y") +
  scale_color_viridis(discrete = TRUE) +
  theme(legend.position = "bottom",
        legend.title = element_blank(),
        legend.key.width = unit(2, "cm"))
```

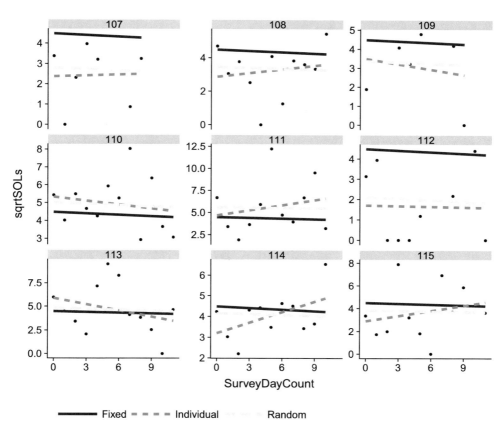

Figure 11-8. *Plots of different model estimated lines against raw data values for nine participants*

To highlight the shrinkage vs. model fit, we can plot the changes in estimated slopes.

```
tmpd <- d[, .(
  Individual = coef(lm(
    sqrtSOLs ~ 1 + SurveyDayCount))[2]),
  by = UserID]

## estimated random slope is deviation + average
tmpd$Random <- ranef(m)$UserID[, "SurveyDayCount"] + fixef(m)[2]
tmpd <- tmpd[order(Individual)]
tmpd[, UserID := factor(UserID, levels = UserID)]

ggplot(tmpd, aes(x = Individual, xend = Random,
                 y = UserID, yend = UserID)) +
```

507

```
geom_segment(
   arrow = arrow(length = unit(0.01, "npc"))) +
geom_vline(xintercept = coef(mreg)[2]) +
xlab("Estimated Slope") +
theme(axis.text.y = element_blank())
```

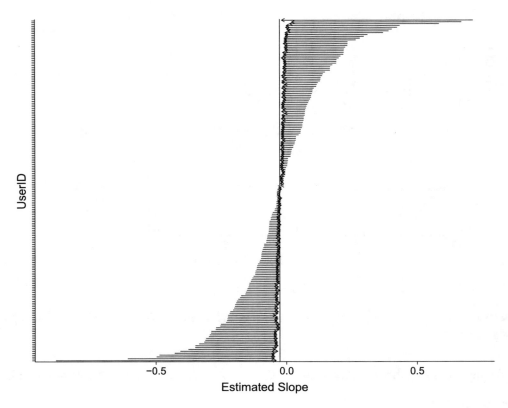

Figure 11-9. *Plot of estimated slope in the individual models and random effects model for each participant, with arrows showing the shrinkage toward the population mean slope*

Intercepts and Slopes as Outcomes

Random intercepts and slopes allow the level (intercept) and association (slope) to differ between each individual (or unit). For a given person (unit), even though there may be repeated measures, they will only have one intercept and one slope value. Thus, although the random intercept and slopes are estimated from repeated measures (within level) data, the intercepts and slopes themselves are between-level variables. That is, an individual's intercept (or slope) does not vary across assessments. Assuming that the intercept and slope values do actually vary across people, there may be interest in identifying predictors of the intercept and slope. To make this more concrete, we can diagram the problem. Figure 11-10 shows a mixed effects model diagram for a two-level model (within and between) with a random intercept and slope. The outcome variable, y, is predicted at the within level by x with a random slope and also a random intercept. The random intercept (i) and random slope (s) are then themselves outcome variables at the between level and are predicted by a between-level predictor, w.

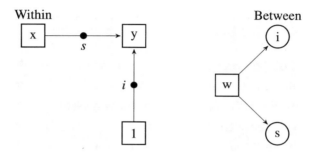

Figure 11-10. *Example diagram at the within and between levels. Squares indicate observed variables (e.g., outcome, predictors). Open circles indicate latent variables (i.e., random effects at the higher level). Filled circles indicate random effects (i.e., random intercept, slope). At the within level, there is a random intercept (i) and random slope (s) for the slope of* y *on* x. *At the between level, there are two latent variables: one for the random intercept (i) and one for the random slope (s) predicted by a between-level variable,* w.

As a specific example, earlier we examined a mixed effects model with transformed minutes to fall asleep (sleep onset latency) predicted by a random intercept and random slope of day in the study. Suppose that we wonder if participants who wake up more on average also take longer to fall asleep (i.e., awakenings predict the random intercept) or if those with fewer awakenings change less over the study period (i.e., flatter random slope). The average number of awakenings does not change, so it is a between-person variable. In terms of a diagram, this is shown in Figure 11-11. In this diagram, the random intercept of sleep onset latency (SOL) and the slope of SOL on study day are both predicted by a between-level predictor, average awakenings.

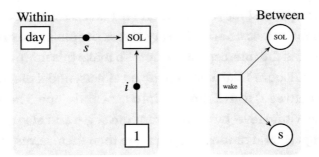

Figure 11-11. *Diagram of the within and between levels for study day predicting sleep onset latency (SOL) with a random intercept and slope and average number of awakenings predicting the random intercept and slope.*

If one thinks about the intercept and slope as new (latent, unobserved) variables, it may be tempting to literally create new variables in a dataset based on the estimates from a mixed effects model. This can be done relatively easily in R by extracting the random coefficient estimate for the intercept and slope. The following code creates a between-level dataset with ID, sex, and age and combines this with the intercept and slope estimates from the mixed effects model, extracted using the `coef()` function.

```
between_data <- cbind(
  d[, .(
  BWASONs = na.omit(BWASONs)[1]),
  by = UserID][order(UserID)],
  coef(m)$UserID)
```

The estimates of the intercept and slope are sometimes called BLUPs for best linear unbiased predictions of an individual's intercept and slope. Using these, we could run a regular generalized linear model at the between level. The results are shown in Table 11-2. They reveal that higher average awakenings predict a higher intercept but do not predict the random slope.

```
between.int <- lm(`(Intercept)` ~ BWASONs,
              data = between_data)
between.slope <- lm(SurveyDayCount ~ BWASONs,
            data = between_data)

texreg(list(
  Intercept = between.int,
```

```
Slope = between.slope),
digits = 3,
label = "tglmml-blups",
float.pos = "!hb")
```

Table 11-2. *Statistical Models*

	Intercept	Slope
(Intercept)	4.093***	−0.022***
	(0.218)	(0.001)
BWASONs	0.394*	0.000
	(0.189)	(0.001)
R²	0.022	0.000
Adj. R²	0.017	−0.005
Num. obs.	191	191
RMSE	1.800	0.010

***$p < 0.001$, **$p < 0.01$, *$p < 0.05$

Although this is an intuitively appealing approach, extracting BLUPs for the random effects and using these in separate analyses is not an optimal strategy. One major limitation is that for each individual, a single estimate for the intercept and slope is extracted, and these *estimates* are then treated in the subsequent model as if they were observed and measured without error. That is, what in reality are *estimates* with some degree of uncertainty are treated as being exactly known with no uncertainty. In reality, there typically is a great deal of uncertainty in these estimates. Although deriving accurate confidence intervals for BLUPs is difficult, approximate ones can be generated easily by specifying the condVar = TRUE argument to the ranef() function. The results are plotted in Figure 11-12, and they show that the model has a high degree of uncertainty about each individual's random intercept and slope estimate. This uncertainty highlights the problem with extracting the single best estimate and treating it as perfect.

```
ggplot(as.data.frame(ranef(m, condVar = TRUE)),
 aes(grp, condval,
     ymin = condval - 2 * condsd,
```

```
      ymax = condval + 2 * condsd)) +
  geom_pointrange(size = .2) +
  facet_wrap(~ term, scales = "free_x") +
  coord_flip() +
  theme(axis.text.y = element_blank(),
        axis.ticks.y = element_blank()) +
   ylab("Random effect + uncertainty") +
   xlab("Participant ID")
```

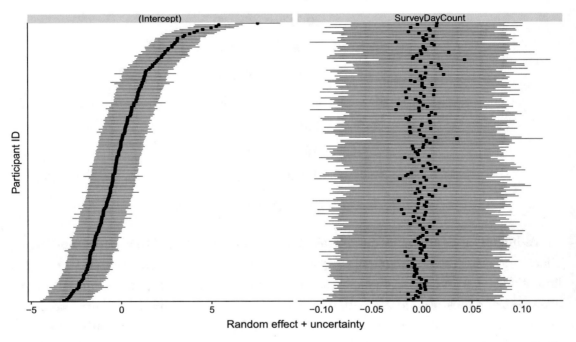

Figure 11-12. *Dot plots of estimated random effects with approximate confidence intervals*

Although extracting BLUPs or other single estimates of each individual's random intercept and slope is not optimal, it often is appropriate and interesting to examine predictors of the random intercept or slope. The best way to including predictors of random effects is to include them as part of the mixed effects model so that the random effects and their predictors are all estimated simultaneously in one model. In this way, uncertainty can be taken into account. Predicting random effects by between-level variables translates into what are termed cross-level interactions, that is, an interaction of a between-level variable and a within-level variable. At first it may seem counterintuitive that predicting the random intercept or slope requires an interaction. However, what is actually meant by predicting the random slope is that the association of study day with sleep onset latency (the slope) depends on the average number of awakenings. Interpreted this way, hopefully it is somewhat clearer how "predicting the slope" and an interaction capture the same question in this instance. Because the random intercept is the regression of the outcome on a constant (by convention, the number one), the interaction aspect of predicting the random intercept is ignorable since any variable times one is just the variable.

In summary, the ideal way to examine whether the average number of awakenings predicts the random intercept and slope in our mixed effects model is to add the average number of awakenings times one as a predictor (this is the prediction of the random intercept) and to add the study day x average number of awakenings interaction as a predictor (predict the random slope). The results of the single mixed effects model and the previous models extracting BLUPs and running a linear regression are shown in Table 11-3. In this table, we can see that there are differences in both the magnitude of estimates and in the standard errors for the predictions of the random intercept (BWASONs) and predicting the random slope (SurveyDayCount:BWASONs). As is often the case, when the uncertainty (measurement error) in the random intercepts and slopes are taken into account, the magnitude of the effect is larger, but there also is greater uncertainty in the estimates, represented by a larger standard error.

```
me.prediction <- lmer(sqrtSOLs ~
   SurveyDayCount + BWASONs +
   SurveyDayCount:BWASONs +
   (1 + SurveyDayCount | UserID),
 data = d)

texreg(list(
  Intercept = extract(between.int),
  Slope = extract(between.slope),
  Random = extract(me.prediction)),
  digits = 3,
  label = "tglmml-blupsme",
  float.pos = "!hb")
```

Table 11-3. *Statistical Models*

	Intercept	Slope	Random
(Intercept)	4.093***	−0.022***	4.014***
	(0.218)	(0.001)	(0.272)
BWASONs	0.394*	0.000	0.480*
	(0.189)	(0.001)	(0.236)
SurveyDayCount			−0.015
			(0.023)
SurveyDayCount:BWASONs			−0.008
			(0.020)
R^2	0.022	0.000	
Adj. R^2	0.017	−0.005	
Num. obs.	191	191	2097
RMSE	1.800	0.010	
AIC			9570.914
BIC			9616.100
Log Likelihood			−4777.457
Num. groups: UserID			191
Var: UserID (Intercept)			3.640
Var: UserID SurveyDayCount			0.002
Cov: UserID (Intercept) SurveyDayCount			−0.004
Var: Residual			4.479

$***p < 0.001, **p < 0.01, *p < 0.05$

11.2 R Examples

Linear Mixed Model with Random Intercept

We previously examined a random intercept-only model for positive affect. Earlier we focused on the contrast of a simple linear model with a random intercept model. Now we will focus on the design and setup of linear mixed models in R.

The lmer() function uses a similar formula interface as does the lm() and other modelling functions in R. The outcome variable is specified followed by a tilde and then all of the predictor variables. What distinguishes linear mixed effects models from linear models is the addition of random effects. The fixed effects are added as identically to fixed effects in a linear model. Random effects are added within parentheses. Inside the parentheses, random effects are written as any other model formula, except that at the end after a vertical bar, the clustering variable is listed.

```
## mixed effects, with random intercept by ID
m.lmm <- lmer(PosAff ~ 1 + (1 | UserID), data = d)
summary(m.lmm)

## Linear mixed model fit by REML. t-tests use Satterthwaite's method [
## lmerModLmerTest]
## Formula: PosAff ~ 1 + (1 | UserID)
##    Data: d
##
## REML criterion at convergence: 14795
##
## Scaled residuals:
##    Min    1Q Median    3Q    Max
## -4.345 -0.647 -0.034  0.617  4.058
##
## Random effects:
##  Groups   Name        Variance Std.Dev.
##  UserID   (Intercept) 0.629    0.793
##  Residual             0.529    0.727
## Number of obs: 6399, groups:  UserID, 191
##
## Fixed effects:
##              Estimate Std. Error       df t value Pr(>|t|)
## (Intercept)    2.6787     0.0581 189.8310    46.1   <2e-16 ***
## ---
## Signif. codes:  0 '***' 0.001 '**' 0.01 '*' 0.05 '.' 0.1 '_' 1
```

In comparison to a linear model only, the standard errors are larger for the fixed effects and there is the addition of a random intercept, as shown in the following summary.

```
## fixed effects only, GLM
m.lm <- lm(PosAff ~ 1, data = d)
summary(m.lm)

##
## Call:
## lm(formula = PosAff ~ 1, data = d)
##
## Residuals:
##     Min      1Q  Median      3Q     Max
## -1.6760 -0.8751 -0.0065  0.7886  2.3240
##
## Coefficients:
##             Estimate Std. Error t value Pr(>|t|)
## (Intercept)   2.6760     0.0134     200   <2e-16 ***
## ---
## Signif. codes:  0 '***' 0.001 '**' 0.01 '*' 0.05 '.' 0.1 '_' 1
##
## Residual standard error: 1.1 on 6398 degrees of freedom
##    (528 observations deleted due to missingness)

## nice side by side comparison
screenreg(list(
  GLM = extract(m.lm),
  GLMM = extract(m.lmm)))

##
## =====================================================
##                          GLM             GLMM
## -----------------------------------------------------
## (Intercept)            2.68 ***        2.68 ***
##                        (0.01)          (0.06)
## -----------------------------------------------------
```

```
## R^2                          0.00
## Adj. R^2                     0.00
## Num. obs.              6399          6399
## RMSE                     1.07
## AIC                                14800.56
## BIC                                14820.85
## Log Likelihood                     -7397.28
## Num. groups: UserID                    191
## Var: UserID (Intercept)               0.63
## Var: Residual                         0.53
## ====================================================
## *** p < 0.001, ** p < 0.01, * p < 0.05
```

Additional fixed effects predictors can easily be added, similar to any other regression modelling function in R. The following example adds average stress as a fixed effects predictor of positive affect. The results from summary() reveal that higher average stress is associated with significantly lower positive affect.

```
## mixed effects, with random intercept by ID
m2.lmm <- lmer(PosAff ~ 1 + BSTRESS + (1 | UserID), data = d)
summary(m2.lmm)
```

```
## Linear mixed model fit by REML. t-tests use Satterthwaite's method [
## lmerModLmerTest]
## Formula: PosAff ~ 1 + BSTRESS + (1 | UserID)
##    Data: d
##
## REML criterion at convergence: 14762
##
## Scaled residuals:
##    Min     1Q Median     3Q    Max
## -4.347 -0.645 -0.034  0.617  4.049
##
## Random effects:
##  Groups    Name        Variance Std.Dev.
##  UserID    (Intercept) 0.517    0.719
##  Residual              0.529    0.727
```

517

```
## Number of obs: 6399, groups:  UserID, 191
##
## Fixed effects:
##              Estimate Std. Error        df t value Pr(>|t|)
## (Intercept)    3.2201     0.0998 188.3579    32.3  < 2e-16 ***
## BSTRESS       -0.2300     0.0359 188.6190    -6.4  1.2e-09 ***
## ---
## Signif. codes:  0 '***' 0.001 '**' 0.01 '*' 0.05 '.' 0.1 '_' 1
##
## Correlation of Fixed Effects:
##         (Intr)
## BSTRESS -0.848
```

Before we can be confident in the model results, it is important to check the assumptions of the model. Several assumptions of the model can be readily tested using the plotDiagnosticsLMER() function. These results are shown in Figure 11-13. They show an approximately symmetrical distribution for the residuals, little evidence of heteroscedasticity, and a normally distributed random intercept. There does appear to be some non-linearity in the association of average stress with positive affect.

```
assumptiontests <- plotDiagnosticsLMER(m2.lmm, plot = FALSE) do.call(plot_
grid, c(
  assumptiontests[c("ResPlot", "ResFittedPlot")],
  assumptiontests$RanefPlot,
  list(ggplot(d, aes(BSTRESS, PosAff)) +
      stat_smooth()),
  ncol = 2))
```

```
## `geom_smooth()` using method = 'gam' and formula 'y_~_s(x,_bs_=_"cs")'
```

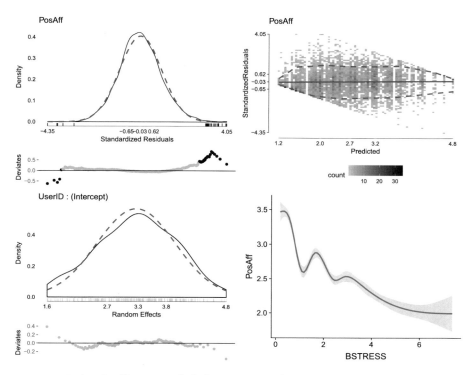

Figure 11-13. *Mixed effects model diagnostic plots showing the distribution of residuals (top left), the residuals vs. fitted values to assess homogeneity of variance (top right), the distribution of the random intercept (bottom left), and a simple, single-level generalized additive model smooth of the association between average stress and positive to assess for non-linearity (bottom right).*

Considering the evidence for non-linear association shown in Figure 11-13, we should consider a more flexible functional form. Previous chapters introduced splines and generalized additive models (GAMs). Although we have not covered GAMs for mixed effects models, we can introduce splines relatively easily. However, first, we need to determine what spline may be approximately an adequate representation of the data. This is easiest to accomplish by visually comparing splines against a GAM. Several B-splines vs. the GAM are shown in Figure 11-14. The results indicate too much flexibility from the 10 degree of freedom model, and while not capturing some of the fluctuations in the GAM, more flexibility than a linear fit and a fairly smooth trend for the 3 degree of freedom B-spline model.

```
ggplot(d, aes(BSTRESS, PosAff)) +
  stat_smooth(method = "lm",
              formula = y ~ bs(x, df = 3),
              colour = viridis(3)[1]) +
  stat_smooth(method = "lm",
              formula = y ~ bs(x, df = 10),
              colour = viridis(3)[2]) +
  stat_smooth(colour = viridis(3)[3])
```

Warning: Removed 528 rows containing non-**finite** values (stat_smooth).
Warning: Removed 528 rows containing non-**finite** values (stat_smooth).

`geom_smooth()` using method = 'gam' and formula 'y_~_s(x,_bs_=_"cs")'

Warning: Removed 528 rows containing non-**finite** values (stat_smooth).

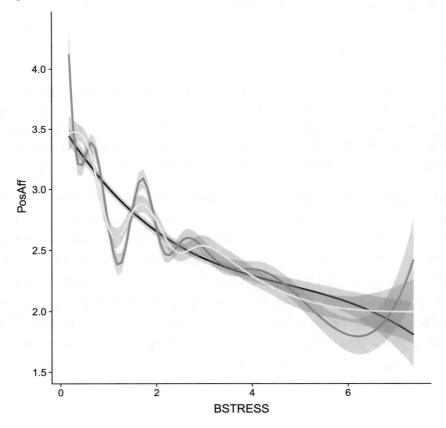

Figure 11-14. *Plot of average stress vs. positive affect using B-splines with 3 degrees of freedom, 10 degrees of freedom, and a generalized additive model.*

We can add the B-splines to the model using the bs() function and then compare the model with B-splines to the one with a linear trend only using the Akaike information criterion (AIC). Note that the AIC relies on a true likelihood; thus, we refit the models using refitML() in place of the default, restricted maximum likelihood which provides a pseudo-likelihood. The AIC suggests a better fit, albeit modestly, for the B-spline vs. the linear model.

```
## mixed effects model
m3.lmm <- lmer(PosAff ~ 1 + bs(BSTRESS, df = 3) +
                 (1 | UserID), data = d)
summary(m3.lmm)

## Linear mixed model fit by REML. t-tests use Satterthwaite's method [
## lmerModLmerTest]
## Formula: PosAff ~ 1 + bs(BSTRESS, df = 3) + (1 | UserID)
##     Data: d
##
## REML criterion at convergence: 14752
##
## Scaled residuals:
##     Min     1Q Median     3Q    Max
## -4.352 -0.644 -0.034  0.618  4.052
##
## Random effects:
##  Groups    Name        Variance Std.Dev.
##  UserID    (Intercept) 0.509    0.713
##  Residual              0.529    0.727
## Number of obs: 6399, groups:  UserID, 191
##
## Fixed effects:
##                        Estimate Std. Error      df t value Pr(>|t|)
## (Intercept)               3.464      0.167 186.213   20.78   <2e-16 ***
## bs(BSTRESS, df = 3)1     -1.536      0.561 186.428   -2.74   0.0068 **
## bs(BSTRESS, df = 3)2     -0.898      0.569 186.628   -1.58   0.1162
## bs(BSTRESS, df = 3)3     -1.706      0.624 186.082   -2.73   0.0069 **
## ---
```

```
## Signif. codes:  0 '***' 0.001 '**' 0.01 '*' 0.05 '.' 0.1 '_' 1
##
## Correlation of Fixed Effects:
##                  (Intr) b(BSTRESS,d=3)1 b(BSTRESS,d=3)2
## b(BSTRESS,d=3)1 -0.857
## b(BSTRESS,d=3)2  0.252 -0.630
## b(BSTRESS,d=3)3 -0.478  0.620          -0.738
```

```
## compare the linear and B-spline models
AIC(refitML(m3.lmm), refitML(m2.lmm))
```

```
##                   df   AIC
## refitML(m3.lmm)   6 14760
## refitML(m2.lmm)   4 14761
```

Next, we will add one more predictor, weekday vs. weekend. First we create the new variable by converting dates using the weekdays() function. Then we can add it to the model. Rather than always re-writing the model, if we simply wish to add or remove predictors, we can use the update() function.

```
## create the new variable in the dataset
d[, Weekend := factor(as.integer(
    weekdays(SurveyDay) %in% c("Saturday", "Sunday")))]
```

```
## update the model adding weekend
m4.lmm <- update(m3.lmm, . ~ . + Weekend)
```

```
## screenreg summary
screenreg(m4.lmm)
```

```
##
## ====================================
##                          Model 1
## ------------------------------------
## (Intercept)                 3.43 ***
##                            (0.17)
## bs(BSTRESS, df = 3)1       -1.54 **
##                            (0.56)
## bs(BSTRESS, df = 3)2       -0.90
```

```
##                                   (0.57)
## bs(BSTRESS, df = 3)3             -1.71 **
##                                   (0.62)
## Weekend1                          0.10 ***
##                                   (0.02)
## ----------------------------------
## AIC                           14745.61
## BIC                           14792.96
## Log Likelihood                -7365.81
## Num. obs.                        6399
## Num. groups: UserID               191
## Var: UserID (Intercept)          0.51
## Var: Residual                    0.53
## ==================================
## *** p < 0.001, ** p < 0.01, * p < 0.05
```

To present the results, especially given the use of B-splines, the clearest approach is to generate predictions and graph the results. The first step is to get predicted values. This is accomplished using the predict() function. One additional aspect of using predict() with LMMs is that there are options to use or not use random effects, specified using the re.form argument. To get population average predictions ignoring the random effects, we can specify the random effects formula for predictions as zero.

```
preddat <- as.data.table(expand.grid(
  BSTRESS = seq(
    from = min(d$BSTRESS, na.rm=TRUE),
    to = max(d$BSTRESS, na.rm=TRUE),
    length.out = 1000),
  Weekend = levels(d$Weekend)))

preddat$yhat <- predict(m4.lmm,
  newdata = preddat,
  re.form = ~ 0)
```

We can plot the predictions to present the model results, capturing the non-linear trends captured by the splines. The results are in Figure 11-15.

```
ggplot(preddat, aes(BSTRESS, yhat, colour = Weekend)) +
  geom_line(size = 1) +
  ylab("Positive Affect") +
  xlab("Average Stress") +
  scale_color_viridis(discrete = TRUE) +
  theme(
    legend.position = c(.75, .8),
    legend.key.width = unit(1, "cm"))
```

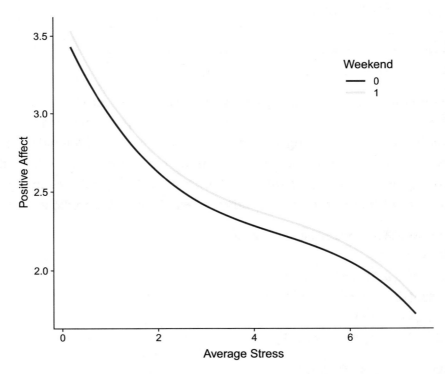

Figure 11-15. *Plot of predicted positive affect from a model of average stress and weekday vs. weekend*

Although standard errors (and by proxy, confidence intervals) can be generated for predictions from regular GLMs, it is more complex for LMMs. Currently, the primary method for generating standard errors or confidence intervals for predictions from LMMs fit by lmer() is to use bootstrapping. To help speed up the bootstraps, we will set up a local cluster for parallel processing. We need to load the relevant packages and export the dataset for predictions.

```
cl <- makeCluster(2)
clusterExport(cl, c("book_directory",
                    "checkpoint_directory",
                    "preddat", "d"))

clusterEvalQ(cl, {
  library(checkpoint)
  checkpoint("2018-09-28", R.version = "3.5.1",
    project = book_directory,
    checkpointLocation = checkpoint_directory,
    scanForPackages = FALSE,
    scan.rnw.with.knitr = TRUE, use.knitr = TRUE)

  library(data.table)
  library(lme4)
  library(lmerTest)
  library(splines)
})
## [[1]]
##  [1] "splines"       "lmerTest"      "lme4"         "Matrix"
##  [5] "data.table"    "checkpoint"    "RevoUtils"    "stats"
##  [9] "graphics"      "grDevices"     "utils"        "datasets"
## [13] "RevoUtilsMath" "methods"       "base"
##
## [[2]]
##  [1] "splines"       "lmerTest"      "lme4"         "Matrix"
##  [5] "data.table"    "checkpoint"    "RevoUtils"    "stats"
##  [9] "graphics"      "grDevices"     "utils"        "datasets"
## [13] "RevoUtilsMath" "methods"       "base"
```

```
genPred <- function(m) {
  predict(m,
    newdata = preddat,
    re.form = ~0)
}
```

The main bootstrapping is a parametric model, and it is conducted using the bootMer() function included in the lme4 package. Lastly, we calculate simple percentile confidence intervals on the results and add them back into our dataset.

```
system.time(
  bootres <- bootMer(m4.lmm,
    FUN = genPred,
    nsim = 1000,
    seed = 12345,
    use.u = FALSE,
    type = "parametric",
    parallel = "snow",
    cl = cl)
)

##      user  system elapsed
##      43.3     0.2    43.8

## calculate percentile bootstrap confidence intervals
## and add to the dataset for plotting
preddat$LL <- apply(bootres$t, 2, quantile, probs = .025)
preddat$UL <- apply(bootres$t, 2, quantile, probs = .975)
```

Now that we have the parametric bootstrapped confidence intervals, we can remake our figure, this time with shaded region to indicate uncertainty in the predictions. The results are in Figure 11-16.

```
ggplot(preddat, aes(BSTRESS, yhat, colour = Weekend,
                    fill = Weekend)) +
  geom_ribbon(aes(ymin = LL, ymax = UL),
              alpha = .25, colour = NA) +
  geom_line(size = 1) +
```

```
ylab("Positive Affect") +
xlab("Average Stress") +
scale_color_viridis(discrete = TRUE) +
scale_fill_viridis(discrete = TRUE) +
theme(
  legend.position = c(.75, .8),
  legend.key.width = unit(1, "cm")) +
coord_cartesian(xlim = c(0, 8), ylim = c(1, 4),
                expand = FALSE)
```

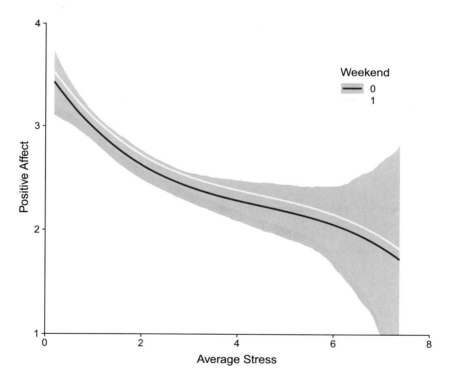

Figure 11-16. *Plot of predicted positive affect from a model of average stress and weekday vs. weekend with bootstrapped confidence intervals*

Linear Mixed Model with Random Intercept and Slope

In addition to random intercepts, LMMs can have random slopes. Random slopes capture individual differences in the association between predictors and the outcome, between people. In order to estimate differences between people in the association of a predictor and outcome, it is necessary that at least some people (ideally all) have more than one value of the predictor and outcome. Consequently, only variables that vary within participants can be used as a random slope.

Before we examine a random slope model, we will see what happens when we add a within-person predictor as a fixed effect only. Previously, we examined the association of average stress with positive affect. Now we will examine the association of within-person stress: deviations from individuals' own average stress levels. Again we rely on the update() function to add this predictor to our previous LMM.

```
## update the model adding within person stress
m5.lmm <- update(m4.lmm, . ~ . + WSTRESS)

## screenreg summary
screenreg(m5.lmm)

##
## ====================================
##                         Model 1
## ------------------------------------
## (Intercept)              3.46 ***
##                         (0.17)
## bs(BSTRESS, df = 3)1    -1.54 **
##                         (0.56)
## bs(BSTRESS, df = 3)2    -0.90
##                         (0.57)
## bs(BSTRESS, df = 3)3    -1.71 **
##                         (0.62)
## Weekend1                 0.01
##                         (0.02)
## WSTRESS                 -0.16 ***
##                         (0.00)
## ------------------------------------
```

```
## AIC                          13389.59
## BIC                          13443.70
## Log Likelihood               -6686.80
## Num. obs.                     6399
## Num. groups: UserID            191
## Var: UserID (Intercept)        0.51
## Var: Residual                  0.42
## =====================================
## *** p < 0.001, ** p < 0.01, * p < 0.05
```

The results show that within-person stress also predicts lower levels of positive affect. Again we can generate predictions and plot them to view the association of within and between stress and weekday vs. weekend with positive affect. Because within deviations from average stress may vary depending whether average stress is low or high, we will calculate the range for within stress separately for low and high average stress. So that we are plotting in a relatively realistic range of the data, we pick two average stress values, the 25th and 75th percentiles. Also for within stress, rather than plotting the entire range, we plot from the 2nd to the 98th percentiles, in the bottom and top quartiles of average stress. This captures the fact that if average stress is very low, it is impossible to go far below average, because participants report stress from 0 to 10.

```
bstress.low <- round(quantile(d[!duplicated(UserID)]$BSTRESS,
                    probs = .25), 1)
bstress.high <- round(quantile(d[!duplicated(UserID)]$BSTRESS,
                    probs = .75), 1)

preddat.low <- as.data.table(expand.grid(
  BSTRESS = bstress.low,
  WSTRESS = seq(
    from = quantile(d[BSTRESS <= bstress.low]$WSTRESS,
             probs = .02, na.rm = TRUE),
    to = quantile(d[BSTRESS <= bstress.low]$WSTRESS,
             probs = .98, na.rm = TRUE),
    length.out = 1000),
  Weekend = factor("1", levels = levels(d$Weekend))))
```

```
preddat.high <- as.data.table(expand.grid(
  BSTRESS = bstress.high,
  WSTRESS = seq(
    from = quantile(d[BSTRESS >= bstress.high]$WSTRESS,
              probs = .02, na.rm = TRUE),
    to = quantile(d[BSTRESS >= bstress.high]$WSTRESS,
              probs = .98, na.rm = TRUE),
    length.out = 1000),
  Weekend = factor("1", levels = levels(d$Weekend))))

preddat <- rbind(
  preddat.low,
  preddat.high)

preddat$yhat <- predict(m5.lmm,
  newdata = preddat,
  re.form = ~ 0)

## convert BSTRESS to factor for plotting
preddat$BSTRESS <- factor(preddat$BSTRESS)
```

Now we can plot the results, which are shown in Figure 11-17. The plot shows that the spread of within stress is smaller for the low levels of average stress than for high levels of average stress.

```
ggplot(preddat, aes(WSTRESS, yhat, colour = BSTRESS)) +
  geom_line(size = 1) +
  ylab("Positive Affect") +
  xlab("Within Stress") +
  scale_color_viridis(discrete = TRUE) +
  theme(
    legend.position = c(.05, .2),
    legend.key.width = unit(1, "cm"))
```

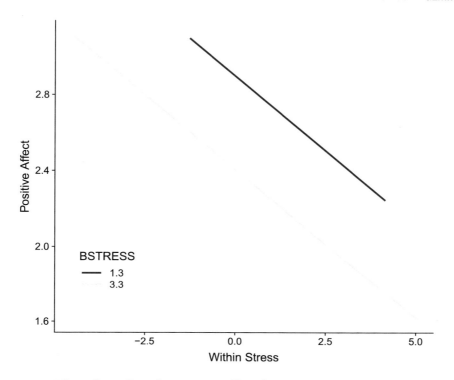

Figure 11-17. *Plot of predicted positive affect from a model of average stress and weekday vs. weekend.*

If we wanted, we could plot predicted lines incorporating the random effects, to show the sort of range in predictions that likely would happen in the population due to individual differences. This requires generating predictions using the random intercept.

```
bstress.low <- round(quantile(d[!duplicated(UserID)]$BSTRESS,
                    probs = .25), 1)
bstress.high <- round(quantile(d[!duplicated(UserID)]$BSTRESS,
                    probs = .75), 1)
```

```
preddat.low <- as.data.table(expand.grid(
  UserID = unique(d$UserID),
  BSTRESS = bstress.low,
  WSTRESS = seq(
    from = quantile(d[BSTRESS <= bstress.low]$WSTRESS,
              probs = .02, na.rm = TRUE),
    to = quantile(d[BSTRESS <= bstress.low]$WSTRESS,
              probs = .98, na.rm = TRUE),
    length.out = 1000),
  Weekend = factor("1", levels = levels(d$Weekend))))

preddat.high <- as.data.table(expand.grid(
  UserID = unique(d$UserID),
  BSTRESS = bstress.high,
  WSTRESS = seq(
    from = quantile(d[BSTRESS >= bstress.high]$WSTRESS,
              probs = .02, na.rm = TRUE),
    to = quantile(d[BSTRESS >= bstress.high]$WSTRESS,
              probs = .98, na.rm = TRUE),
    length.out = 1000),
  Weekend = factor("1", levels = levels(d$Weekend))))

preddat <- rbind(
  preddat.low,
  preddat.high)

preddat$yhat <- predict(m5.lmm,
  newdata = preddat,
  re.form = NULL)

## convert BSTRESS to factor for plotting
preddat$BSTRESS <- factor(preddat$BSTRESS)
```

Now we can plot the results, which are shown in Figure 11-18. The plot shows the large variation in level of positive affect between individuals. Although the specific lines are based on the random effects from our sample, it gives a sense of the sort of variability that likely occurs in the population.

```
ggplot(preddat, aes(WSTRESS, yhat, group = UserID)) +
  geom_line(alpha = .2) +
  ylab("Positive Affect") +
  xlab("Within Stress") +
  facet_wrap(~ BSTRESS, ncol = 2) +
  coord_cartesian(
    xlim = c(-4, 5),
    ylim = c(1, 5),
    expand = FALSE)
```

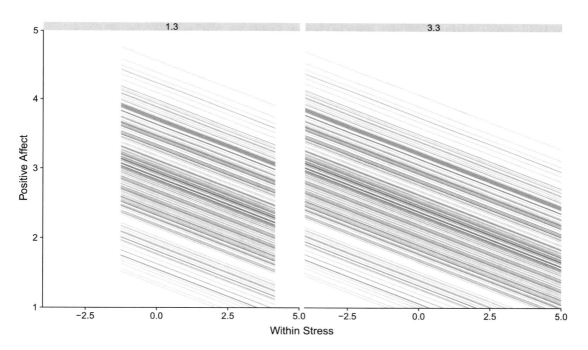

Figure 11-18. *Plot of predicted positive affect from a model of average stress and weekday vs. weekend*

Next we examine a random slope for within stress. We update our previous model again. This update is more complex, however, because of how random effects are specified in lmer(). Our previous models all included a random intercept. It might seem natural to try to add a random slope by writing: (WSTRESS | UserID). However, as with fixed effects formulae, R automatically assumes an intercept should be included. Thus (WSTRESS | UserID) expands to (1 + WSTRESS | UserID). This is not appropriate as there already is a random intercept included. Another apparently logical alternative would be to explicitly exclude the intercept, which works the same as for fixed effects: (0 + WSTRESS | UserID). However, this approach would result in a random intercept and random slope, but the two would be forced to be uncorrelated. lmer() only includes correlations among random effects when they are in the same block. Thus, what we really need is (WSTRESS | UserID), but we need to remove the initial random intercept. The following code accomplishes this by first dropping the random intercept from the old model and then adding a new random effects block that includes the random intercept and random slope. If we were writing a model from scratch, we could simply write the one random effects line, but since we are updating an existing model, we need to drop the old random intercept.

```
m6.lmm <- update(m5.lmm, . ~ . - (1 | UserID) +
  (1 + WSTRESS | UserID))

screenreg(m6.lmm)

##
## ===============================================
##                                   Model 1
## -----------------------------------------------
## (Intercept)                        3.45 ***
##                                   (0.16)
## bs(BSTRESS, df = 3)1              -1.59 **
##                                   (0.54)
## bs(BSTRESS, df = 3)2              -0.75
##                                   (0.54)
## bs(BSTRESS, df = 3)3              -1.72 **
##                                   (0.59)
## Weekend1                           0.02
##                                   (0.02)
## WSTRESS                           -0.16 ***
##                                   (0.01)
```

```
## ----------------------------------------------
## AIC                               13196.60
## BIC                               13264.23
## Log Likelihood                    -6588.30
## Num. obs.                             6399
## Num. groups: UserID                    191
## Var: UserID (Intercept)               0.51
## Var: UserID WSTRESS                   0.01
## Cov: UserID (Intercept) WSTRESS      -0.02
## Var: Residual                         0.40
## ================================================
## *** p < 0.001, ** p < 0.01, * p < 0.05
```

Again we can generate predictions. This time we focus on including both the random intercept and slope to highlight the variation that occurs between individuals. As the model uses the same variables as our previous examples, we only need to generate new predictions and then plot them. We do not need to recreate the data to be used for prediction.

```
## convert BSTRESS  from factor  to numeric for prediction
preddat$BSTRESS  <- as.numeric(as.character(
  preddat$BSTRESS))

preddat$yhat2 <- predict(m6.lmm,
  newdata = preddat,
  re.form =  NULL)

##  convert BSTRESS  to factor for plotting
preddat$BSTRESS  <-  factor(preddat$BSTRESS)
```

Now we can plot the results, which are shown in Figure 11-19. The plot shows the large variation in level and some variation in the slope of the stress and positive affect association between individuals.

```
ggplot(preddat, aes(WSTRESS, yhat2, group = UserID)) +
  geom_line(alpha = .2) +
  ylab("Positive Affect") +
  xlab("Within Stress") +
  facet_wrap(~ BSTRESS, ncol = 2) +
  coord_cartesian(
    xlim = c(-4, 5),
    ylim = c(1, 5),
    expand = FALSE)
```

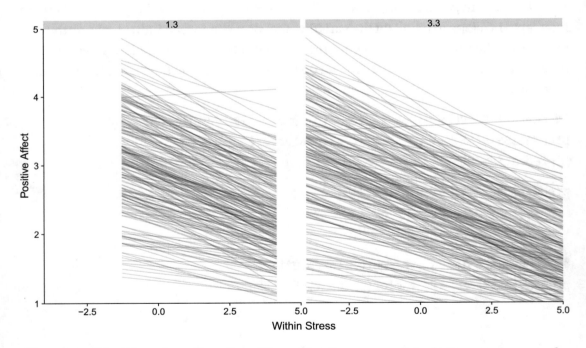

Figure 11-19. *Plot of predicted positive affect from a model of average stress and weekday vs. weekend*

As usual, it is a good idea to check the diagnostics. These are shown in Figure 11-20.

```
assumptiontests <- plotDiagnosticsLMER(m6.lmm, plot = FALSE)
do.call(plot_grid, c(
  assumptiontests[c("ResPlot", "ResFittedPlot")],
  assumptiontests$RanefPlot,
  list(ggplot(d, aes(WSTRESS, PosAff)) +
       stat_smooth()),
  ncol = 2))

## `geom_smooth()` using method = 'gam' and formula 'y_~_s(x,_bs_=_"cs")'
```

The diagnostics in Figure 11-20 suggest there is one multivariate outlier, but otherwise diagnostics appear fairly appropriate, although there again appears to be some degree of non-linear trend for within-person stress. Before proceeding, we will exclude the multivariate outlier. First we look at the extreme values from the assumption tests to identify the ID of the multivariate extreme value.

```
assumptiontests$ExtremeValues[
  EffectType == "Multivariate Random Effect UserID"]

##       PosAff UserID                       EffectType
##  1:     4.7    123 Multivariate Random Effect UserID
##  2:     3.9    123 Multivariate Random Effect UserID
##  3:     3.8    123 Multivariate Random Effect UserID
## ---
## 20:     3.7    123 Multivariate Random Effect UserID
## 21:     4.9    123 Multivariate Random Effect UserID
## 22:     4.6    123 Multivariate Random Effect UserID
```

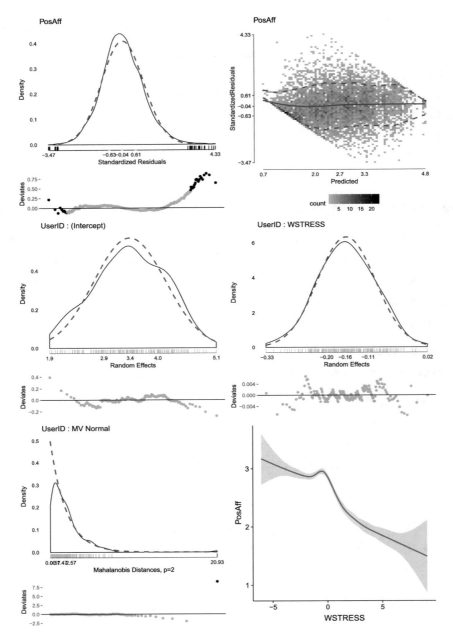

Figure 11-20. *Mixed effects model diagnostic plots showing the distribution of residuals (top left), the residuals vs. fitted values to assess homogeneity of variance (top right), the distribution of the random intercept (bottom left), and a simple, single-level generalized additive model smooth of the association between within stress and positive affect to assess for non-linearity (bottom right).*

Next we can update our model, this time not changing the formula but changing the dataset. The results appear similar, in this case, although the B-splines change somewhat. As we will later compare models based on fit indices to identify which trend for within stress is optimal, we change from restricted maximum likelihood to maximum likelihood by setting REML = FALSE.

```
m7.lmm <- update(m6.lmm,
  data = d[UserID != 123],
  REML = FALSE)

screenreg(list(m6.lmm, m7.lmm))
```

```
##
## ============================================================
##                              Model 1        Model 2
## ------------------------------------------------------------
## (Intercept)                   3.45 ***       3.43 ***
##                              (0.16)         (0.16)
## bs(BSTRESS, df = 3)1         -1.59 **       -1.50 **
##                              (0.54)         (0.52)
## bs(BSTRESS, df = 3)2         -0.75          -0.91
##                              (0.54)         (0.52)
## bs(BSTRESS, df = 3)3         -1.72 **       -1.63 **
##                              (0.59)         (0.57)
## Weekend1                      0.02           0.02
##                              (0.02)         (0.02)
## WSTRESS                      -0.16 ***      -0.16 ***
##                              (0.01)         (0.01)
## ------------------------------------------------------------
```

```
## AIC                                        13196.60      13083.77
## BIC                                        13264.23      13151.38
## Log Likelihood                             -6588.30      -6531.89
## Num. obs.                                      6399          6377
## Num. groups: UserID                           191           190
## Var: UserID (Intercept)                       0.51          0.50
## Var: UserID WSTRESS                           0.01          0.01
## Cov: UserID (Intercept) WSTRESS              -0.02         -0.03
## Var: Residual                                 0.40          0.40
## ==============================================================
## *** p < 0.001, ** p < 0.01, * p < 0.05
```

Next we recheck the diagnostic plots to evaluate the model with the multivariate outlier removed. These are shown in Figure 11-21.

```
assumptiontests <- plotDiagnosticsLMER(m7.lmm, plot = FALSE)
do.call(plot_grid, c(
  assumptiontests[c("ResPlot", "ResFittedPlot")],
  assumptiontests$RanefPlot,
  list(ggplot(d[UserID != 123], aes(WSTRESS, PosAff)) +
       stat_smooth()),
  ncol = 2))

## `geom_smooth()` using method = 'gam' and formula 'y_~_s(x,_bs_=_"cs")'
```

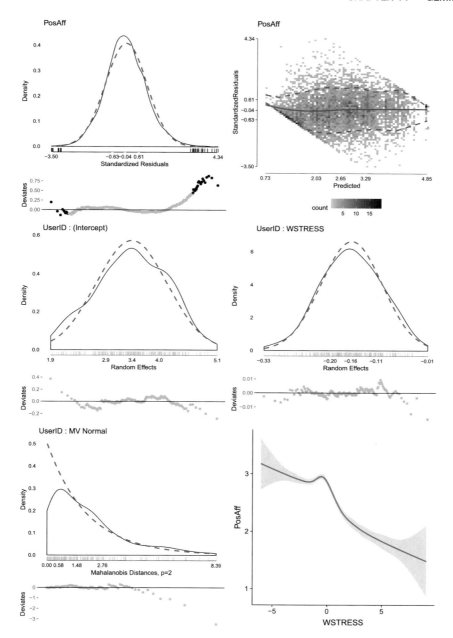

Figure 11-21. *Mixed effects model diagnostic plots showing the distribution of residuals (top left), the residuals vs. fitted values to assess homogeneity of variance (top right), the distribution of the random intercept (bottom left), and a simple, single-level generalized additive model smooth of the association between within stress and positive affect to assess for non-linearity (bottom right).*

The updated diagnostics are relatively good for the model, but the issue of potential non-linearity for within stress remains. One way to explore this is to fit several possible models and use the Akaike information criterion (AIC) or Bayesian information criterion (BIC) to select the optimal model. We will contrast four models: (a) linear, (b) quadratic, (c) linear piecewise with the bend at 0, and (d) B-splines. Note that for all trends, we include both the fixed and random effects. Further, all models are based on the random linear slope model with the multivariate outlier removed. At what point to include or exclude an outlier is somewhat debatable. Potentially a multivariate outlier in a linear trend would not be with some of the other trends, which might argue for excluding from the final selected model. However, an outlier also can alter which type of model is selected as optimal, which would argue for excluding outliers prior to comparing models. In this case, that is what we did.

```
m7.lmmb <- update(m7.lmm, . ~ . - (1 + WSTRESS | UserID) +
  WSTRESS + I(WSTRESS^2) +
  (1 + WSTRESS + I(WSTRESS^2) | UserID))

m7.lmmc <- update(m7.lmm, . ~ . - (1 + WSTRESS | UserID)
  - WSTRESS +
  pmin(WSTRESS, 0) + pmax(WSTRESS, 0) +
  (1 + pmin(WSTRESS, 0) + pmax(WSTRESS, 0) | UserID))

m7.lmmd <- update(update(m7.lmm, . ~ . - WSTRESS), . ~ .
  - (1 + WSTRESS | UserID)  +
  bs(WSTRESS, df = 3) + (1 + bs(WSTRESS, df = 3) | UserID))
```

Once all the models have been fit, we can compare the AICs using `AIC()`. The results show the linear is clearly suboptimal. The AICs for the quadratic, piecewise linear, and B-splines are much closer to each other. Because of how much easier it is to present and interpret a two-piece linear model than quadratic or B-splines, we will proceed using the piecewise linear model.

```
AIC(
  m7.lmm,
  m7.lmmb,
  m7.lmmc,
  m7.lmmd)
```

```
##           df   AIC
## m7.lmm   10 13084
## m7.lmmb  14 13007
## m7.lmmc  14 13014
## m7.lmmd  19 13004
```

We will recheck the diagnostic plots one last time to ensure that everything looks fine for our piecewise linear model. These are shown in Figure 11-22. The diagnostics suggest there is little evidence any assumptions were violated.

```
assumptiontests <- plotDiagnosticsLMER(m7.lmmc, plot = FALSE)
do.call(plot_grid, c(
  assumptiontests[c("ResPlot", "ResFittedPlot")],
  assumptiontests$RanefPlot,
  ncol = 2))
```

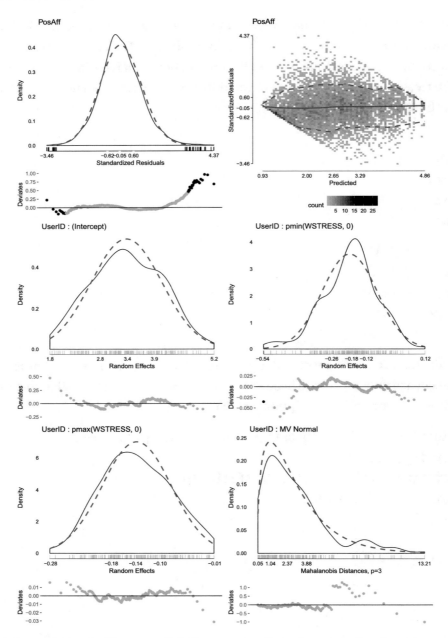

Figure 11-22. *Mixed effects model diagnostic plots showing the distribution of residuals (top left), the residuals vs. fitted values to assess homogeneity of variance (top right), the distribution of the random effects (middle row and bottom left), and a test for multivariate normality of the random effects (bottom right).*

To proceed, we will examine a summary of the model with effect sizes, including the marginal and conditional R^2 and Cohen's f^2 values for each predictor. This can be accomplished using the `detailedTests()` function from the `JWileymisc` package. Unfortunately, the model frame extracted from models is not quite right when there are B-splines included, which there still are for the average effects of stress.

To get around this issue at the time of writing (by the time you are reading this, `JWileymisc` will be updated so it is not required), we do a bit of hacking where we re-define the built-in `model.frame()` function. After we get our results saved in the object, `test.m7.lmmc`, we remove all the extra functions and copies we made so that our hack does not impact other functions or regular use of R.

```
## hack
model.frame <- function(obj) {
  d[UserID != 123][
    !is.na(PosAff) & !is.na(BSTRESS) &
    !is.na(WSTRESS) & !is.na(Weekend)]
}
detailedTests <- detailedTests
environment(detailedTests) <- environment()
.detailedTestsLMER <- .detailedTestsLMER
environment(.detailedTestsLMER) <- environment()

## calculate the detailed tests
test.m7.lmmc <- detailedTests(m7.lmmc,
  method = "Wald")

## remove our hack
rm(model.frame, detailedTests,
   .detailedTestsLMER)
```

Now we can get all of the tests and get a nicely formatted set of results, using the `formatLMER()` function. Note that this may take a few seconds, even when using the simplest Wald method for confidence intervals. The final set of results is shown in Table 11-4. Under the effect sizes, there are two for each variable with the first being the marginal Cohen's f^2 and the second, after the slash, being the conditional Cohen's f^2, depending whether the calculations were based on the marginal or conditional R^2 values.

```
effecttable <- formatLMER(list(test.m7.lmmc))

xtable(effecttable,
    caption = paste("Final random intercept and slope model",
        "with a B-spline for average stress and linear piecewise",
        "model for within person stress."),
    label = "tglmml-effecttable")
```

Table 11-4. *Final Random Intercept and Slope Model with a B-Spline for Average Stress and Linear Piecewise Model for Within-Person Stress*

	Term	Model 1
1	Fixed Effects	
2	(Intercept)	3.35*** [3.03, 3.66]
3	Weekend1	0.02 [−0.01, 0.06]
4	bs(BSTRESS, df = 3)1	−1.47** [−2.48, −0.46]
5	bs(BSTRESS, df = 3)2	−0.75 [−1.74, 0.23]
6	bs(BSTRESS, df = 3)3	−1.53** [−2.65, −0.40]
7	pmax(WSTRESS, 0)	−0.14*** [−0.16, −0.12]
8	pmin(WSTRESS, 0)	−0.19*** [−0.22, −0.16]
9	Random Effects	
10	cor_pmax(WSTRESS, 0).(Intercept)|UserID	−0.83
11	cor_pmax(WSTRESS, 0).pmin(WSTRESS, 0)|UserID	0.02
12	cor_pmin(WSTRESS, 0).(Intercept)|UserID	0.17
13	sd_(Intercept)|UserID	0.76
14	sd_pmax(WSTRESS, 0)|UserID	0.07
15	sd_pmin(WSTRESS, 0)|UserID	0.16
16	sigma	0.62
17	Overall Model	
18	Model DF	14

(continued)

Table 11-4. (*continued*)

	Term	Model 1
19	N (UserID)	190
20	N (Observations)	6377
21	logLik	−6492.77
22	AIC	13013.53
23	BIC	13108.18
24	Marginal R2	0.18
25	Conditional R2	0.65
26	Effect Sizes	
27	bs(BSTRESS, df = 3) (Fixed)	0.11/−0.03, p < .001
28	Weekend (Fixed)	0.00/ 0.00, p = .191
29	pmin(WSTRESS, 0) (Fixed + Random)	−0.01/ 0.06, p < .001
30	pmin(WSTRESS, 0) (Random)	−0.01/ 0.03, p < .001
31	pmax(WSTRESS, 0) (Fixed + Random)	0.00/ 0.03, p < .001
32	pmax(WSTRESS, 0) (Random)	−0.03/−0.02, p < .001

Also note that it is possible to format the results differently. The coefficients, p-values, confidence intervals, and effect sizes are written out using the `sprintf()` function, and we can specify different labels or formatting around them. The following example prints actual p-values instead of stars for the fixed effects coefficients and uses parentheses instead of brackets for confidence intervals, plus adding some labels to all the effects. The results are in Table 11-5.

```
xtable(
formatLMER(
  list(test.m7.lmmc),
  format = list(
    FixedEffects = c("b = %s, %s, CI = (%s; %s)"),
    RandomEffects = c("%s", "%s (%s; %s)"),
    EffectSizes = c("Marg f2 = %s; Cond f2 = %s, %s")),
```

```
pcontrol = list(
  digits = 3,
  stars = FALSE,
  includeP = TRUE,
  includeSign = TRUE,
  dropLeadingZero = TRUE)),
caption = paste("Different formatting for the final",
  "random intercept and slope model."),
label = "tglmml-effecttablealt")
```

Table 11-5. *Different Formatting for the Final Random Intercept and Slope Model*

	Term	Model 1
1	Fixed Effects	
2	(Intercept)	b = 3.35, p < .001, CI = (3.03; 3.66)
3	Weekend1	b = 0.02, p = .191, CI = (–0.01; 0.06)
4	bs(BSTRESS, df = 3)1	b = –1.47, p = .005, CI = (–2.48; –0.46)
5	bs(BSTRESS, df = 3)2	b = –0.75, p = .136, CI = (–1.74; 0.23)
6	bs(BSTRESS, df = 3)3	b = –1.53, p = .009, CI = (–2.65; –0.40)
7	pmax(WSTRESS, 0)	b = –0.14, p < .001, CI = (–0.16; –0.12)
8	pmin(WSTRESS, 0)	b = –0.19, p < .001, CI = (–0.22; –0.16)
9	Random Effects	
10	cor_pmax(WSTRESS, 0).(Intercept)|UserID	–0.83
11	cor_pmax(WSTRESS, 0).pmin(WSTRESS, 0)|UserID	0.02
12	cor_pmin(WSTRESS, 0).(Intercept)|UserID	0.17
13	sd_(Intercept)|UserID	0.76
14	sd_pmax(WSTRESS, 0)|UserID	0.07
15	sd_pmin(WSTRESS, 0)|UserID	0.16
16	sigma	0.62
17	Overall Model	

(continued)

Table 11-5. (*continued*)

Term	Model 1
18 Model DF	14
19 N (UserID)	190
20 N (Observations)	6377
21 logLik	−6492.77
22 AIC	13013.53
23 BIC	13108.18
24 Marginal R2	0.18
25 Conditional R2	0.65
26 Effect Sizes	
27 bs(BSTRESS, df = 3) (Fixed)	Marg f2 = 0.11; Cond f2 = −0.03, p < .001
28 Weekend (Fixed)	Marg f2 = 0.00; Cond f2 = 0.00, p = .191
29 pmin(WSTRESS, 0) (Fixed + Random)	Marg f2 = −0.01; Cond f2 = 0.06, p < .001
30 pmin(WSTRESS, 0) (Random)	Marg f2 = −0.01; Cond f2 = 0.03, p < .001
31 pmax(WSTRESS, 0) (Fixed + Random)	Marg f2 = 0.00; Cond f2 = 0.03, p < .001
32 pmax(WSTRESS, 0) (Random)	Marg f2 = −0.03; Cond f2 = −0.02, p < .001

11.3 Summary

This chapter introduced linear mixed models (LMMs) as a special case of generalized linear mixed models (GLMMs). A unique feature of GLMMs is the introduction of random effects in addition to the usual fixed effects present in most statistical models. Random effects allow the association of variables to differ between different individuals or whatever the higher grouping variable is (e.g., people, schools, hospitals, etc.). Random effects provide an elegant solution to the problem of non-independent assessments without conducting models on each individual's data. This chapter also covered challenges on how to determine effect sizes and generate predictions from GLMMs, specifically showing how bootstrapping can be used to accurately capture uncertainty in the predictions. A summary of the functions used throughout the chapter along with a brief description of each is shown in Table 11-6.

Table 11-6. *Listing of Key Functions Described in This Chapter and Summary of What They Do*

Function	What It Does
aes()	Controls which variables affect which aesthetics (e.g., x/y axis and color).
AIC()	Akaike information criterion.
apply()	Takes a given function and applies it over the specified variables.
bootMer()	Bootstrap sampling from a lmer() model to generate confidence intervals for coefficients or predictions.
cat()	Concatenates and prints entered strings.
clusterEvalQ()	Function from parallel that copies given environment to each cluster instance.
clusterExport()	Exports values from global environment for use by each cluster environment.
confint()	CONFidence INTerval generation function.
coord_cartesian()	Sets limits on a graph in the Descartes' fashion.
data()	Places data in memory (in this case from JWileymisc package).
detailedTests()	Calculates confidence intervals, overall model effect sizes, and effect sizes for individual coefficients for lmer() type models.
element_blank()	Ensures an element of ggplot is not drawn.
facet_wrap()	Creates a copy of the overall graph for each facet (or factor) named (e.g., individual vs. random).
factor()	Designates a particular collection of data elements to be factors.
fitted()	Similar to predict(), except the original data used to build the model is the input. In either case, model output yhat values are returned.
fixef()	Extracts fixed effects, takes model as argument.
formatLMER()	Formats model outputs and ensures consistency, alignment, and standard formatting.

(continued)

Table 11-6. (*continued*)

Function	What It Does
geom_line()	Draws a line.
geom_ribbon()	Draws a shaded region on either side of the y values (e.g., for confidence intervals).
geom_segment()	Draws a line segment between given points, which is useful for discrete data.
geom_vline()	Draws a vertical line.
ggplot()	Grammar of graphics plot object (as opposed to base R graphics).
iccMixed()	Calculates the intraclass correlation coefficient for a variable using mixed effects models behind the scenes. Requires the name of the variable, the name of the clustering or ID variable, and the dataset as arguments.
image()	Base graphics function displaying squares.
is.na()	Returns Boolean value(s) indicating NA elements.
lm()	Fits a linear model.
lmer()	Estimates linear mixed effects models.
makeCluster()	Makes a parallel compute cluster.
melt()	Takes wide data and melts it down to long data.
offset()	Fixes a coefficient in a model to a specific value (rather than allowing the model to compute a coefficient algorithmically).
plotDiagnosticsLMER()	Makes various diagnostic plots for lmer() class model.
predict()	Similar to fitted(), except requires a data argument. In either case, model output yhat values are returned.
print()	Prints the specified strings to the console.
R2LMER()	Calculates both marginal and conditional R^2 from linear mixed models.
rbind()	Binds data together by rows.
ranef()	Extracts random effects from a fitted lmer() model object.

(*continued*)

Table 11-6. (*continued*)

Function	What It Does
refitML()	Model refit function in place of default, restricted maximum likelihood.
round()	Rounds values to the specified number of decimal places.
scale_color_viridis()	Provides color on a scale using the viridis color package.
sprintf()	Prints characters subject to a given formatting.
stat_smooth()	Smooths a graph to avoid plotting overfit.
update()	Updates and refits a model.
VarCorr()	Extracts the random effects variances and covariances or standard deviations and correlations from a lmer() class model. Takes the fitted model as its argument.

CHAPTER 12

GLMMs: Advanced

This chapter on generalized linear *mixed* models (GLMMs) builds on the foundation of working with multilevel data from the GLMMs Introduction chapter and the GLMMs Linear chapter that focused strictly on continuous, normally distributed outcomes. This chapter focuses on GLMMs for other types of outcomes, specifically for binary outcomes and count outcomes.

We do use the optimx package [70] and the dfoptim package [97]. While not used directly (they are dependencies), they do need to be installed.

```
library(checkpoint)
checkpoint("2018-09-28", R.version = "3.5.1",
  project = book_directory,
  checkpointLocation = checkpoint_directory,
  scanForPackages = FALSE,
  scan.rnw.with.knitr = TRUE, use.knitr = TRUE)

library(knitr)
library(ggplot2)
library(cowplot)
library(viridis)
library(JWileymisc)
library(data.table)
library(lme4)
library(lmerTest)
library(chron)
library(zoo)
library(pander)
library(texreg)
library(xtable)
```

© Matt Wiley and Joshua F. Wiley 2019
M. Wiley and J. F. Wiley, *Advanced R Statistical Programming and Data Models*,
https://doi.org/10.1007/978-1-4842-2872-2_12

```
library(splines)
library(parallel)
library(boot)
library(optimx)
library(dfoptim)

options(width = 70, digits = 2)
```

12.1 Conceptual Background

This chapter does not introduce any substantially new conceptual content. Rather, it is a synthesis of pieces covered in several earlier chapters. If you have not read them already, the relevant chapters are the previous two chapters: GLMMs Introduction and GLMMs Linear. Together, these two chapters provided some coverage of the unique aspects of mixed effects or multilevel models. The other aspect needed is familiarity with different distribution families and link functions. These concepts were introduced in earlier chapters, specifically GLM 1 and GLM 2. In these two chapters, we explored how linear regression models could be extended to logistic regression and Poisson regression models to analyze binary and count outcome data. In this chapter, we will examine the same extensions, but instead of extending linear regression, we are extending linear mixed models that include both fixed and random effects. In practice, however, there are few differences conceptually. If you understand mixed effects models and you understand logistic and Poisson regression, you will find the same ideas and concepts applied throughout this chapter.

12.2 Logistic GLMM

Random Intercept

The first set of models we will examine are random intercept logistic regression models. First we load the data, including the raw data and that processed in the GLMMs Introduction chapter.

```
data(aces_daily)
draw <- as.data.table(aces_daily)
d <- readRDS("aces_daily_sim_processed.RDS")
```

Technically, there are no binary outcomes in the dataset we have been using. However, we can create a binary outcome by categorizing a continuous outcome. Each day, participants reported how many minutes it took them to fall asleep. Taking more than 30 minutes to fall asleep often is considered a clinically significant length.

```
d[, SOLs30 := as.integer(SOLs >= 30)]
```

GLMMs are set up very similarly to linear mixed models that we ran in the GLMMs Linear chapter. The primary difference is the use of the `glmer()` function in place of the `lmer()` function, and the need to specify the distribution and optionally link function. For binary outcomes, we use the binomial distribution with a logit link function, as discussed in the GLM 2 chapter. We also add two predictors: average number of awakenings after sleep onset and average use of disengagement coping. One additional challenge with GLMMs aside from normal outcomes is that there are no closed solutions and instead the solutions must be approximated through numerical integration. R defaults to what is known as the Laplace approximation, one integration point, but using additional integration points can improve the accuracy. We use 9 points by setting nAGQ = 9.

```
m1.glmm <- glmer(SOLs30 ~ BCOPEDis + BWASONs + (1 | UserID),
                 family = binomial(link = logit),
                 data = d, nAGQ = 9)
summary(m1.glmm)

## Generalized linear mixed model fit by maximum likelihood (Adaptive
##    Gauss-Hermite Quadrature, nAGQ = 9) [glmerMod]
##   Family: binomial  ( logit )
## Formula: SOLs30 ~ BCOPEDis + BWASONs + (1 | UserID)
##    Data: d
##
##       AIC       BIC    logLik deviance df.resid
##      1969      1991      -980     1961     2093
##
## Scaled residuals:
##     Min      1Q  Median      3Q     Max
## -2.558  -0.453  -0.226   0.346   3.343
##
```

```
## Random effects:
##  Groups Name          Variance Std.Dev.
##  UserID (Intercept) 3.55      1.88
## Number of obs: 2097, groups:  UserID, 191
##
## Fixed effects:
##              Estimate Std. Error z value Pr(>|z|)
## (Intercept)   -3.606      0.684    -5.27  1.4e-07 ***
## BCOPEDis       0.777      0.295     2.63   0.0085 **
## BWASONs        0.520      0.228     2.29   0.0223 *
## ---
## Signif. codes:  0 '***' 0.001 '**' 0.01 '*' 0.05 '.' 0.1 '_' 1
##
## Correlation of Fixed Effects:
##           (Intr) BCOPED
## BCOPEDis -0.918
## BWASONs  -0.410  0.101
```

Looking at the results, we can say that someone with zero average use of disengagement coping and zero average awakenings after sleep onset will have -3.6 log odds of taking 30 or more minutes to fall asleep. We can use the inverse link function, $\frac{1}{1+e^{-\mu}}$, to translate this into a probability. In R we can do it using the `plogis()` function.

```
plogis(fixef(m1.glmm)[["(Intercept)"]])
```

```
## [1] 0.026
```

Examining the individual coefficients, we see that for each additional unit higher average disengagement coping people use, they are expected to have 0.8 higher log odds of long sleep onset latency (30 or more minutes). Likewise, for each additional awakening on average people have, they are expected to have 0.5 higher log odds of long sleep onset latency. Put more simply, the results show that both higher average use of disengagement coping and more awakenings after sleep onset predict a higher probability of taking 30 minutes or longer to fall asleep.

Another way to interpret the fixed effects is to convert them to odds ratios. Odds ratios for the fixed effects can be calculated similarly to regular logistic regression, by exponentiating them. The following code does this for the fixed effects and the confidence intervals for the fixed effects, which are selected using the `fixef()` function and the `parm = "beta_"` argument to the `confint()` function so that only confidence intervals on the fixed effects and not the random effects are returned. These are combined using `cbind()` and then the whole result is exponentiated to give estimates and confidence intervals all on the odds ratio scale.

```
exp(cbind(
  B = fixef(m1.glmm),
  confint(m1.glmm, parm = "beta_", method = "Wald")))

##                   B   2.5 % 97.5 %
## (Intercept) 0.027 0.0071    0.1
## BCOPEDis    2.176 1.2199    3.9
## BWASONs     1.682 1.0768    2.6
```

Examining the odds ratios, we can say that for each additional unit higher average disengagement coping people use, they are expected to have 2.2 times the odds of long sleep onset latency. Likewise, for each additional awakening on average people have, they are expected to have 1.7 times the odds of long sleep onset latency.

To further aid interpretation, we can generate predicted probabilities. However, generating predicted probabilities from a GLMM is more complicated than it is from a regular logistic regression due to the random effects. Although the random intercept will always average to zero on the logit scale, it will not average to zero on the probability scale.

First we generate predictions using the average random effects (setting them to zero, essentially).

```
preddat <- as.data.table(expand.grid(
  BCOPEDis = seq(
    from = min(d$BCOPEDis, na.rm=TRUE),
    to = max(d$BCOPEDis, na.rm = TRUE),
    length.out = 1000),
  BWASONs = quantile(d$BWASONs, probs = c(.2, .8),
                     na.rm = TRUE)))
```

```
## predictions based on average random effects
preddat$yhat <- predict(m1.glmm,
  newdata = preddat,
  type = "response",
  re.form = ~ 0)
```

Next we generate predictions for each person in the sample.

```
preddat2 <- as.data.table(expand.grid(
  UserID = unique(d$UserID),
  BCOPEDis = seq(
    from = min(d$BCOPEDis, na.rm=TRUE),
    to = max(d$BCOPEDis, na.rm = TRUE),
    length.out = 1000),
  BWASONs = quantile(d$BWASONs, probs = c(.2, .8),
                     na.rm = TRUE)))

## predictions based on average random effects
preddat2$yhat <- predict(m1.glmm,
  newdata = preddat2,
  type = "response",
  re.form = NULL)
```

Now we average the predicted probabilities across people. That is, this is averaging after generating predicted probabilities, rather than averaging before generating predicted probabilities.

```
## calculate predicted probabilities
## averaging across participants
preddat3 <- preddat2[, .(yhat = mean(yhat)),
        by = .(BCOPEDis, BWASONs)]
```

Now we can plot the various results, shown in Figure 12-1. The plots highlight how large the difference in predicted probabilities can be when averaging across the random effects on the probability scale vs. averaging across them on the logit scale and generating a single set of predicted probabilities. In general, generating multiple predicted probabilities for each participant and averaging across these is more appropriate than is averaging across values used to generate predictions, although it takes more effort to generate.

```
ggplot(rbind(
  cbind(preddat, Type = "Zero"),
  cbind(preddat3, Type = "Average")),
  aes(BCOPEDis, yhat, colour = Type)) +
  geom_line(size = 1) +
  scale_color_viridis(discrete = TRUE) +
  facet_wrap(~ round(BWASONs, 1)) +
  theme(
    legend.key.width = unit(1, "cm"),
    legend.position = c(.1, .9)) +
  xlab("Average disengagement coping") +
  ylab("Probability of sleep onset latency 30+ min") +
  coord_cartesian(
    xlim = c(1, 4),
    ylim = c(0, .6),
    expand = FALSE)
```

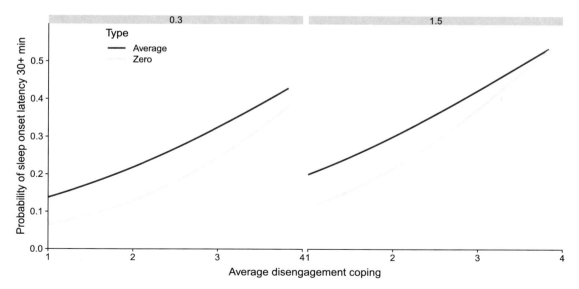

Figure 12-1. *Graph of the predicted probabilities setting random effects to zero (average on the logit scale) and averaging all of them*

We can also see the impact of the random intercept on the probability scale by plotting the individual predicted probabilities. These results are shown in Figure 12-2.

```
ggplot(preddat2,
  aes(BCOPEDis, yhat, group = UserID)) +
  geom_line(alpha = .2) +
  facet_wrap(~ round(BWASONs, 1))+
  xlab("Average disengagement coping") +
  ylab("Probability of sleep onset latency 30+ min") +
  coord_cartesian(
    xlim = c(1, 4),
    ylim = c(0, 1),
    expand = FALSE)
```

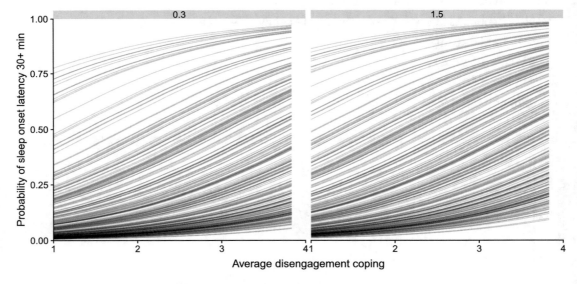

Figure 12-2. *Graph of the individual predicted probabilities across levels of disengagement coping and average number of awakenings*

Random Intercept and Slope

Just as with linear mixed models, GLMMs can have both random intercepts and random slopes. Random slopes can capture individual differences in the association of within-person predictors with the probability of the event occurring, here taking 30 or more minutes to fall asleep. Here we examine a model with average and within-person positive affect predicting time to fall asleep, including a random intercept and a random slope for within-person positive affect.

```
m2.glmm <- glmer(SOLs30 ~ BPosAff + WPosAff +
  (1 + WPosAff | UserID),
  family = binomial(link = logit),
  data = d, nAGQ = 1)

summary(m2.glmm)

## Generalized linear mixed model fit by maximum likelihood (Laplace
##   Approximation) [glmerMod]
##  Family: binomial  ( logit )
## Formula: SOLs30 ~ BPosAff + WPosAff + (1 + WPosAff | UserID)
##    Data: d
##
##      AIC      BIC   logLik deviance df.resid
##     1813     1846     -900     1801     1894
##
## Scaled residuals:
##    Min     1Q Median     3Q    Max
## -2.334 -0.451 -0.242  0.341  3.415
##
## Random effects:
##  Groups Name         Variance Std.Dev. Corr
##  UserID (Intercept) 3.7615    1.94
##         WPosAff      0.0897    0.30     1.00
## Number of obs: 1900, groups:  UserID, 191
##
## Fixed effects:
##             Estimate Std. Error z value Pr(>|z|)
## (Intercept)   -1.877      0.561   -3.34  0.00082 ***
## BPosAff        0.117      0.198    0.59  0.55573
## WPosAff       -0.337      0.123   -2.73  0.00627 **
## ---
## Signif. codes:  0 '***' 0.001 '**' 0.01 '*' 0.05 '.' 0.1 '_' 1
##
```

```
## Correlation of Fixed Effects:
##          (Intr) BPsAff
## BPosAff -0.952
## WPosAff  0.092 -0.001
```

These results reveal that when positive affect is one unit above an individual's own mean positive affect, they are expected to have -0.34 lower log odds of taking a long time to fall asleep that night. Note that this is not simply saying happier people fall asleep faster, because the within-person variable is a deviation from an individual's own mean. These results indicate that on days when people are happier relative to how they normally feel, they are less likely to take a long time to fall asleep.

As before, we can easily calculate the odds ratios. These show that for each unit above average evening positive affect, participants had about 0.71 times the odds of taking 30 or more minutes to fall asleep.

```
exp(cbind(
  B = fixef(m2.glmm),
  confint(m2.glmm, parm = "beta_", method = "Wald")))

##                   B 2.5 % 97.5 %
## (Intercept) 0.15 0.051   0.46
## BPosAff     1.12 0.762   1.66
## WPosAff     0.71 0.560   0.91
```

Converting the impact of within-person positive affect to probabilities requires taking into account other predictors and both the random intercept and the slope, and averaging across both of these. Due to the possibility of different ranges for within-person positive affect depending on the average level of positive affect, we calculate these separately.

```
bpa.low <- quantile(d$BPosAff, probs = .2, na.rm=TRUE)
bpa.high <- quantile(d$BPosAff, probs = .8, na.rm=TRUE)

preddat4.low <- as.data.table(expand.grid(
  UserID = unique(d$UserID),
  WPosAff = seq(
    from = min(d[BPosAff <= bpa.low]$WPosAff,
               na.rm = TRUE),
```

```
    to = max(d[BPosAff <= bpa.low]$WPosAff,
             na.rm = TRUE),
    length.out = 1000),
  BPosAff = bpa.low))

preddat4.high <- as.data.table(expand.grid(
  UserID = unique(d$UserID),
  WPosAff = seq(
    from = min(d[BPosAff >= bpa.high]$WPosAff,
               na.rm = TRUE),
    to = max(d[BPosAff >= bpa.high]$WPosAff,
             na.rm = TRUE),
    length.out = 1000),
  BPosAff = bpa.high))

preddat4 <- rbind(
  preddat4.low,
  preddat4.high)

## predictions including random effects
preddat4$yhat <- predict(m2.glmm,
  newdata = preddat4,
  type = "response",
  re.form = NULL)

## calculate predicted probabilities
## averaging across participants
preddat4b <- preddat4[, .(yhat = mean(yhat)),
        by = .(WPosAff, BPosAff)]
```

Now we can plot the various results, in Figure 12-3, which shows that indeed the range of within-person variation is quite different in those with "low" average positive affect (the 20th percentile) compared to those with "high" average positive affect (the 80th percentile).

```
ggplot(preddat4b,
  aes(WPosAff, yhat, colour = factor(round(BPosAff, 1)))) +
  geom_line(size = 1) +
```

```
scale_color_viridis("Average\nPositive Affect",
                    discrete = TRUE) +
theme(
  legend.key.width = unit(1.5, "cm"),
  legend.position = c(.7, .9)) +
coord_cartesian(
  xlim = c(-4, 4),
  ylim = c(0, .45),
  expand = FALSE) +
xlab(paste0("Within person positive affect\n",
            "(deviations from own mean)")) +
ylab("Probability of sleep onset latency 30+ min")
```

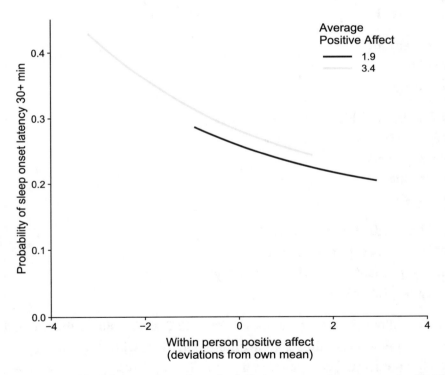

Figure 12-3. *Graph of the predicted probabilities averaging across individuals accounting for random intercept and slope of within-person positive affect*

12.3 Poisson and Negative Binomial GLMMs
Random Intercept

For outcomes that are counts, it is common to assume a Poisson or negative binomial distribution as a normal distribution typically is a poor representation of count data. We introduced the Poisson model in the GLM 2 chapter for fixed effects–only (single-level) models. This section expands that by allowing a random intercept. One feature of the Poisson distribution is that the variance is expected to be equal to the mean. If there is more variability than the mean, there is overdispersion. Although this often is an issue for Poisson models, it may be less of a concern in mixed effects Poisson models, because with a random intercept for each person, there is a greater chance that the model can adjust to each individual and that the assumption of variance equal to the mean may be more tenable.

Running a Poisson GLMM with random intercept in R can be done using the glmer() function and specifying the distribution and link function using the argument family = poisson(link = log). Otherwise, the model is identical in structure to other GLMMs. For this analysis, we will use the number of awakenings at night as the outcome variable and age (years) and whether participants were born in Australia (coded 1) or not born in Australia (coded 0) as two predictor variables.

```
m3.glmm <- glmer(WASONs ~ Age + BornAUS +
  (1 | UserID),
  family = poisson(link = log),
  data = d, nAGQ = 9)

summary(m3.glmm)

## Generalized linear mixed model fit by maximum likelihood (Adaptive
##   Gauss-Hermite Quadrature, nAGQ = 9) [glmerMod]
## Family: poisson  ( log )
## Formula: WASONs ~ Age + BornAUS + (1 | UserID)
##    Data: d
##
##    AIC      BIC    logLik deviance df.resid
##   2070     2092    -1031     2062     1910
##
```

```
## Scaled residuals:
##    Min    1Q Median    3Q    Max
## -1.763 -0.673 -0.360  0.477  3.983
##
## Random effects:
##  Groups Name         Variance Std.Dev.
##  UserID (Intercept) 0.52      0.721
## Number of obs: 1914, groups:  UserID, 190
##
## Fixed effects:
##              Estimate Std. Error z value Pr(>|z|)
## (Intercept)  -1.7574     0.5810    -3.02  0.00249 **
## Age           0.0588     0.0266     2.21  0.02681 *
## BornAUS       0.4250     0.1243     3.42  0.00063 ***
## ---
## Signif. codes:  0 '***' 0.001 '**' 0.01 '*' 0.05 '.' 0.1 '_' 1
##
## Correlation of Fixed Effects:
##          (Intr) Age
## Age       -0.991
## BornAUS  -0.061 -0.019
```

Because Poisson GLMMs use a log link, the results are on the log scale. Examining the individual coefficients, we can interpret them as follows. The intercept indicates that a zero-year-old not born in Australia is expected to have -1.8 log awakenings per night. For each year older someone is, they are expected to have 0.1 more log awakenings per night. Finally compared to people born outside Australia, those born in Australia are expected to have 0.4 higher log awakenings.

As with the logistic GLMMs, we can use the inverse link function, in this case simply exp(), to make the results easier to understand. Transforming the intercept, we can say that a zero-year-old not born in Australia is expected to have 0.2 awakenings per night.

We can also exponentiate the coefficients to get the results as a multiplier of the baseline count. This results in the interpretation that for each year older someone is, they are expected to have 1.1 times the awakenings per night. Compared to people born outside Australia, those born in Australia are expected to have 1.5 times the awakenings. When the Poisson coefficients are exponentiated, they are referred to as incident rate ratios (IRRs). IRRs provide a relative measure of how many times higher the rate is in one group over another or how many times higher the rate is as a continuous predictor changes.

To get the IRRs and confidence intervals, we follow the same approach as we did to get ORs for logistic GLMMs. First we extract the coefficients and confidence intervals and then exponentiate the final results to place them on the IRR scale instead of the log scale.

```
exp(cbind(
  B = fixef(m3.glmm),
  confint(m3.glmm, parm = "beta_", method = "Wald")))
```

```
##                   B 2.5 % 97.5 %
## (Intercept) 0.17 0.055   0.54
## Age             1.06 1.007   1.12
## BornAUS         1.53 1.199   1.95
```

To see the absolute effects, we generate predictions for each person in the sample and average the results.

```
preddat5 <- as.data.table(expand.grid(
  UserID = unique(d[!is.na(BornAUS) & !is.na(Age)]$UserID),
  Age = seq(
    from = min(d$Age, na.rm=TRUE),
    to = max(d$Age, na.rm = TRUE),
    length.out = 1000),
  BornAUS = 0:1))

## predictions based on average random effects
preddat5$yhat <- predict(m3.glmm,
  newdata = preddat5,
  type = "response",
  re.form = NULL)

## calculate predicted counts
## averaging across participants
preddat5 <- preddat5[, .(yhat = mean(yhat)),
        by = .(Age, BornAUS)]
```

We can also see the impact of the random intercept on the probability scale by plotting the individual predicted probabilities. These results are shown in Figure 12-4.

```
ggplot(preddat5,
  aes(Age, yhat, colour = factor(BornAUS))) +
  geom_line(size = 2) +
  scale_colour_viridis("Born in Australia", discrete = TRUE) +
  xlab("Age (years)") +
  ylab("Predicted # wakenings after sleep onset") +
  theme(
    legend.key.width = unit(1.5, "cm"),
    legend.position = c(.1, .9)) +
  coord_cartesian(
    xlim = c(18, 26.5),
    ylim = c(0, 2),
    expand = FALSE)
```

If there are concerns of overdispersion, an alternative for count outcomes is to use a negative binomial model. Support for this is being added in the lme4 package and such models can be fit using the glmer.nb() function. Note that these models are considerably slower than fitting Poisson models. Next, we refit our Poisson random intercept model as a negative binomial model.

```
m3.glmm.nb <- glmer.nb(formula(m3.glmm),
  data = d)

## Warning in checkConv(attr(opt, "derivs"), opt$par, ctrl =
control$checkConv, : Model failed to converge with max|grad| = 0.00224463
(tol = 0.001, component 1)

## Warning in theta.ml(Y, mu, weights = object@resp$weights, limit = limit,
: iteration limit reached

## Warning in checkConv(attr(opt, "derivs"), opt$par, ctrl =
control$checkConv, : Model failed to converge with max|grad| = 0.00115603
(tol = 0.001, component 1)

## Warning in checkConv(attr(opt, "derivs"), opt$par, ctrl =
control$checkConv, : Model failed to converge with max|grad| = 0.00162663
(tol = 0.001, component 1)
```

```
## Warning in checkConv(attr(opt, "derivs"), opt$par, ctrl =
control$checkConv, : Model failed to converge with max|grad| = 0.00194197
(tol = 0.001, component 1)

## Warning in checkConv(attr(opt, "derivs"), opt$par, ctrl =
control$checkConv, : Model failed to converge with max|grad| = 0.00147706
(tol = 0.001, component 1)

## Warning in checkConv(attr(opt, "derivs"), opt$par, ctrl =
control$checkConv, : Model failed to converge with max|grad| = 0.0016048
(tol = 0.001, component 1)

## Warning in checkConv(attr(opt, "derivs"), opt$par, ctrl =
control$checkConv, : Model failed to converge with max|grad| = 0.0014179
(tol = 0.001, component 1)

## Warning in checkConv(attr(opt, "derivs"), opt$par, ctrl =
control$checkConv, : Model failed to converge with max|grad| = 0.00177251
(tol = 0.001, component 1)

## Warning in checkConv(attr(opt, "derivs"), opt$par, ctrl =
control$checkConv, : Model failed to converge with max|grad| = 0.0012395
(tol = 0.001, component 1)

## Warning in checkConv(attr(opt, "derivs"), opt$par, ctrl =
control$checkConv, : Model failed to converge with max|grad| = 0.00139838
(tol = 0.001, component 1)

## Warning in checkConv(attr(opt, "derivs"), opt$par, ctrl =
control$checkConv, : Model failed to converge with max|grad| = 0.00142382
(tol = 0.001, component 1)

## Warning in checkConv(attr(opt, "derivs"), opt$par, ctrl =
control$checkConv, : Model failed to converge with max|grad| = 0.00144676
(tol = 0.001, component 1)

## Warning in checkConv(attr(opt, "derivs"), opt$par, ctrl =
control$checkConv, : Model failed to converge with max|grad| = 0.00138734
(tol = 0.001, component 1)
```

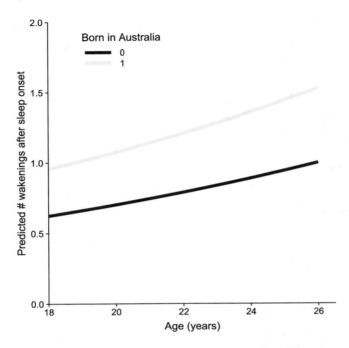

Figure 12-4. *Graph of the average predicted counts of awakenings after sleep onset by age and whether people were born in Australia or not*

The negative binomial GLMM generates warnings about failures to converge. To investigate these convergence warnings, we could try fitting the model using different optimizers to see whether different optimizers converge to the same results. The lme4 package does not directly include a function for this, but it ships with an R script that can be loaded to provide a function. Note that this also requires loading a few additional R packages that provide additional optimizers. Once we have sourced the code from the lme4 package, we can use the allFit() function to fit our model using different optimizers.

```
## load R code shipped with lme4 to provide the allFit()
source(system.file("utils", "allFit.R", package="lme4"))
m3.all <- allFit(m3.glmm.nb)

## bobyqa :

## Warning in checkConv(attr(opt, "derivs"), opt$par, ctrl =
control$checkConv, : Model failed to converge with max|grad| = 0.00130544
(tol = 0.001, component 1)
```

```
## [OK]
## Nelder_Mead :

## Warning in checkConv(attr(opt, "derivs"), opt$par, ctrl =
control$checkConv, : Model failed to converge with max|grad| = 0.00199142
(tol = 0.001, component 1)

## [OK]
## nlminbw : [OK]
## nmkbw :

## Warning in checkConv(attr(opt, "derivs"), opt$par, ctrl =
control$checkConv, : Model failed to converge with max|grad| = 0.00894524
(tol = 0.001, component 1)

## [OK]
## optimx.L-BFGS-B :

## Warning in checkConv(attr(opt, "derivs"), opt$par, ctrl =
control$checkConv, : Model failed to converge with max|grad| = 0.361426
(tol = 0.001, component 1)


## [OK]
## nloptwrap.NLOPT_LN_NELDERMEAD :

## Warning in checkConv(attr(opt, "derivs"), opt$par, ctrl =
control$checkConv, : Model failed to converge with max|grad| = 0.00541331
(tol = 0.001, component 1)

## [OK]
## nloptwrap.NLOPT_LN_BOBYQA :

## Warning in checkConv(attr(opt, "derivs"), opt$par, ctrl =
control$checkConv, : Model failed to converge with max|grad| = 0.00541331
(tol = 0.001, component 1)

## [OK]
```

Again we get warnings about a failure to converge, although in many cases we can see the absolute gradient is near the default tolerance of .001. Next we can make a summary of all the fits and look at the results for the fixed effects, log likelihood, and theta, which in this model is the random intercept variance.

```
m3.all.sum <- summary(m3.all)
```

```
m3.all.sum$fixef
```

```
##                                (Intercept)    Age  BornAUS
## bobyqa                                -1.8  0.059     0.42
## Nelder_Mead                           -1.8  0.059     0.42
## nlminbw                               -1.8  0.059     0.42
## nmkbw                                 -1.8  0.059     0.42
## optimx.L-BFGS-B                       -1.7  0.055     0.42
## nloptwrap.NLOPT_LN_NELDERMEAD         -1.8  0.059     0.42
## nloptwrap.NLOPT_LN_BOBYQA             -1.8  0.059     0.42
```

```
m3.all.sum$llik
```

```
##                        bobyqa                    Nelder_Mead
##                         -2270                          -2270
##                       nlminbw                          nmkbw
##                         -2270                          -2270
##             optimx.L-BFGS-B nloptwrap.NLOPT_LN_NELDERMEAD
##                         -2270                          -2270
##     nloptwrap.NLOPT_LN_BOBYQA
##                         -2270
```

```
m3.all.sum$theta
```

```
##                                UserID.(Intercept)
## bobyqa                                        0.72
## Nelder_Mead                                   0.72
## nlminbw                                       0.72
## nmkbw                                         0.72
## optimx.L-BFGS-B                               0.72
## nloptwrap.NLOPT_LN_NELDERMEAD                 0.72
## nloptwrap.NLOPT_LN_BOBYQA                     0.72
```

The results show that all the different optimizers converge to the same estimates. Thus despite the convergence warnings, we may feel relatively confident that we have in fact found the optimal solution and the model has converged.

Next we use the `screenreg()` function to print the results of the two models side by side to see how the Poisson differs from the negative binomial. In this case we can see that they are quite similar, except that the standard errors for age and the intercept are somewhat smaller in the negative binomial model than in the Poisson model.

```
screenreg(
  list(Poisson = m3.glmm,
       NegBin = m3.glmm.nb))
```

```
##
## ========================================================
##                            Poisson        NegBin
## --------------------------------------------------------
## (Intercept)               -1.76 **       -1.76 ***
##                           (0.58)         (0.48)
## Age                        0.06 *         0.06 **
##                           (0.03)         (0.02)
## BornAUS                    0.42 ***       0.42 ***
##                           (0.12)         (0.12)
## --------------------------------------------------------
## AIC                     2070.12         4549.55
## BIC                     2092.34         4577.34
## Log Likelihood         -1031.06        -2269.78
## Num. obs.               1914            1914
## Num. groups: UserID      190             190
## Var: UserID (Intercept)    0.52            0.51
## ========================================================
## *** p < 0.001, ** p < 0.01, * p < 0.05
```

As before, we can make a table of the incident rate ratios (IRRs) and their confidence intervals by first extracting the fixed effects and confidence intervals and then exponentiating them. The following code does this with a side-by-side comparison of the Poisson and negative binomial GLMMs.

```
exp(cbind(
  fixef(m3.glmm),
  confint(m3.glmm, parm = "beta_", method = "Wald"),
  fixef(m3.glmm.nb),
  confint(m3.glmm.nb, parm = "beta_", method = "Wald")))
```

```
##                       2.5 % 97.5 %       2.5 % 97.5 %
## (Intercept) 0.17 0.055    0.54 0.17 0.068    0.44
## Age         1.06 1.007    1.12 1.06 1.016    1.11
## BornAUS     1.53 1.199    1.95 1.53 1.202    1.95
```

It also can be helpful to compare the expected and observed distributions to see how closely they align. The following code extracts the estimate of overdispersion from the negative binomial GLMM, called theta, and then calculates the observed and expected densities of each number of awakening and stores these in a dataset for plotting.

```
theta <- getME(m3.glmm.nb, "glmer.nb.theta")

density <- data.table(
  X = as.integer(names(table(d$WASONs))),
  Observed = as.vector(prop.table(table(d$WASONs))))

density$NegBin <- colMeans(do.call(rbind, lapply(fitted(m3.glmm.nb),
function(mu) {
  dnbinom(density$X, size = theta, mu = mu)
})))

density$Poisson <- colMeans(do.call(rbind, lapply(fitted(m3.glmm),
function(mu) {
  dpois(density$X, lambda = mu)
})))
```

Now we can plot the densities to visualize how close our models match the observed data distribution. The results are in Figure 12-5. The graph shows that overall there is fairly good agreement between the observed and expected distributions. This supports that our distribution expectations are not completely unreasonable. We also see that there are no differences in this case between the Poisson and the negative binomial model.

```
ggplot(melt(density, id.vars = "X"),
  aes(X, value, fill = variable)) +
  geom_col(position = "dodge") +
  scale_fill_viridis("Type", discrete = TRUE) +
  theme(legend.position = c(.8, .8)) +
  xlab("Number of awakenings") +
  ylab("Density") +
  coord_cartesian(
    xlim = c(-.5, 4.5),
    ylim = c(0, .5),
    expand = FALSE)
```

To better understand why the Poisson and negative binomial GLMMs are giving the same end results in terms of distribution, we can look at the model estimate for theta. It is a very high estimate, and in these cases the negative binomial tends to the Poisson.

```
getME(m3.glmm.nb, "glmer.nb.theta")
```

```
## [1] 40993
```

Taking into account the estimate of theta, the similarity in results and expected distributions from the Poisson and the negative binomial model, in this case we most likely would choose to stay with the simpler Poisson model.

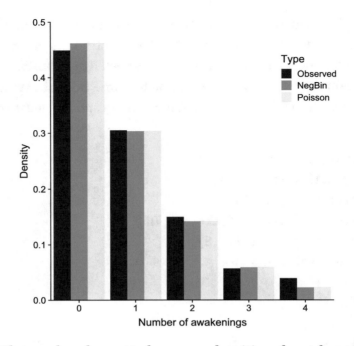

Figure 12-5. *Observed and expected average densities of number of awakenings at night based on a Poisson and negative binomial GLMM.*

Random Intercept and Slope

As with logistic GLMMs, we can include not only random intercepts but also random slopes. Here we will continue attempting to predict awakenings after sleep onset. Given the results from the random intercept model, we will only explore a Poisson GLMM here, rather than also including a negative binomial GLMM.

In addition to age and whether participants were born in Australia or not, a good predictor of awakenings may be previous number of awakenings. As part of our GLMMs Introduction chapter, we created lagged variables. The variable, WWASONsLag1, captures within-person deviations from an individual's own mean, on the previous night. That is, it tells us how many more (or less) awakenings did a participant have compared to their usual number last night. Although the number of awakenings is a discrete count, deviations from average awakenings across 12 days take on a more continuous distribution as people can be above or below their own average. A plot of the distribution of lagged within-person deviations of awakenings after sleep onset is shown in Figure 12-6. The graph shows a distribution that is not normal, but is more continuous and relatively symmetrical.

```
testdistr(d[, WWASONsLag1],
          varlab = "Within WASONs lag 1")
```

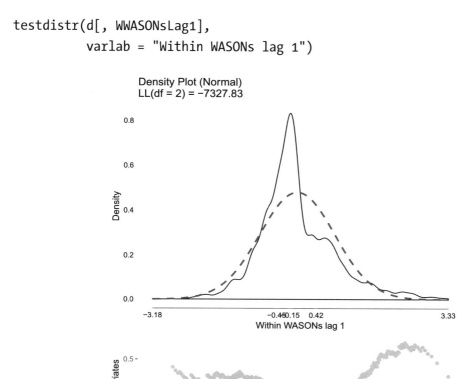

Figure 12-6. *Distribution of within-person deviations from typical number of awakenings after sleep onset lagged to the previous day.*

In the following model, we include WWASONsLag1 as both a fixed effect and a random effect. Because we added the random effect with the intercept, the model will allow the random effects to be correlated.

```
m4.glmm <- glmer(WASONs ~ Age + BornAUS +
   WWASONsLag1 +
  (1 + WWASONsLag1  | UserID),
  family = poisson(link = log),
  data = d, nAGQ = 1)
```

summary(m4.glmm)

```
## Generalized linear mixed model fit by maximum likelihood (Laplace
##   Approximation) [glmerMod]
##  Family: poisson  ( log )
## Formula:
## WASONs ~ Age + BornAUS + WWASONsLag1 + (1 + WWASONsLag1 | UserID)
##    Data: d
##
##      AIC      BIC   logLik deviance df.resid
##     4246     4284    -2116     4232     1770
##
## Scaled residuals:
##    Min     1Q Median     3Q    Max
## -1.768 -0.655 -0.333  0.485  3.892
##
## Random effects:
##  Groups Name          Variance Std.Dev. Corr
##  UserID (Intercept) 0.5043    0.7101
##         WWASONsLag1 0.0044    0.0663   1.00
## Number of obs: 1777, groups:  UserID, 189
##
## Fixed effects:
##             Estimate Std. Error z value Pr(>|z|)
## (Intercept)  -1.6039     0.5728   -2.80  0.00511 **
## Age           0.0527     0.0262    2.01  0.04472 *
## BornAUS       0.4045     0.1218    3.32  0.00089 ***
## WWASONsLag1  -0.0986     0.0457   -2.16  0.03093 *
## ---
## Signif. codes:  0 '***' 0.001 '**' 0.01 '*' 0.05 '.' 0.1 '_' 1
##
## Correlation of Fixed Effects:
##             (Intr) Age    BrnAUS
## Age         -0.991
## BornAUS     -0.059 -0.020
## WWASONsLag1 -0.039  0.043  0.112
```

The model summary shows a negative association on average (fixed effects) between lagged awakenings after sleep onset and number of awakenings the following night. Specifically, for each additional awakening above an individual's own average, they are expected to have -0.1 lower log awakenings the following night. We can exponentiate this value to get the IRR. The IRR indicates that for each additional awakening above an individual's own average, they are expected to have 0.91 times as many awakenings the following night.

```
exp(cbind(
  B = fixef(m4.glmm),
  confint(m4.glmm, parm = "beta_", method = "Wald")))
```

```
##                 B 2.5 % 97.5 %
## (Intercept) 0.20 0.065   0.62
## Age         1.05 1.001   1.11
## BornAUS     1.50 1.180   1.90
## WWASONsLag1 0.91 0.828   0.99
```

Although not huge, there was some variation in the slope, as indicated by the random variance and standard deviation. We can extract the slopes for each participant, which incorporates both the fixed and random effects using the coef() function. Then, we can exponentiate these slopes and plot them to show the distribution of IRRs. The distribution shows that nearly all participants are predicted to have an IRR below 1, indicating that for nearly all people, when they have more than usual awakenings one night, they tend to have fewer awakenings the following night (Figure 12-7).

```
testdistr(exp(coef(m4.glmm)$UserID$WWASONsLag1))
```

Finally, we could generate predictions for the number of awakenings. Often it is desirable to include a measure of uncertainty in predictions. However, producing confidence intervals around predictions from GLMMs is complex. Approximate confidence intervals can be obtained through the use of bootstrapping. However, it is worth noting that even bootstrapped confidence intervals are currently limited to a parametric bootstrap, and thus assumptions are still made about the distributions. First we set up a new dataset for prediction and generate the overall predictions, on the link scale.

```
preddat.boot <- as.data.table(expand.grid(
  UserID = unique(model.frame(m4.glmm)$UserID),
  WWASONsLag1 = seq(
    from = min(d$WWASONsLag1, na.rm = TRUE),
    to = max(d$WWASONsLag1, na.rm = TRUE),
    length.out = 100),
  Age = quantile(d[!duplicated(UserID)]$Age,
                  probs = c(.2, .8), na.rm = TRUE),
  BornAUS = 0:1))

preddat.boot$yhat <- predict(m4.glmm,
  newdata = preddat.boot)
```

To help speed up the bootstraps, we will set up a local cluster for parallel processing. We need to load the relevant packages and export the dataset for predictions.

```
genPred <- function(m) {
  predict(m,
    newdata = preddat.boot)
}

cl <- makeCluster(4)
clusterExport(cl, c("book_directory",
                    "checkpoint_directory",
                    "preddat.boot", "d", "genPred"))

clusterEvalQ(cl, {
  library(checkpoint)
  checkpoint("2018-09-28", R.version = "3.5.1",
    project = book_directory,
    checkpointLocation = checkpoint_directory,
    scanForPackages = FALSE,
    scan.rnw.with.knitr = TRUE, use.knitr = TRUE)

  library(data.table)
  library(lme4)
  library(lmerTest)
})
```

```
## [[1]]
##  [1] "lmerTest"     "lme4"        "Matrix"       "data.table"
##  [5] "checkpoint"   "RevoUtils"   "stats"        "graphics"
##  [9] "grDevices"    "utils"       "datasets"     "RevoUtilsMath"
## [13] "methods"      "base"
##
## [[2]]
##  [1] "lmerTest"     "lme4"        "Matrix"       "data.table"
##  [5] "checkpoint"   "RevoUtils"   "stats"        "graphics"
##  [9] "grDevices"    "utils"       "datasets"     "RevoUtilsMath"
## [13] "methods"      "base"
##
## [[3]]
##  [1] "lmerTest"     "lme4"        "Matrix"       "data.table"
##  [5] "checkpoint"   "RevoUtils"   "stats"        "graphics"
##  [9] "grDevices"    "utils"       "datasets"     "RevoUtilsMath"
## [13] "methods"      "base"
##
## [[4]]
##  [1] "lmerTest"     "lme4"        "Matrix"       "data.table"
##  [5] "checkpoint"   "RevoUtils"   "stats"        "graphics"
##  [9] "grDevices"    "utils"       "datasets"     "RevoUtilsMath"
## [13] "methods"      "base"
```

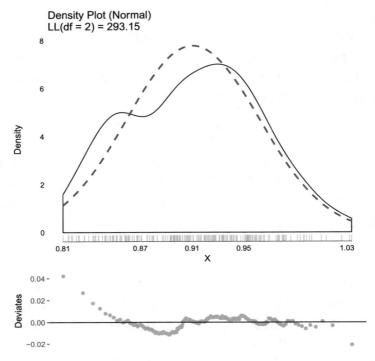

Figure 12-7. *For nearly all people, when they have more than usual awakenings one night, they tend to have fewer awakenings the following night.*

The main bootstrapping is a parametric model, and it is conducted using the bootMer() function included in the lme4 package. Note that we do few bootstraps here, as even with a parallel cluster using four cores, it is relatively slow. More common would be to run 1,000, 5,000, or 10,000 bootstrap samples, but that may take many minutes or hours.

```
system.time(bootres <- bootMer(m4.glmm,
    FUN = genPred,
    nsim = 100,
    seed = 12345,
    use.u = FALSE,
    type = "parametric",
    parallel = "snow",
    ncpus = 4,
    cl = cl))

##    user  system elapsed
##     3.3     1.7   169.4
```

Lastly, we calculate simple percentile confidence intervals on the results and add them back into our dataset. This is somewhat more complicated than in the linear mixed model case, as we collapse across ID. This is because it makes a difference whether one averages random effects on the link scale (log, where they average to zero) or the response scale (counts, where random effects do not average to zero). In our prediction dataset, there are 75,600 rows, corresponding to 189 unique IDs, each repeated 400 times, for different predictor values. To find all the rows with the same predictors but across IDs, we can create an index for the first, second, etc. combination of predictor values. We can do this by simply repeating an index value by the length of unique IDs, as long as the dataset is sorted by predictor values.

```
preddat.boot[, Index := rep(1L:400L,
  each = length(unique(UserID)))]
```

First, we will make a new small predicted dataset that has just the average predicted values and required set of predictor values. Note that we exponentiate the predicted log counts and then average across IDs, not average then exponentiate.

```
preddat.boot.avg <- preddat.boot[, .(yhat = mean(exp(yhat))),
  by = .(WWASONsLag1, Age, BornAUS)]
```

Now we can loop by the new index to get the confidence intervals from the bootstrap samples for the average counts. If we were to directly take the confidence intervals based on percentiles, this would incorporate uncertainty due to differences between people and in the average estimate. Instead, we exponentiate, average across all IDs, and then take the percentiles across the bootstrap samples. The averaging across people (IDs) after exponentiating, but prior to calculating the percentile confidence intervals, means that we are averaging out variability across people and only considering variability in the mean estimates across bootstraps. We could also include the variability due to IDs, but that would answer a different question.

To actually do this, note that boot results have different predictions on the columns (i.e., columns are what differentiate different IDs and various values of our predictor variables) and each new bootstrap result is a different row. Thus we use our index on the columns and take the row means, after exponentiating, in order to average out variability across IDs for a particular set of predictor values. Then we calculate percentiles for confidence intervals and add these back into our prediction dataset that was averaged across IDs.

```
dim(bootres$t)
```

```
## [1]    100 75600
```

```
for (i in 1:400) {
  ## find which indices to use
  ok <- which(preddat.boot$Index == i)

  ## now average across people
  tmp_avg <- rowMeans(exp(bootres$t[, ok]))

  ## lower confidence interval
  preddat.boot.avg[i,
    LL := quantile(tmp_avg, probs = .025, na.rm = TRUE)]
  preddat.boot.avg[i,
    UL := quantile(tmp_avg, probs = .975, na.rm = TRUE)]
}
```

Now that we have the parametric bootstrapped confidence intervals, we can plot a figure showing predicted number of awakenings by previous night awakenings, age, and whether participants were born in or out of Australia. The results are in Figure 12-8. The graph shows the decline in next night awakenings when people have higher than their own average number of awakenings. We can also see the main effect of age, with the "older" young adults having more awakenings predicted than the "younger" young adults. Also by colors, we see that those born in Australia tend to report more awakenings than those not born in Australia. The confidence intervals are not strictly symmetrical, which is normal for confidence intervals on the response scale (i.e., after exponentiating). There is some jaggedness in the confidence intervals. They would likely be more smooth had we generated 5,000 bootstrap samples instead of just 100.

```
ggplot(preddat.boot.avg, aes(WWASONsLag1, yhat,
  colour = factor(BornAUS), fill = factor(BornAUS))) +
  geom_ribbon(aes(ymin = LL, ymax = UL),
              alpha = .25, colour = NA) +
  geom_line(size = 1) +
  ylab("Predicted Awakenings") +
  xlab("Within person awakenings lag 1") +
  scale_color_viridis("Born in Australia", discrete = TRUE) +
```

```
scale_fill_viridis("Born in Australia", discrete = TRUE) +
theme(
   legend.position = "bottom",
   legend.key.width = unit(1, "cm")) +
facet_wrap(~ Age) +
coord_cartesian(
  xlim = c(-3, 3),
  ylim = c(0, 2.5),
  expand = FALSE)
```

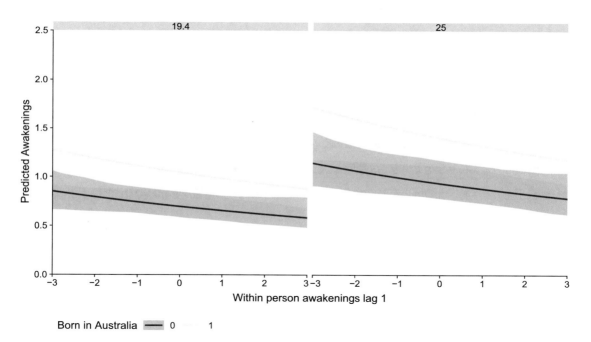

Figure 12-8. *Plot of predicted number of awakenings by previous night awakenings (relative to own average), separated by age (years) at the 20th and 80th percentiles (19.4y and 25y, respectively) and born in Australia (0 = no, 1 = yes). Bootstrap confidence intervals around the mean predicted count are shown through shading.*

12.4 Summary

This chapter builds on the linear mixed effects models from earlier chapters to mixed effects models for binary and count outcomes. Specifically, the chapter covers logistic mixed effects models and Poisson and negative binomial mixed effects models. The chapter also introduced unique challenges in generating predictions from GLMMs, specifically around how to account for the random effects in predictions on the original scale. Finally, we show how confidence intervals for predictions on the original scale can be computed accounting for random effects using bootstrapping. A summary of some of the key functions is in Table 12-1.

Table 12-1. *Listing of Key Functions Described in This Chapter and Summary of What They Do*

Function	What It Does
glmer()	Estimates generalized linear mixed effects models.
glmer.nb()	Estimates negative binomial generalized linear mixed effects models.
bootMer()	Bootstrap linear or generalized linear mixed effects models.
binomial()	Distribution family function for logistic generalized linear mixed models. Normally used with the canonical logit link.
poisson()	Distribution family function for Poisson generalized linear mixed models for count outcomes. Normally used with the canonical log link.
summary()	Provides a summary of the data input.
fixef()	Extracts fixed effects, takes model as argument.
coef()	Extracts model coefficients from a generalized linear mixed model. Note that unlike for single-level models, with mixed effects models fit using glmer() or lmer(), this function returns coefficients for each participant or cluster level that incorporate both fixed and random effects.
confint()	CONFidence INTerval generation function.
predict()	Similar to fitted(), except requires a data argument. In either case, model output yhat values are returned.
quantile()	Computes the quantiles of the given data.

CHAPTER 13

Modelling IIV

Up until this point, we have focused exclusively on statistical models of the location (or mean) of a distribution. This chapter focuses on something new, the scale or variability of a distribution. Specifically, this chapter introduces the concept of intra-individual variability (IIV), the variability within individual units across repeated assessments. Although a relatively niche area of study, IIV provides additional information about an individual unit and allows new types of research or practical questions to be evaluated, such as do people (schools, factories, etc.) with greater variability have different outcomes? This chapter makes use of the package varian which was developed by one of the authors specifically for variability analysis.

```
library(checkpoint)
checkpoint("2018-09-28", R.version = "3.5.1",
  project = book_directory,
  checkpointLocation = checkpoint_directory,
  scanForPackages = FALSE,
  scan.rnw.with.knitr = TRUE, use.knitr = TRUE)

library(knitr)
library(ggplot2)
library(cowplot)
library(viridis)
library(data.table)
library(JWileymisc)
library(varian)
library(mice)
library(parallel)

options(width = 70, digits = 2)
```

© Matt Wiley and Joshua F. Wiley 2019
M. Wiley and J. F. Wiley, *Advanced R Statistical Programming and Data Models*,
https://doi.org/10.1007/978-1-4842-2872-2_13

13.1 Conceptual Background

Bayesian Inference

An important note is on the use of Bayesian methods. Coverage of Bayesian thinking is beyond the scope of this book. However, there is an excellent book on Bayesian data analysis in extensive detail by Gelman and colleagues (2013) [36]. Although an in-depth knowledge of Bayesian methods is not required, a familiarity with the basics both theoretically and practically of how sampling is performed is necessary to fully utilize this chapter. We provide a *very* brief overview of a few aspects of Bayesian inference here, but readers who have not encountered these methods are strongly recommended to read other works.

Bayesian inferences use Bayes' rule to calculate the posterior distribution of the parameters, *p(parameters — data, prior)*, which is the probability of the parameter(s), conditioned on the data and the prior distribution. Markov chain Monte Carlo (MCMC) is used to draw samples from the posterior probability distribution of the parameters. Point estimates are obtained by summarizing the MCMC samples from the posterior distribution, such as by calculating the mean or median of the samples; uncertainty can be characterized by calculating the standard deviation or presenting the percentiles (e.g., the 2.5th and 97.5th percentiles for a 95% confidence interval; CI) of the MCMC samples. If p-values are desired, two-tailed empirical p-values can be calculated as two times the smaller of the proportion of samples falling above or below 0, that is: $2*min(prop(\theta \leq 0),$ $prop(\theta > 0))$. Convergence can be checked by calculating the percent scale reduction factors (PSRFs) [17] for each parameter in the model. PSRFs, also referred to as Rhats, estimate the percent scale reduction possible by running the MCMC chains longer. A value of one indicates convergence, although typically values sufficiently close to one are considered indicative of convergence (e.g., ¡ 1.1).

Bayesian inference relies on summarizing the posterior distribution. Therefore, in order to have stable summaries, it is important to have an adequate posterior sample size. However, if there is high autocorrelation in the samples, many posterior samples may not sufficiently characterize the entire posterior parameter distribution. Often, information about this can be drawn from the effective sample size for each parameter adjusted for autocorrelation estimates based on variograms and the multichain variance. It also is possible to evaluate the effective posterior sample size for each individual parameter. If the estimated posterior effective sample size is insufficient, it may indicate that additional iterations are required, or that some other approach such as rescaling the data, simplifying the model, or using stronger priors may be required.

What Is IIV?

In most applications, people focus on mean (also known as "location") differences. For example, common questions are whether two groups have different means, whether means can be predicted by other factors (i.e., most regression models), and how means change over time. However, researchers also have acknowledged that inter- and intra-individual variabilities (also known as "scale") are important [30]. For example, empirical research by Russell, Moskowitz, Zuroff, Sookman, and Paris (2007) [84] demonstrated that patients with borderline personality disorder, which is characterized by unstable relationships and affective instability [1], exhibited significantly higher intra-individual variability in affect than did a comparison group. In the context of aging and developmental processes, Ram and Gerstorf (2009) [77] overviewed important conceptual, methodological, and research design considerations for studying intra-individual variability. Conceptual and empirical interest IIV also exists in the field of sleep (e.g., Buysse et al., 2010 [18]; Suh et al., 2012 [91]), driven by findings that no two nights' sleep is the same and many people vary considerably in when they go to bed each night, how long they sleep, and the quality of that sleep. Thus although interest in scale and IIV is small compared to interest in understanding means, there is interest in IIV.

This chapter aims to introduce a rigorous and accessible method for estimating intra-individual variability. We also will highlight the utility of the estimated index of variability when intra-individual differences are of theoretical and empirical relevance. We will start by introducing some straightforward methods for quantifying IIV before introducing more complex statistical location and scale models.

Methods for Quantifying and Modelling Variability

Before introducing our primary Bayesian location scale model, we review some existing methods for quantifying and modelling variability.

Perhaps the most common measure of variability is the variance or standard deviation. Using the standard deviation (or squared, the variance), intra-individual variability is calculated as the standard deviation of observations for each subject, also known as individual standard deviation (ISD). After calculating the ISD, it can be used as an outcome or predictor in another statistical model. Because the ISD quantifies deviations from individual units' mean, systematic time effects (e.g., a linear increase over time) will increase the ISD. This may or may not be appropriate depending whether you are interested in capturing all variability within a person or only non-systematic variability.

For example, in sleep, seasonal variation in daylight influences individuals' sleep/ wake behaviors, and such variations may not be of direct research or clinical relevance. In these cases, the ISD may overestimate the type of intra-individual variability that is of interest. Such biases can be addressed by calculating the ISD on the residuals, after adjusting for time or other relevant factors (i.e., detrending).

Another traditional approach to quantifying variability is the root mean square of successive differences (RMSSD) [103]. Because the RMSSD is based on successive differences, it naturally removes the effects of systematic trends over time. For example, the following code shows the same data first ordered from smallest to largest and then randomized. Although the standard deviation does not change when the data are reordered, the RMSSD is much smaller in the ordered than unordered data. This highlights how with systematic trends (the ordered data) the RMSSD can be smaller than the standard deviation, but with unordered data it may be the same size or even greater.

```
## ordered
sd(c(1, 3, 5, 7, 9))

## [1] 3.2

rmssd(c(1, 3, 5, 7, 9))

## [1] 2

## randomized
sd(c(3, 1, 9, 5, 7))

## [1] 3.2

rmssd(c(3, 1, 9, 5, 7))

## [1] 4.7
```

Other measures of variability are the variance (i.e., ISD^2), mean square of successive differences (i.e., $RMSSD^2$), median absolute difference, the range, interquartile range, and coefficient of variation.

To get a better sense of how these various measures relate to each other, we will load the ACES data we worked with from the chapters on generalized linear mixed models (GLMMs). We also load the processed data we made and saved in the "GLMMs: Introduction" chapter. Next we define a function, `variability_measures()`, to make it easier to calculate all of these measures on data.

```
data(aces_daily)
draw <- as.data.table(aces_daily)
d <- readRDS("aces_daily_sim_processed.RDS")

variability_measures <- function(x) {
  x <- na.omit(x)
  list(
    SD = sd(x),
    VAR = sd(x)^2,
    RMSSD = rmssd(x),
    MSSD = rmssd(x)^2,
    MAD = median(abs(x - median(x))),
    RANGE = range(x),
    IQR = abs(diff(quantile(x, probs = c(.25, .75))))),
    CV = sd(x) / mean(x))
}
```

Now we can calculate the variability measures by participant ID to make a new dataset and then estimate and plot the correlation matrices for four different variables in the ACES data.

```
plot_grid(
  plot(SEMSummary(~ .,
    data = d[, variability_measures(PosAff), by = UserID][,-1]),
    order = "asis") +
    ggtitle("PosAff"),
  plot(SEMSummary(~ .,
    data = d[, variability_measures(NegAff), by = UserID][,-1]),
    order = "asis") +
    ggtitle("NegAff"),
  plot(SEMSummary(~ .,
    data = d[, variability_measures(COPEPrc), by = UserID][,-1]),
    order = "asis") +
    ggtitle("COPEPrc"),
  plot(SEMSummary(~ .,
    data = d[, variability_measures(SOLs), by = UserID][,-1]),
    order = "asis") +
    ggtitle("SOLs"),
ncol = 2)
```

591

The plots in Figure 13-1 help show empirically what also can be gleaned based on how they are calculated: the standard deviation (SD), variance (VAR), root mean squared successive differences (RMSSD), mean squared successive differences (MSSD), median absolute deviations (MAD), and interquartile range (IQR) all tend to be strongly correlated. The range and the coefficient of variation are more substantially different, in general.

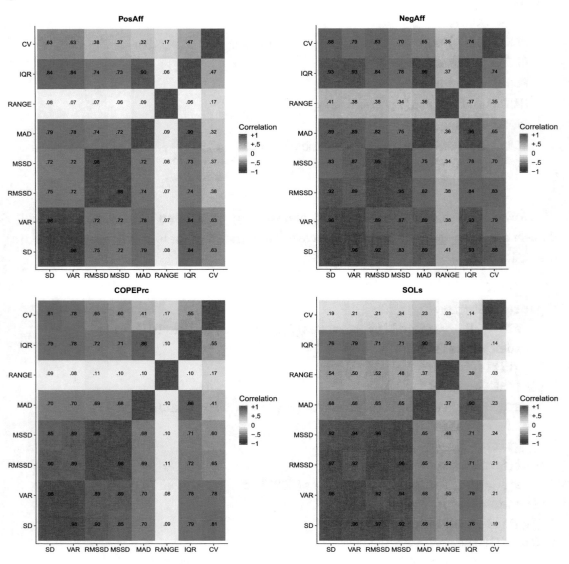

Figure 13-1. *Correlations between variability measures applied to individuals across four different measures*

However, all of these approaches are limited by measurement error. Although this criticism is true of any calculated statistic, in research focused on means, it is less of a practical issue. Means demonstrate good reliability, thus low measurement error, with few repeated measures. In contrast, the ISD has very poor reliability with few measures, particularly when the number of repeated observations is small and when individual difference in ISDs is small [31]. With analytically derived reliabilities of ISD and ISD2, as many as 50 repeated observations were required for a reliable questionnaire (reliability 0.9) to estimate intra-individual variability with reasonable reliability (reliability 0.8) [105]. In many cases, that many measurements may not be feasible to collect due to participant burden and cost.

The field of psychometrics has conducted extensive work into how to assess and account for measurement error. For example, latent variable models can be used to account for measurement error [13]. However, most statistical models are designed to test location (mean), not scale (variability) effects. One notable exception is for independent observations, where flexible generalized additive models (GAMs) for location and scale have been developed (see [79] and [89]). However, location scale GAMs are not applicable to repeated measures data (dependent observations) and thus to the study of intra-individual variability.

Technological advances in assessments, including mobile apps for self-reports and the explosion of wearables providing a wealth of intensive measurements, have resulted in a growing number of studies including many repeated measures per person. Subsequently, recent methodological literature has focused on developing new techniques for quantifying characteristics of individuals over time, including IIV.

Hedeker, Mermelstein, and Demirtas (2008) [42] developed mixed effects location and scale models in a maximum likelihood framework. Hedeker's models allow explanatory variables to predict both between- and within-subject factors, including random intercepts and random IIVs. It has been applied to analyze ecological momentary assessment data (for an excellent introduction to the application, see Hedeker, Mermelstein, and Demirtas, 2012) [43]. The model for continuous outcomes was extended by Hedeker, Demirtas, and Mermelstein (2009) [41] to ordinal data, and Li and Hedeker (2012) [56] extended the location scale model to three-level models.

Another approach is to first calculate the squared successive differences and then model these using a generalized linear mixed model. This approach uses successive differences and is therefore less sensitive to systematic intra-individual changes. The intercept, effectively the mean squared successive difference (RMSSD2), can be predicted by entering each squared successive difference into a generalized linear mixed model. One limitation of this approach is its ability to handle missing data. For example, if data are collected across three days, there are two successive differences, $t_2 - t_1$ and $t_3 - t_2$, and if the second day (t_2) is missing, both successive differences are undefined.

Wang, Hamaker, and Bergeman (2012) [106] proposed a Bayesian multilevel model that directly uses the raw data. This model incorporated both temporal dependency (captured via an autocorrelation coefficient) and magnitude of variability, whereas Hedeker and colleagues' (2008) [42] method models only the magnitude. Accounting for temporal dependence is particularly useful for many repeated measures. Autocorrelations also are sometimes referred to as inertia, as they indicate how difficult it is to change direction. Despite the benefits, accounting for individual difference in level (intercept), temporal dependence (autocorrelation), and variability (IIV) results in a more complex model that will tend to require a larger sample size. Wang and colleagues (2012) [106] observed that in over 200 participants, their method converged when using all 56 repeated measures per participant, but did not converge when using 7 or even 14 repeated measures per participant.

To summarize, unless there are many repeated measures (likely more than 50), simple methods to quantify variability such as the ISD or RMSSD are not optimal due to low reliability and an inability to account for measurement error. Explicit location scale models exist and are likely superior choices for fewer repeated measures. Even when there are many repeated measures, explicit location scale models have advantages in that they can begin separating sources of variability (e.g., by removing temporal dependence) [42, 106]. So far, we have focused on IIV as an outcome. In the next section, we introduce a Bayesian variability model that provides estimates of IIV and uses the IIV as a predictor.

Intra-individual Variability as a Predictor

Considering IIV as a predictor has essentially two aspects. First, we must obtain either a reliable measure of IIV or a measure of IIV along with an estimate of our uncertainty in the measure (i.e., accounting for the measurement error). Next, this measure can be used as a predictor in another model. The details of the second model are not particularly important as the IIV estimates may be entered into virtually any model.

The foundation of estimating IIVs is multilevel or mixed effects models. We have introduced these models in previous chapters. Mixed effects models are an ideal starting point because of their flexibility. Specifically, depending on the research question, there are different ways one might define IIV. For example, suppose that two people, A and B, were given an intervention and assessed weekly. The trajectories over time are shown in Figure 13-2.

```
iivdat <- data.table(
  Assessment = 0:15,
  PersonA = c(1, 3, 2, 4, 3, 5, 4, 6, 5, 7, 6, 8, 7, 9, 8, 10),
  PersonB = c(2, 5, 2, 6, 3, 7, 4, 8, 5, 9, 6, 10, 7, 11, 8, 12))

ggplot(iivdat, aes(Assessment)) +
  stat_smooth(aes(y = PersonA), method = "lm", se=FALSE,
              colour = viridis(2)[1], linetype = 2) +
  geom_line(aes(y = PersonA),
            colour = viridis(2)[1], size = 1) +
  stat_smooth(aes(y = PersonB), method = "lm", se=FALSE,
              colour = viridis(2)[2], linetype = 2) +
  geom_line(aes(y = PersonB),
            colour = viridis(2)[2], size = 1) +
  ylab("Outcome Scores")
```

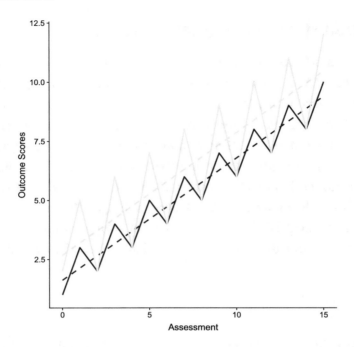

Figure 13-2. *Two hypothetical people given an intervention, Person A (purple) and Person B (yellow), both improve at about the same rate, but Person B is less consistent*

The figure shows that although both people respond about the same to the intervention, one improves more steadily than another. These differences are evident in the ISDs.

```
## ISD
sd(iivdat$PersonA)
```

```
## [1] 2.6
```

```
sd(iivdat$PersonB)
```

```
## [1] 3
```

However, without any adjustment, these ISDs incorporate variability in outcome scores due to systematic changes over time and the fluctuations around these systematic trends. If the research aims to use the total IIV, then these may be ideal. However, if the question surrounds how steady or variable the improvements over time are, then we should first remove these systematic changes. The results are shown as follows. With the systematic trends gone, the ISDs are much smaller than before.

```
## ISD, after removing systematic improvements
sd(resid(lm(PersonA ~ Assessment, data = iivdat)))
```

```
## [1] 0.77
```

```
sd(resid(lm(PersonB ~ Assessment, data = iivdat)))
```

```
## [1] 1.8
```

Although we do not advocate directly extracting and operating on the residuals, conceptually, this is the goal. Mixed effects models are an ideal choice as they allow the analyst the flexibility to adjust or not adjust for systematic trends over time. Mixed effects models also allow any other relevant predictors to be taken into account so that the desired source of IIV can be isolated. Depending on one's goals, this may involve not adding any predictors, adding time, or adding many different potential predictors into the model and modelling IIV on the residuals.

For practical reasons, it is convenient to estimate the model using Markov chain Monte Carlo (MCMC) simulation in a Bayesian framework. A Bayesian framework is helpful because

- Bayesian methods allow flexibility in specifying the model and distributions without needing to explicitly derive the likelihood function

- Maximum likelihood methods are prone to difficulty in convergence when estimating IIVs, whereas Bayesian methods may be slow, but typically if run long enough will mix and converge

- Through MCMC simulation, the multiple draws from the posterior distribution allow the uncertainty in estimation of IIVs to be captured as a distribution of probable estimates sampled, rather than a single "best estimate"

The next section introduces a Bayesian variability model (BVM) in more detail.

Bayesian Variability Model

We have introduced mixed effects models in previous chapters. If those are unfamiliar, it would be best to review earlier chapters where location-only mixed effects models are explained prior to continuing. A more comprehensive presentation and evaluation of the model is presented in an earlier technical report of ours [116].

For reference, let Y be a between-person outcome, and V a within-person variable, such that V is measured repeatedly on each person. IIVs will be estimated from V and used to predict their score on Y. Additionally, the model can estimate each individual's mean of V as an additional predictor of Y. In the literature on IIV, we have argued that it is important to at least statistically adjust for the mean of a variable when testing whether IIV predicts an outcome to demonstrate the added value of examining IIV [116, 8, 6, 7]. First, consider a multilevel model for V; for simplicity and without loss of generality, we start with an unconditional model:

$$V_{ij} \sim \mathrm{N}\left(\mu_j, \sigma\right) \tag{13.1}$$

In this example, V_{ij} is the value for the ith ($i = 1, 2, ..., I_j$) assessment for the jth ($j = 1, 2, ..., N$) subject. Each individual has their own estimated mean, μ_j, and the distribution of means is assumed to follow a normal distribution:

$$\mu_j \sim N\left(\mu_\mu, \sigma_\mu\right)$$

This basic mixed effects model has a single estimate of the residual variability that is assumed to apply to everyone. We can extend this model by allowing the standard deviation to also vary by subjects. The extended model still assumes a normal distribution, but now allows both the location, μ, and the residual standard deviation, σ, to vary. The new model is specified as

$$V_{ij} \sim N\left(\mu_j, \sigma_j\right) \tag{13.2}$$

This equation is the same as before, but μ_j represents individual standard deviations or, in the case of a conditional model, individual residual standard deviations. Estimating ISDs based on observed residuals cannot account for the fact that uncertainty in model estimates may lead to differences in the residuals. However, as part of a model, the uncertainty in the location parameters will propagate to the residuals and thus into the ISD estimates.

Like the individual means, which are assumed to come from a normal distribution, the individual (residual) standard deviations are assumed to come from a distribution, specifically a gamma distribution with scale and shape parameters α and β:

$$\sigma_j \sim -\left(\alpha, \beta\right) \tag{13.3}$$

Estimates of the individual (residual) standard deviations, σ_j, then are used as predictors of the outcome, Y, using any standard model. As an example, we could use a multiple linear regression, that is

$$Y_j \sim N\left(\mu 2_j, \sigma\right) \tag{13.4}$$

In this model, the predicted value, μ, is

$$\mu 2_j = \beta_0 + \beta_1 Covariate_1 + \ldots + \beta_k Covariate_k + \alpha_1 \sigma_j + \alpha_2 \mu_j \tag{13.5}$$

To highlight the difference between any other predictors or covariates in the model and the individual means and ISDs from the mixed effects model, we use a separate parameter vector. We indicate β for the regular predictors and α for the latent means and ISDs.

Note that although we used linear regression here, we could easily replace a linear regression model with nearly any statistical model.

The ISD approach for estimating IIVs has been criticized as it does not take the order of observations into account. However, this is addressed by estimating IIVs in the context of a mixed effects model. As part of the mixed effects model, systematic trends over time and other relevant variables can be added as predictors. By including lagged outcome measures, it is possible to include auto-regressive effects as well. This provides the flexibility to have the ISD include all the variability or only the parts of interest. Additionally by including the detrending in a statistical model, uncertainty is again captured and propagated to the IIV estimates.

Software Implementation: VARIAN

The Bayesian variability model (BVM) can be estimated in any general purpose Bayesian framework, such as JAGS, BUGS, or Stan. We will use Stan [35, 19], a general purpose programming language for Bayesian inference using Markov chain Monte Carlo (MCMC) and sampling using the No-U-Turn Sampler which is an extension of Hamiltonian Monte Carlo (Hoffman & Gelman, 2014) [44].

Although it is possible, and most flexible, to specify each model manually in Stan, to make it easier for analysts less familiar with Bayesian methods, we also will make use of the R package, `varian` [116], available from CRAN or at GitHub: `https://github.com/ElkhartGroup/VARIAN`. `varian` links to Stan and allows the BVM to be estimated in just a few lines of code. At the moment, `varian` only supports modelling variability of continuous, normally distributed variables.

In `varian` by default the priors are weakly informative, assuming the standard deviation of variables is approximately ≤ 10. Specifically, means and regression coefficients use a normal prior with mean zero and standard deviation of 1,000. The scale and shape parameters for the gamma distribution and the residual variance from the second stage outcome use the half-Cauchy prior, which has been recommended as a better weakly informative prior than either the uniform or inverse-gamma families for variance components [34, 74]. Specifically, `varian` uses location and scale parameters of zero and ten, respectively. Depending on the scale of the data, it may be necessary to specify alternate parameters for the priors or to rescale the data in order for the default priors to be weakly informative.

Convergence is estimated using the percent scale reduction factor (PSRF). Because each individual estimate of σ_j and μ_j is a parameter, there are many individual PSRFs. In `varian` as a diagnostic plot, we present a histogram of the Rhats to allow easy visual inspection of convergence for all parameters. A simulation study suggests that the BVM as described and implemented in `varian` produces minimally biased estimates as long as there are five or more repeated measures per person.

13.2 R Examples
IIV Predicting a Continuous Outcome

To use the BVM described earlier to predict a continuous outcome, we can use the `varian()` function. The `varian` package and `Stan` compile models in C++. Thus for these to work, you will need to have a C++ compiler installed on your system and accessible to R. On machines running Windows OS, the easiest way to do this is to install R tools, available from `https://cran.r-project.org/bin/windows/Rtools/`. On machines running Mac OS, the most common way to get the necessary compilers is to install `Xcode` from the app store. On different variants of linux or unix, it should be straightforward to use their package managers to install GCC or another C++ compiler. If you have errors when the models compile, try updating R and your compiler (either by installing the latest version of R tools or `Xcode` or specific compilers). Finally, note that during compilation, it is normal for there to be a number of messages and warnings. If there are no errors, the model is typically okay, and most of the messages and warnings can be ignored.

The `varian()` function takes several arguments. First, the model formula for the between-person outcome is specified to the `y.formula` argument. The IIV estimates are automatically included, so all that need be specified is the outcome variable and any additional covariates or predictors. Next, the formula for the IIVs is specified to `v.formula`. Because the IIVs require a repeated measures outcome, there must be an ID variable. The data are specified, as usual. In addition, we specify the `design`, which indicates whether IIVs alone should be estimated, they should predict an outcome, or they should predict a mediator and an outcome. In the following example model, we estimate IIVs and use them to predict one continuous outcome. Additionally, there are several arguments pertaining to the MCMC sampling, including the total number of iterations, `totaliter`; the number of warmup iterations, `warmup`; the thinning interval, `thin`; and the number of independent chains to use, `chains`. To make the example run faster, we use a relatively low number of iterations and a low thin. To ensure good convergence and stable estimates in final models, one would likely choose a larger value so that the ultimate effective sample size was several thousand. In this instance, we test whether the IIV of positive affect predicts average negative affect beyond average positive affect. That is, does instability of positive affect uniquely predict someone's typical level of negative affect?

```
cl <- makeCluster(2)
clusterExport(cl, c("book_directory", "checkpoint_directory" ))

clusterEvalQ(cl, {
  library(checkpoint)
  checkpoint("2018-09-28", R.version = "3.5.1",
  project = book_directory,
  checkpointLocation = checkpoint_directory,
  scanForPackages = FALSE,
  scan.rnw.with.knitr = TRUE, use.knitr = TRUE)

  library(varian)
})
## [[1]]
##  [1] "varian"       "rstan"       "StanHeaders"  "ggplot2"
##  [5] "checkpoint"   "RevoUtils"   "stats"        "graphics"
##  [9] "grDevices"    "utils"       "datasets"     "RevoUtilsMath"
## [13] "methods"      "base"
##
## [[2]]
```

```
##  [1] "varian"       "rstan"        "StanHeaders"  "ggplot2"
##  [5] "checkpoint"   "RevoUtils"    "stats"        "graphics"
##  [9] "grDevices"    "utils"        "datasets"     "RevoUtilsMath"
## [13] "methods"      "base"

system.time(m <- varian(
  y.formula = BNegAff ~ 1,
  v.formula = PosAff ~ 1 | UserID,
  data = d,
  design = "V -> Y",
  useU = TRUE,
  totaliter = 10000,
  warmup = 500, thin = 5,
  chains = 2, verbose=TRUE,
  cl = cl))

##    user  system elapsed
##     1.3     1.5   510.2
```

Prior to examining the model estimates, we can examine some basic model convergence diagnostics using the vm_diagnostics() function. The results are shown in Figure 13-3. The range of Rhat values suggests good convergence. However, the effective sample size for all parameters varies and some are relatively low. For now we will proceed, but in practice one may desire to increase iterations and use stronger priors or other means to ensure the smallest effective sample size is larger.

```
## check diagnostics
vm_diagnostics(m)
```

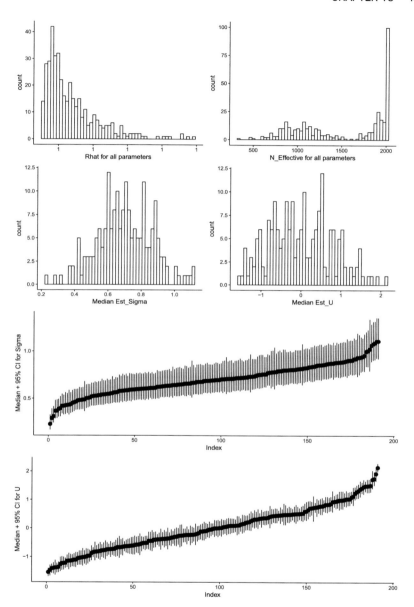

Figure 13-3. *Diagnostics including percent scale reduction factor (Rhat), effective sample size, distribution of the individual standard deviations, distribution of the individual means, and individual estimates of individual standard deviations and means with credible intervals.*

Although there could be other parameters of interest, often the main interest surrounds whether the IIVs indeed predict the outcome. To examine this, we extract the MCMC samples and plot them using `vmp_plot()`. We plot specifically `Yalpha` which

603

is the *alpha* coefficient vector for the outcome, Y. The results are in Figure 13-4. The graphs show the distribution of the individual ISDs, the individual means, and their joint distribution through a scatter plot. The bar graphs show the proportion of MCMC samples that fall above and below 0, and these are used to generate p-values. In this instance, both the IIV and individual means are significant predictors, but in opposite directions. Higher average positive affect predicts significantly lower negative affect. However, independent of mean positive affect, people with less stable positive affect (i.e., more variable) are expected to have higher typical negative affect.

```
## extract MCMC samples
mcmc.samples <- extract(m$results,
  permute = TRUE)

## examine MCMC samples of
## the alpha regression coefficients
vmp_plot(mcmc.samples$Yalpha)
```

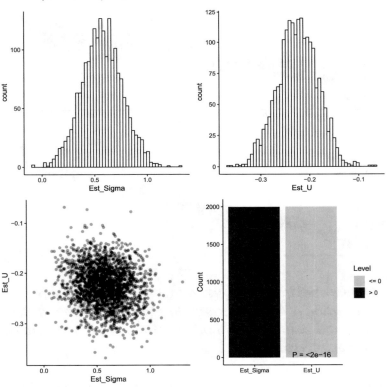

Figure 13-4. *Plots of the distributions, bivariate scatter plot, and proportion of cases on each side of zero for empirical p-values*

Finally, we can get summaries of any parameter we might want using the `param_summary()` function. The following code shows summaries for each component: the IIV model and the model predicting average negative affect. The results show that the intercept for average negative affect is quite low, near one (on a one to five possible range). The results for IIVs indicate that higher IIV of positive affect is associated with higher negative affect. Your results may be somewhat different from this book due to random sampling inherent in MCMC sampling, which particularly impacts results when the effective sample size is small. However, in these examples, a one unit higher IIV is associated with about half a point higher average negative affect. We can also see the results for the average level of positive affect, which has a negative association with average negative affect: the more positive people are on average, the less negative they are. Finally, for average negative affect, we can get a summary of the residual error, not explained by either positive affect IIV or positive affect individual means.

```
## intercept of average negative affect
param_summary(mcmc.samples$YB[, 1])

##    Mean Median   SE LL2.5 UL97.5 p-value
## 1   1.1    1.1 0.14  0.87    1.4  < .001

## IIV on average negative affect
param_summary(mcmc.samples$Yalpha[, 1])

##   Mean Median   SE LL2.5 UL97.5 p-value
## 1 0.57   0.57 0.19  0.19   0.95    .002

## individual mean on average negative affect
param_summary(mcmc.samples$Yalpha[, 2])

##     Mean Median   SE LL2.5 UL97.5 p-value
## 1 -0.22  -0.22 0.04 -0.31  -0.15  < .001

## residual error of average negative affect
param_summary(mcmc.samples$sigma_Y)

##   Mean Median   SE LL2.5 UL97.5 p-value
## 1 0.43   0.43 0.02  0.38   0.48  < .001
```

We also can get parameter summaries for the IIV model, including the intercept, effects of any predictors included, and summaries of the random effects and gamma distribution parameters for the IIVs. These are shown in the following.

```
## intercept of positive affect
param_summary(mcmc.samples$VB[, 1])

##    Mean Median   SE LL2.5 UL97.5 p-value
## 1   2.7    2.7 0.06   2.6    2.8  < .001

## positive affect random intercept standard deviation
param_summary(mcmc.samples$sigma_U)

##    Mean Median   SE LL2.5 UL97.5 p-value
## 1   0.8    0.8 0.04  0.73   0.89  < .001

## estimate of the gamma rate parameter for IIVs
param_summary(mcmc.samples$rate)

##    Mean Median  SE LL2.5 UL97.5 p-value
## 1    19     19 2.4    15     24  < .001

## estimate of the gamma shape parameter for IIVs
param_summary(mcmc.samples$shape)

##    Mean Median  SE LL2.5 UL97.5 p-value
## 1    14     13 1.6    11     17  < .001
```

Finally, although not needed if the goal is to use the IIV estimates directly through the varian() function, to use the estimates in other models, they must first be extracted. Extracting the Bayesian estimates of the IIVs can be helpful to use them in a wide variety of models. For example, they could be used as predictors in a generalized additive model, machine learning models, etc. The IIVs are named Sigma_V. This will be a matrix with one column for each unique ID in the dataset that is not missing, and one row for each MCMC sample, after thinning. In this case, it is a 2,000 x 191 dimension matrix. When used as part of one Bayesian model, the uncertainty in IIV estimates is automatically propagated to uncertainty in later parameter estimates. If the results are extracted, however, care must be taken to ensure that the uncertainty is still propagated. One way to do this is to treat the MCMC samples as multiple imputations of a missing variable. In essence, we can think of the IIVs and individual means as missing values because we do not observe them. Our model imputes our best estimates, but it does so with some error. Note that the individual means are also available, named U.

```
dim(mcmc.samples$Sigma_V)
```

```
## [1] 2000   191
```

```
str(mcmc.samples$Sigma_V)
```

```
##   num [1:2000, 1:191] 0.558 0.478 0.525 0.6 0.44 ...
##   - attr(*, "dimnames")=List of 2
##    ..$ iterations: NULL
##    ..$           : NULL
```

The simplest, but not optimal, way of using the IIV estimates is to simply average across the MCMC samples. The following code extracts these, averages them, merges with average negative affect, and estimates a regression model. The results are similar, but rather different, both in the average estimates and in the confidence intervals, to those from the one-step Bayesian model.

```
avg_dataset <- cbind(
  d[!duplicated(UserID), .(BNegAff)],
  IIV = colMeans(mcmc.samples$Sigma_V),
  IIM = colMeans(mcmc.samples$U))

avg_model <- lm(BNegAff ~ IIV + IIM, data = avg_dataset)

summary(avg_model)
```

```
##
## Call:
## lm(formula = BNegAff ~ IIV + IIM, data = avg_dataset)
##
## Residuals:
##     Min      1Q  Median      3Q     Max
## -0.7862 -0.2762 -0.0538  0.1933  1.5671
##
## Coefficients:
##              Estimate Std. Error t value Pr(>|t|)
## (Intercept)    1.0518     0.1334    7.88  2.5e-13 ***
```

```
## IIV              0.7119      0.1841      3.87   0.00015 ***
## IIM             -0.2321      0.0389     -5.97   1.2e-08 ***
## ---
## Signif. codes:  0 '***' 0.001 '**' 0.01 '*' 0.05 '.' 0.1 '_' 1
##
## Residual standard error: 0.42 on 188 degrees of freedom
## Multiple R-squared:  0.2,    Adjusted R-squared:  0.192
## F-statistic: 23.5 on 2 and 188 DF,  p-value: 7.72e-10
```

Next, we can make many datasets by using the different MCMC samples. We do not use all 1,000, but take only every 10. On each of these, we can estimate a regression model and then combine and pool the results by converting to a multiply imputed analysis object using the as.mira() function and then combining the results using the pool() function. Working with multiply imputed data is covered in greater detail in our chapter on missing data.

```
ind_dataset <- lapply(seq(1, 1000, by = 10), function(i) {
  cbind(
  d[!duplicated(UserID), .(BNegAff)],
  IIV = mcmc.samples$Sigma_V[i, ],
  IIM = mcmc.samples$U[i, ])
})

ind_model <- lapply(ind_dataset, function(tmpdat) {
  lm(BNegAff ~ IIV + IIM, data = tmpdat)
})

ind_model_pooled <- pool(as.mira(ind_model))
```

As one final comparison, we can fit a model using simple ISD estimates.

```
raw_model <- lm(BNegAff ~ IIV + IIM,
 data = d[, .(BNegAff = BNegAff[1],
             IIV = sd(PosAff, na.rm = TRUE),
             IIM = mean(PosAff, na.rm = TRUE)),
         by = UserID])
```

For comparison, we show the results from the Bayesian model and calculate the regression coefficients and confidence intervals from the model that uses the average IIV estimates, the raw ISD model, and the model that treats them as multiply imputed. Comparing across these different models, we can see that although none of them match exactly, the results are much closer for the Bayesian model and treating the estimates as multiply imputed.

```
## Bayesian Results
param_summary(mcmc.samples$YB[, 1]) ## intercept

##    Mean Median   SE LL2.5 UL97.5 p-value
## 1   1.1    1.1 0.14  0.87    1.4  < .001

param_summary(mcmc.samples$Yalpha[, 1]) ## IIV

##    Mean Median   SE LL2.5 UL97.5 p-value
## 1 0.57   0.57 0.19  0.19   0.95    .002

param_summary(mcmc.samples$Yalpha[, 2]) ## IIM

##     Mean Median   SE LL2.5 UL97.5 p-value
## 1 -0.22  -0.22 0.04 -0.31  -0.15  < .001

## using averages only
cbind(B = coef(avg_model), confint(avg_model))

##                   B 2.5 % 97.5 %
## (Intercept)    1.05  0.79   1.31
## IIV            0.71  0.35   1.08
## IIM           -0.23 -0.31  -0.16

## using raw ISDs
cbind(B = coef(raw_model), confint(raw_model))

##                   B 2.5 % 97.5 %
## (Intercept)    1.82  1.52   2.11
## IIV            0.46  0.15   0.77
## IIM           -0.22 -0.29  -0.14
```

```
## treating as multiply imputed
summary(ind_model_pooled, conf.int = TRUE)

##              estimate std.error statistic  df p.value 2.5 % 97.5 %
## (Intercept)     1.14      0.14       8.2 129 5.3e-14  0.87   1.42
## IIV             0.58      0.19       3.0 127 3.0e-03  0.20   0.96
## IIM            -0.22      0.04      -5.6 175 7.2e-08 -0.30  -0.15
```

These examples show how the Bayesian method can help to estimate IIVs. Further, they highlight the bias that arises from calculating raw ISDs or from treating IIV estimates as measured without error. These results are all the more notable as each participant had on average just over 30 positive affect assessments available. Thus, the reliability of the ISDs would be much higher than if, say, only 5 or 14 assessments were available. Nevertheless, there were large differences in the estimates and uncertainty estimates for each model, with only the single-step Bayesian solution and the one treating the IIV estimates as multiply imputed yielding fairly close results.

13.3 Summary

This chapter introduced the concept of intra-individual variability (IIV) and how IIV can provide an additional dimension of information about repeated measures. The chapter summarized limitations of calculating individual standard deviations and other simple methods of quantifying IIV. It also introduced a Bayesian variability model (BVM) that uses mixed effects models to control for any variables of interest, such as time effects and temporal dependence, and calculates IIVs on the residuals. By doing this in one step and in a Bayesian framework, the BVM can accommodate uncertainty in IIV estimates, and this reduces bias in the results and provides more accurate statistical inference. A summary of functions used in the chapter is in Table 13-1.

Table 13-1. *Listing of Key Functions Described in This Chapter and Summary of What They Do*

Function	What It Does
sd()	Returns the standard deviation of sample data
rmssd()	Returns the root mean square of successive differences of sample data
varian()	Estimates a Bayesian variability model based on a location scale mixed effects model estimated in a Bayesian framework. Can be used to have intra-individual variabilities (IIVs) predict other outcomes or simply to estimate IIVs accounting for measurement error and extract them
vm_ diagnostics()	Plots diagnostics for the Bayesian variability model estimated using the varian() function
extract()	Extracts the Markov chain Monte Carlo samples for plotting or making summaries or using in other models
vmp_plot()	Plots parameters from a Bayesian variability model to view their individual and joint distributions
param_ summary()	Creates a summary of parameters from a Bayesian variability model including the average, median, credible intervals, and empirical p-values
as.mira()	Converts a list of model results to a multiply imputed object class to allow models to be pooled. Used in IIV analyses when multiple IIV estimates are extracted and treated as multiply imputed values
pool()	Pools model results from a list of analyses repeated on different multiply imputed datasets

Bibliography

[1] American Psychiatric Association. *The Diagnostic and Statistical Manual of Mental Disorders: DSM-5*. 2013.

[2] The World Bank. *Gender Statistics*, July 30, 2018. The Gender Statistics database is a comprehensive source for the latest sex-disaggregated data and gender statistics covering demography, education, health, access to economic opportunities, public life and decision-making, and agency. licencsed under CC-BY 4.0.

[3] J. W. Bartlett, S. R. Seaman, I. R. White, and J. R. Carpenter. Multiple imputation of covariates by fully conditional specification: Accommodating the substantive model. *Statistical Methods in Medical Research*, 24(4):462–487, 2015.

[4] D. Bates and D. Eddelbuettel. Fast and elegant numerical linear algebra using the RcppEigen package. *Journal of Statistical Software*, 52(5):1–24, 2013.

[5] B. Bei, J. F. Wiley, N. B. Allen, and J. Trinder. A cognitive vulnerability model of sleep and mood in adolescents under restricted and extended sleep opportunities. *Sleep*, 38(3):453–461, 2015.

[6] Bei Bei, Rachel Manber, Nicholas B Allen, John Trinder, and Joshua F Wiley. Too long, too short, or too variable? sleep intraindividual variability and its associations with perceived sleep quality and mood in adolescents during naturalistically unconstrained sleep. *Sleep*, 40(2), 2017.

[7] Bei Bei, Teresa E Seeman, Judith E Carroll, and Joshua F Wiley. Sleep and physiological dysregulation: a closer look at sleep intraindividual variability. *Sleep*, 40(9), 2017.

[8]   Bei Bei, Joshua F Wiley, John Trinder, and Rachel Manber. Beyond the mean: A systematic review on the correlates of daily intraindividual variability of sleep/wake patterns. *Sleep medicine reviews*, 28:108–124, 2016.

[9]   Henrik Bengtsson. *matrixStats: Functions that Apply to Rows and Columns of Matrices (and to Vectors)*, 2018. R package version 0.54.0.

[10] Christoph Bergmeir and Jos e´ M. Ben´ıtez. Neural networks in R using the stuttgart neural network simulator: RSNNS. *Journal of Statistical Software*, 46(7):1–26, 2012.

[11] Przemyslaw Biecek. *DALEX: Descriptive mAchine Learning EXplanations*, 2018. R package version 0.2.3.

[12] Roger Bivand, Jan Hauke, and Tomasz Kossowski. Computing the jacobian in gaussian spatial autoregressive models: An illustrated comparison of available methods. *Geographical Analysis*, 45(2):150–179, 2013.

[13] K. A Bollen. *Structural Equations with Latent Variables*. Wiley, 1989.

[14] J. E. Bower, J. F. Wiley, L Petersen, M. Irwin, S. Cole, and P. Ganz. Fatigue after breast cancer treatment: Biobehavioral predictors of fatigue trajectories. *Health Psychology*, in press.

[15] Patrick Breheny and Woodrow Burchett. *visreg: Visualization of Regression Models*, 2017. R package version 2.4-1.

[16] L. Breiman. Random forests. *Machine Learning*, 45(1):5–32, 2001.

[17] Stephen P Brooks and Andrew Gelman. General methods for monitoring convergence of iterative simulations. *Journal of computational and graphical statistics*, 7(4):434–455, 1998.

[18] D. J. Buysse, Y. Cheng, A. Germain, D. E. Moul, P. L. Franzen, M. Fletcher, and T. H. Monk. Night-to-night sleep variability in older adults with and without chronic insomnia. *Sleep medicine*, 11(1):56–64, 2010.

[19] Bob Carpenter, Andrew Gelman, Matthew D Hoffman, Daniel
      Lee, Ben Goodrich, Michael Betancourt, Marcus Brubaker,
      Jiqiang Guo, Peter Li, and Allen Riddell. Stan: A probabilistic
      programming language. *Journal of statistical software*, 76(1), 2017.

[20] W. Chang. *R Graphics Cookbook*. O'Reilly Media, Inc., Sebastopol,
      CA, 2012.

[21] W. S. Cleveland. Robust locally weighted regression and
      smoothing scatterplots. *Journal of the American Statistical
      Association*, 74(368):829–836, 1979.

[22] Stephen R Cole and Miguel A Hern a´n. Constructing inverse
      probability weights for marginal structural models. *American
      journal of epidemiology*, 168(6):656–664, 2008.

[23] Microsoft Corporation. *checkpoint: Install Packages from
      Snapshots on the Checkpoint Server for Reproducibility*, 2017. R
      package version 0.4.3.

[24] Microsoft Corporation and Steve Weston. *doParallel: Foreach
      Parallel Adaptor for the 'parallel' Package*. R package version 1.0.12.

[25] H. A. David. Gini's mean difference rediscovered. *Biometrika*,
      pages 573–575, 1968.

[26] A. C. Davison and D. V. Hinkley. *Bootstrap Methods and Their
      Applications*. Cambridge University Press, Cambridge, 1997.
      ISBN 0-521-57391-2.

[27] R. de Jong, S. van Buuren, and M. Spiess. Multiple imputation
      of predictor variables using generalized additive models.
      Communications in Statistics - Simulation and Computation,
      45(3):968–985, 2016.

[28] Lisa L Doove, Stef Van Buuren, and Elise Dusseldorp. Recursive
      partitioning for missing data imputation in the presence of
      interaction effects. *Computational Statistics & Data Analysis*,
      72:92–104, 2014.

[29] M. Dowle and A. Srinivasan. *data.table: Extension of 'data.frame'*,
      2016. R package version 1.10.0.

[30] Dara R. Eizenman, John R. Nesselroade, David L. Featherman, and John W. Rowe. Intraindividual variability in perceived control in a older sample: The macarthur successful aging studies. *Psychology and Aging*, 12(3):489–502, 1997.

[31] R. Estabrook, K. J. Grimm, and R. P. Bowles. A monte carlo simulation study of the reliability of intraindividual variability. *Psychology and Aging*, 27(3):560, 2012.

[32] Max Kuhn. Contributions from Jed Wing, Steve Weston, Andre Williams, Chris Keefer, Allan Engelhardt, Tony Cooper, Zachary Mayer, Brenton Kenkel, the R Core Team, Michael Benesty, Reynald Lescarbeau, Andrew Ziem, Luca Scrucca, Yuan Tang, Can Candan, and Tyler Hunt. *caret: Classification and Regression Training*, 2018. R package version 6.0-80.

[33] Simon Garnier. *viridis: Default Color Maps from 'matplotlib'*, 2018. R package version 0.5.1.

[34] Andrew Gelman et al. Prior distributions for variance parameters in hierarchical models (comment on article by browne and draper). *Bayesian analysis*, 1(3):515–534, 2006.

[35] Andrew Gelman, Daniel Lee, and Jiqiang Guo. Stan: A probabilistic programming language for bayesian inference and optimization. *Journal of Educational and Behavioral Statistics*, 40(5):530–543, 2015.

[36] Andrew Gelman, Hal S Stern, John B Carlin, David B Dunson, Aki Vehtari, and Donald B Rubin. *Bayesian data analysis*. Chapman and Hall/CRC, 2013.

[37] H. Goldstein, J. R. Carpenter, and W. J. Browne. Fitting multilevel multivariate models with missing data in responses and covariates that may include interactions and non-linear terms. *Journal of the Royal Statistical Society: Series A (Statistics in Society)*, 177(2): 553–564, 2014.

[38] J. W. Graham. Missing data analysis: Making it work in the real world. *Annual Review of Psychology*, 60:549– 576, 2009.

[39]   Peter J Green and Bernard W Silverman. *Nonparametric regression and generalized linear models: a roughness penalty approach.* CRC Press, 1993.

[40]   T. J. Hastie and R. J. Tibshirani. *Generalized Additive Models*, volume 43. CRC Press, Boca Raton, FL, 1990.

[41]   D. Hedeker, H. Demirtas, and R. J. Mermelstein. A mixed ordinal location scale model for analysis of ecological momentary assessment (ema) data. *Statistics and Its Interface*, 2(4):391–401, 2009.

[42]   D. Hedeker, R. J. Mermelstein, and H. Demirtas. An application of a mixed-effects location scale model for analysis of ecological momentary assessment (ema) data. *Biometrics*, 64(2):627–634, 2008.

[43]   D. Hedeker, R. J. Mermelstein, and H. Demirtas. Modeling between-subject and within-subject variances in ecological momentary assessment data using mixed-effects location scale models. *Statistics in Medicine*, 31(27):3328–3336, 2012.

[44]   Matthew D Hoffman and Andrew Gelman. The no-u-turn sampler: adaptively setting path lengths in hamiltonian monte carlo. *Journal of Machine Learning Research*, 15(1):1593–1623, 2014.

[45]   J. S. House. *Americans' Changing Lives: Waves I, II, III, IV, and V, 1986, 1989, 1994, 2002, and 2011*, ICPSR04690-v7; 2014-09-09. Ann Arbor, MI: Inter-university Consortium for Political and Social Research [distributor].

[46]   R. A. Hughes, I. R. White, S. R. Seaman, J. R. Carpenter, K. Tilling, and J. A. C. Sterne. Joint modelling rationale for chained equations. *BMC Medical Research Methodology*, 14(1):28, 2014.

[47]   G. James, D. Witten, T. Hastie, and Tibshirani R. *An Introduction to Statistical Learning.* Springer, New York, NY, 1st edition, 2013.

[48]   P. C. D. Johnson. Extension of nakagawa & schielzeth's r2glmm to random slopes models. *Methods in Ecology and Evolution*, 5(9):944–946, 2014.

[49]   Alexandros Karatzoglou, Alex Smola, Kurt Hornik, and Achim
        Zeileis. kernlab – an S4 package for kernel methods in R. *Journal
        of Statistical Software*, 11(9):1–20, 2004.

[50]   R. Koenker. *Quantile Regression*. Cambridge University Press,
        New York, NY, 2005.

[51]   R. Koenker and G. Bassett Jr. Regression quantiles. *Econometrica*,
        pages 33–50, 1978.

[52]   Max Kuhn and Hadley Wickham. *rsample: General Resampling
        Infrastructure*, 2017. R package version 0.0.2.

[53]   V. Kuperman, Z. Estes, M. Brysbaert, and A. B. Warriner. Emotion
        and language: valence and arousal affect word recognition.
        Journal of Experimental Psychology: General, 143(3):1065, 2014.

[54]   Alexandra Kuznetsova, Per B Brockhoff, and Rune HB
        Christensen. lmertest package: Tests in linear mixed effects
        models. *Journal of Statistical Software*, 82(13):1–26, 2017.

[55]   O. Langsrud. Anova for unbalanced data: Use type ii instead of type
        iii sums of squares. *Statistics and Computing*, 13(2):163–167, 2003.

[56]   X. Li and D. Hedeker. A three-level mixed-effects location scale
        model with an application to ecological momentary assessment
        data. *Statistics in Medicine*, 31(26):3192–3210, 2012.

[57]   X. Lin and D. Zhang. Inference in generalized additive mixed
        modelsby using smoothing splines. *Journal of the Royal Statistical
        Society: Series B (Statistical Methodology)*, 61(2):381–400, 1999.

[58]   Roderick J. A. Little and Donald B. Rubin. *Statistical Analysis with
        Missing Data*, volume 2. Wiley, Hoboken, New Jersey, 2002.

[59]   M. Maechler, P. J. Rousseeuw, C. Croux, V. Todorov, A. Ruckstuhl,
        M. Salibian-Barrera, T. Verbeke, M. Koller, E. L. T. Conceicao, and
        M. A. di Palma. *robustbase: Basic Robust Statistics*, 2016. R package
        version 0.92-7.

[60]   P. C. Mahalanobis. On the generalized distance in statistics.
        Proceedings of the National Institute of Sciences, 2:49–55, 1936.

[61] P. McCullagh and J. A. Nelder. *Generalized Linear Models.* Monograph on Statistics and Applied Probability. Chapman and Hall/CRC, Boca Raton, Florida, 2nd edition, 1989.

[62] David Meyer, Evgenia Dimitriadou, Kurt Hornik, Andreas Weingessel, and Friedrich Leisch. *e1071: Misc Functions of the Department of Statistics, Probability Theory Group (Formerly: E1071), TU Wien*, 2018. R package version 1.7-0.

[63] Microsoft and Steve Weston. *foreach: Provides Foreach Looping Construct for R*. R package version 1.4.5.

[64] K. Mohan and J. Pearl. Graphical models for recovering probabilistic and causal queries from missing data. In Z. Ghahramani, M. Welling, C. Cortes, N. D. Lawrence, and K. Q. Weinberger, editors, *Advances in Neural Information Processing Systems 27*, pages 1520–1528. Curran Associates, Inc., 2014.

[65] K. Mohan and J. Pearl. *On the testability of models with missing data*, pages 643–650. Journal of Machine Learning Research, 2014.

[66] K. Mohan, J. Pearl, and J. Tian. Graphical models for inference with missing data. In C. J. C. Burges, L. Bottou, M. Welling, Z. Ghahramani, and K. Q. Weinberger, editors, *Advances in Neural Information Processing Systems 26*, pages 1277–1285. Curran Associates, Inc., 2013.

[67] K. G. Moons, R. A. Donders, T. Stijnen, and Jr. Harrell, F. E. Using the outcome for imputation of missing predictor values was preferred. *Journal of Clinical Epidemiology*, 59(10):1092–101, 2006.

[68] Ashley I Naimi, Erica EM Moodie, Nathalie Auger, and Jay S Kaufman. Constructing inverse probability weights for continuous exposures: a comparison of methods. *Epidemiology*, 25(2):292–299, 2014.

[69] S. Nakagawa and H. Schielzeth. A general and simple method for obtaining r2 from generalized linear mixed-effects models. *Methods in Ecology and Evolution*, 4(2):133–142, 2013.

[70] John C. Nash. On best practice optimization methods in R. *Journal of Statistical Software*, 60(2):1–14, 2014.

[71] J. A. Nelder and R. W. M. Wedderburn. Generalized linear models. *Journal of the Royal Statistical Society. Series A (General)*, 135(3):370–384, 1972.

[72] E. Paradis, J. Claude, and K. Strimmer. APE: analyses of phylogenetics and evolution in R language. *Bioinformatics*, 20:289–290, 2004. R package version 5.1.

[73] J. Pearl. *Causality: Models, Reasoning and Inference*. Cambridge University Press, 2nd edition, 2009.

[74] Nicholas G Polson, James G Scott, et al. On the half-cauchy prior for a global scale parameter. *Bayesian Analysis*, 7(4):887–902, 2012.

[75] C. A. Pope III, R. T. Burnett, M. J. Thun, E. E. Calle, D. Krewski, K. Ito, and G. D. Thurston. Lung cancer, cardiopulmonary mortality, and long-term exposure to fine particulate air pollution. *Jama*, 287(9):1132–1141, 2002.

[76] R Core Team. *R: A Language and Environment for Statistical Computing*. R Foundation for Statistical Computing, Vienna, Austria, 2017.

[77] N. Ram and D. Gerstorf. Time-structured and net intraindividual variability: Tools for examining the development of dynamic characteristics and processes. *Psychology and Aging*, 24(4): 778–791, 2009.

[78] R. A. Rigby and D. M. Stasinopoulos. Generalized additive models for location, scale and shape. *Journal of the Royal Statistical Society: Series C (Applied Statistics)*, 54(3):507–554, 2005.

[79] R. A. Rigby and D. M. Stasinopoulos. Generalized additive models for location, scale and shape. *Journal of the Royal Statistical Society: Series C (Applied Statistics)*, 54(3):507–554, 2005.

[80] James M Robins, Miguel Angel Hernan, and Babette Brumback. Marginal structural models and causal inference in epidemiology, 2000.

[81]   Julia M Rohrer. Thinking clearly about correlations and causation: Graphical causal models for observational data. *Advances in Methods and Practices in Psychological Science*, 1(1):27–42, 2018.

[82]   P. J. Rousseeuw. *Multivariate estimation with high breakdown point*, pages 283–297. Reidel Publishing Company, Dordrecht, 1985.

[83]   P. J. Rousseeuw and K. Van Driessen. A fast algorithm for the minimum covariance determinant estimator. *Technometrics*, 41(3):212–223, 1999.

[84]   J. J. Russell, D. Moskowitz, D. C. Zuroff, D. Sookman, and J. Paris. Stability and variability of affective experience and interpersonal behavior in borderline personality disorder. *Journal of Abnormal Psychology*, 116(3):578–588, 2007.

[85]   Joseph L Schafer. *Analysis of incomplete multivariate data*. CRC press, 1997.

[86]   A. D. Shah, J. W. Bartlett, J. Carpenter, O. Nicholas, and H. Hemingway. Comparison of random forest and parametric imputation models for imputing missing data using mice: a caliber study. *American Journal of Epidemiology*, 179(6):764–774, 2014.

[87]   Annette L Stanton, Joshua F Wiley, Jennifer L Krull, Catherine M Crespi, Constance Hammen, John JB Allen, Martha L Barro´n, Alexandra Jorge, and Karen L Weihs. Depressive episodes, symptoms, and trajectories in women recently diagnosed with breast cancer. *Breast cancer research and treatment*, 154(1):105–115, 2015.

[88]   D. M. Stasinopoulos and R. A. Rigby. Generalized additive models for location scale and shape (gamlss) in r. *Journal of Statistical Software*, 23(7):1–46, 2007.

[89]   D. M. Stasinopoulos and R. A. Rigby. Generalized additive models for location scale and shape (gamlss) in r. *Journal of Statistical Software*, 23(7):1–46, 2007.

[90]   Daniel J Stekhoven and Peter B u¨hlmann. Missforest—non-parametric missing value imputation for mixed-type data. *Bioinformatics*, 28(1):112–118, 2011.

[91] S. Suh, S. Nowakowski, R. A. Bernert, J. C. Ong, A. T. Siebern, C. L. Dowdle, and R. Manber. Clinical significance of night-to-night sleep variability in insomnia. *Sleep medicine*, 13(5):469–475, 2012.

[92] J. W. Tukey. The future of data analysis. *Annals of Mathematical Statistics*, 33(1):1–67, 1962.

[93] S. van Buuren. *Multiple imputation of multilevel data*, pages 173–196. Taylor and Francis Group, New York, NY, 2011.

[94] S. van Buuren, J. P. L. Brand, C. G. M. Groothuis-Oudshoorn, and D. B. Rubin. Fully conditional specification in multivariate imputation. *Journal of Statistical Computation and Simulation*, 76(12):1049–1064, 2006.

[95] S. van Buuren and K. Groothuis-Oudshoorn. Mice: Multivariate imputation by chained equations in r. *Journal of Statistical Software*, 45(3), 2011.

[96] Willem M van der Wal, Ronald B Geskus, et al. Ipw: an r package for inverse probability weighting. *Journal of Statistical Software*, 43(13):1–23, 2011.

[97] Ravi Varadhan, Johns Hopkins University, Hans W. Borchers, and ABB Corporate Research. *dfoptim: Derivative-Free Optimization*, 2018. R package version 2018.2-1.

[98] W. N. Venables and B. D. Ripley. *Modern Applied Statistics with S*. Statistics and Computing. Springer Science and Business Media, New York, NY, 4th edition, 2002.

[99] W. N. Venables and B. D. Ripley. *Modern Applied Statistics with S*. Springer, New York, fourth edition, 2002. ISBN 0-387-95457-0.

[100] G. Vink and S. van Buuren. Multiple imputation of squared terms. *Sociological Methods and Research*, 42(4):598–607, 2013.

[101] P. T. von Hippel. How to impute interactions, squares, and other transformed variables. *Sociological Methodology*, 39(1):265–291, 2009.

[102] P. T. von Hippel. Should a normal imputation model be modified to impute skewed variables? *Sociological Methods and Research*, 42(1):105–138, 2013.

[103] J. von Neumann, R. H. Kent, H. R. Bellinson, and B. I. Hart. The mean square successive difference. *Annals of Mathematical Statistics*, 12(4):153–162, 1941.

[104] Grace Wahba. *Spline models for observational data*, volume 59. Siam, 1990.

[105] L. Wang and K. J. Grimm. Investigating reliabilities of intraindividual variability indicators. *Multivariate Behavioral Research*, 47(5):771–802, 2012.

[106] Lijuan Peggy Wang, Ellen Hamaker, and CS Bergeman. Investigating inter-individual differences in short-term intra-individual variability. *Psychological Methods*, 17(4):567, 2012.

[107] Y. Wei, A. Pere, R. Koenker, and X. He. Quantile regression methods for reference growth charts. *Statistics in Medicine*, 25(8):1369–1382, 2006.

[108] I. R. White and P. Royston. Imputing missing covariate values for the cox model. *Statistics in Medicine*, 28(15):1982–1998, 2009.

[109] H. Wickham. *ggplot2: elegant graphics for data analysis*. Springer New York, 2nd edition, 2016.

[110] Hadley Wickham. The split-apply-combine strategy for data analysis. *Journal of Statistical Software*, 40(1):1–29, 2011.

[111] Hadley Wickham. *tidyverse: Easily Install and Load 'Tidyverse' Packages*, 2017. R package version 1.1.1.

[112] Hadley Wickham and Jennifer Bryan. *readxl: Read Excel Files*, 2018. R package version 1.1.0.

[113] Hadley Wickham et al. Tidy data. *Journal of Statistical Software*, 59(10):1–23, 2014.

[114] J. F. Wiley. *JWileymisc: Miscellaneous Utilities and Functions*, 2017. R package version 0.3.0.

[115] J. F. Wiley, E. H. Cleary, A. Karan, and A. L. Stanton. Disease controllability moderates the effect of coping efficacy on positive affect. *Psychology and Health*, 31(4):498–508, 2016.

[116] Joshua F Wiley, Bei Bei, John Trinder, and Rachel Manber. Variability as a predictor: A bayesian variability model for small samples and few repeated measures. *arXiv preprint arXiv:1411.2961*, 2014.

[117] C. O. Wilke. *cowplot: Streamlined Plot Theme and Plot Annotations for 'ggplot2'*, 2016. R package version 0.7.0.

[118] L. Wilkinson. Dot plots. *The American Statistician*, 53(3):276–281, 1999.

[119] Graham J. Williams. *Data Mining with Rattle and R: The art of excavating data for knowledge discovery*. Use R! Springer, 2011.

[120] Greg Ridgeway with contributions from others. *gbm: Generalized Boosted Regression Models*, 2017. R package version 2.1.3.

[121] S. N. Wood. Stable and efficient multiple smoothing parameter estimation for generalized additive models. *Journal of the American Statistical Association*, 99(467):673–686, 2004.

[122] S. N. Wood. *Generalized Additive Models: An Introduction with R*. CRC Press, Boca Raton, FL, 2006.

[123] S. N Wood. Fast stable direct fitting and smoothness selection for generalized additive models. *Journal of the Royal Statistical Society: Series B (Statistical Methodology)*, 70(3):495–518, 2008.

[124] Marvin N. Wright and Andreas Ziegler. ranger: A fast implementation of random forests for high dimensional data in C++ and R. *Journal of Statistical Software*, 77(1):1–17, 2017.

[125] Thomas W Yee. *Vector generalized linear and additive models: with an implementation in R*. Springer, 2015.

Index

© Matt Wiley and Joshua F. Wiley 2019
M. Wiley and J. F. Wiley, *Advanced R Statistical Programming and Data Models*,
https://doi.org/10.1007/978-1-4842-2872-2

D

E

N

Printed in the United States
By Bookmasters